the ideas of
PHYSICS
second edition

the ideas of
PHYSICS

second edition

DOUGLAS C. GIANCOLI

California State Polytechnic University,
Pomona

HARCOURT BRACE JOVANOVICH, INC.

New York   San Diego   Chicago   San Francisco   Atlanta

ISBN: 0-15-540559-4

Library of Congress Catalog Card Number: 77-93866

Printed in the United States of America

Cover from Paul Klee, "Eros," Private Collection, Switzerland. Photographed by Fred Wirz, Lucerne.

CREDITS AND ACKNOWLEDGMENTS: Fig. 1–1: *La Grande Danseuse d'Avignon*, Pablo Picasso (1907). Oil on canvas, 59¼″ × 39½″. Private Collection, Switzerland. Photo courtesy of Aquavella Galleries, Inc., New York; 1–3: Douglas Giancoli; 2–1: American Museum of Natural History; 2–2: Yerkes Observatory photograph; 2–8: New York Public Library; 2–12: AIP, Niels Bohr Library; 2–17: Coleco Industries, Inc.; 2–19: Douglas Giancoli; 3–1: National Portrait Gallery, London; 3–9: NASA; 3–21: PSSC PHYSICS, 2nd ed., © 1961 by D. C. Heath and Co., Lexington, Mass.; 3–31: © 1962 by the Regents of the University of California; reprinted by permission of the University of California Press; 4–19: Douglas Giancoli; 4–24: Douglas Giancoli; 4–26: *The Interior of the Pantheon*, Giovanni Paolo Panini. National Gallery of Art, Washington. Samuel H. Kree Collection; 4–27: Douglas Giancoli; 5–21: HBJ Photo; 6–1: AIP, Niels Bohr Library, Burndy Library; 7–17(B): Taylor Instrument, Consumer Product Division, Sybron Corp.; 7–23: Douglas Giancoli; 8–6: Douglas Giancoli; 9–13: Courtesy Coca-Cola, USA; 10–1: Douglas Giancoli; 10–11: Dravo Corporation; 10–13: Douglas Giancoli; 10–27: Rotkin, PFI; 11–7: Joanne Pinkowitz; 11–10: Douglas Giancoli; 11–12: Douglas Giancoli; 11–16: Courtesy of Poly-Optics, Inc.; 12–2: PSSC PHYSICS, © 1965 by D. C. Heath & Co., Lexington, Mass.; 12–14: Douglas Giancoli; 12–16: Brent Brolin; 12–20: Fritz Goro, Life Magazine, © Time Inc.; 12–22: Fritz Goro, Life Magazine, © Time Inc.; 12–24: Georges Seurat, *Sunday Afternoon on the Island of La Grande Jatte*, detail, courtesy of The Art Institute of Chicago; 12–25: American Optical Corporation; 12–26: American Optical Corporation; 12–27: Courtesy Bausch & Lomb; 13–11: Joanne Pinkowitz; 13–12: Wide World Photos; 13–22: Sylvania; 13–27: RCA; 13–29: Intel Corporation; 13–35: Donald N. Sharp Memorial Community Hospital, San Diego; 13–37: Masterson Photography; 13–38: RCA; 14–2: HBJ Photo; 15–9: Courtesy of Bausch & Lomb; 15–20: *Autumn Rhythm*, Jackson Pollock (1950). Oil on canvas, 105″ × 207″. The Metropolitan Museum of Art. George A. Hearn Fund; 15–27: Lee Bobker, Vision Associates; 15–29: JEDL; 15–32: Holoconcepts of America; 16–10: Lawrence Radiation Lab; 16–11: Lawrence Radiation Lab; 16–12: AIP Neils Bohr Library, Megger Gallery of Nobel Laureates; 16–16: University of Chicago; 16–20: Lawrence Radiation Lab.

preface

In this, the second edition of *The Ideas of Physics,* a number of significant changes have been made, which are described below. The basic objective of the book remains the same: to present physics in an interesting, understandable, and enjoyable manner to readers who have little or no background in science. An appreciation rather than a working knowledge of physics is intended. Therefore, the ideas, or concepts, of physics are stressed rather than the mathematical aspects. In fact, very little mathematics is used.

Since I believe that a major barrier to understanding physics is language, I have avoided jargon and kept the language as simple as possible. I am convinced that even the deepest ideas of physics can be expressed using uncomplicated language.

Although the coverage of topics is more or less traditional, I have avoided the conventional approach in which general principles are stated and various results are then deduced from them. That is not the way science has developed nor is it the way scientists themselves work. The approach I have used is more in keeping with how science actually is practiced; that is, I begin with commonplace observations and concrete examples and *then* proceed to generalizations or hypotheses that unify these observations.

Merely reading about common observations and experiences, however, is often not sufficient. If readers are to understand and enjoy physics, they must have some of these experiences themselves while they are trying to understand the related concepts. To encourage this kind of direct experience, I have presented easily performed experiments in the main body of the text. These experiments require only the simplest materials, which most people will already have. Some of the experiments require only a minute or two to perform and should be done when the reader encounters them. Others require more time and might be considered "projects," which students could write up. (Incidentally, I have found that doing such a project is often more stimulating to nonscience students than answering a set of questions or problems.)

Since no scientific theory is born in a vacuum and each scientist builds on the work of his predecessors and contemporaries, I have presented the historical development of ideas related to a given topic rather than simply stating the latest viewpoint. I believe that students can better understand the motion of celestial bodies or our present view of the atom, for example, if they have traced the development of ideas that have led to our present view. It is through this approach, too, that a student can get the flavor of what physics is about and come to realize that it is indeed a human endeavor.

I have included a reasonable amount of philosophic material related to physics, for I feel it is important to the subject and can be of considerable interest. Likewise, the practical side of physics, "how things work," is also interesting and important. Therefore, many practical applications are discussed, such as the camera, loudspeakers, transistors, radio and television sets, nuclear power, and the internal combustion engine.

The book is fairly comprehensive. The order of topics is for the most part conventional—the one exception being that relativity is presented rather early (Chapter 6), as the culmination of classical mechanics. This means that some of the excitement of twentieth-century physics occurs early in the book instead of all at the end.

In order to lead readers into the unfamiliar (and sometimes frightening) subject known as physics as gently as possible, the first chapter is mostly of a philosophic nature. In addition, the second chapter is largely historical until the familiar notion of speed is introduced. The book continues through classical mechanics, including the concepts of energy and momentum, and then to the theory of relativity. Chapter 7 begins what might be considered the second part of the book, which deals in large part with the structure of matter and includes such topics as heat, kinetic theory, waves, sound, light, and electricity and magnetism, and culminates in the quantum theory of atomic structure and nuclear physics.

The Ideas of Physics can readily be used in a one-semester or one-quarter introductory course. Chapters are written so that some sections and even whole chapters can be omitted without losing continuity. If there is not time to cover the entire book, a number of alternatives are feasible. For example, if as much modern physics as possible is desired, a course could include Chapters 2 and 3, 5 and 6, and 10 through 15, with Chapter 16 (Nuclear Physics) as optional. If necessary, Chapter 14 and all but the first section of Chapter 13 can be omitted without losing the essentials necessary for understanding Chapter 15 on the quantum theory and atomic structure. If only classical physics is to be covered, Chapters 2 through 5 and 7 through 14, in whole or in part, can be used.

A number of important changes and additions have been made in this second edition. First of all, there is more emphasis on the metric system, and a section is devoted to an explanation of its use. Each chapter is now followed by a set of multiple-choice questions for review and then a series of exercises requiring word answers or very simple calculations. The total number of questions has been greatly expanded; there are now three times as many as in the first edition. Many new student experiments have also been added. A new appendix contains a brief summary of each chapter with important terms to help students when reviewing. The section on powers of ten notation has been relegated to another appendix.

Many new topics have been added to the main text. These include sections on: fluid flow and Bernoulli's principle, heat transfer, solar heating, polarization, total internal reflection and fiber optics, color mixing and color vision, noise pollution, series and parallel resistance circuits, solenoids, transmission of radio and television signals; a lengthy section on energy sources and power plants; and a whole new chapter on rotational motion and bodies in equilibrium (both topics use the concept of torque). In addition, the chapters on heat and light have each been split into two chapters.

For those courses in which a more mathematical treatment is desired, there is now available an *Optional Mathematical Supplement.* It contains worked-out examples for each chapter, a small amount of additional material, and many exercises for the student that involve arithmetic or simple algebra. As its name implies, this supplement is optional, to be used at the discretion of the instructor. Also available is an *Instructor's Manual* with answers to exercises in the main text, solutions to exercises in the supplement, plus suggested test questions.

I would like to thank the reviewers and the many users of the first edition of this book who have provided many valuable comments. To the many persons on the staff of Harcourt Brace Jovanovich who worked so hard in the production of this book, I offer my sincere appreciation. In particular I wish to thank Roger Dunn for his continual encouragement and help. Finally, I express my grateful appreciation to Andrea, Francesca, and Mary Teresa, my children, for their unceasing patience and support, which were so important during the writing of this book.

DOUGLAS C. GIANCOLI

contents

the ideas of
PHYSICS
second edition

1

PHYSICS: A HUMAN ENDEAVOR

Many people feel that science, and especially physics, is beyond their understanding. Even people in important decision-making positions often have little knowledge of science and how it works. Yet science and its offspring—technology—are everywhere. To deal with them rationally requires some understanding. It is thus the aim of this book to bring the ideas of physics to nonscientists in an uncomplicated way.

This is one reason for studying physics. Another reason is that after reading this book you may feel more confident about making minor repairs on a toaster or bicycle. And perhaps the most important reason for studying physics is an aesthetic one—physics is an intellectually stimulating activity; it is an exciting field of human endeavor. Like artists and poets, physicists and other scientists seek to comprehend and interpret the world around them. In their own fields they each attempt to increase our degree of consciousness. Great scientists, like great artists, poets, or musicians, make us aware as we have not been aware before. They extend certain aspects of our experience and deepen our appreciation of life.

1. SCIENCE: A CREATIVE ACTIVITY

Science is a human endeavor. The practice of science is built on the unprovable assumption that a real physical world exists independently of humans. The existence of a real physical world is taken for granted by most scientists, artists, and for that matter, most everyone.

The aim of science is to find order in the physical world around us. Scientists seek to order the great variety of phenomena they perceive in the natural world, in part by seeking relationships among the various things that occur in the world. Perhaps people were motivated initially by the very practical need to survive. By finding order in nature they would better know what to expect in their daily affairs. But eventually people found that examining nature, finding whatever order they could and thereby bringing a unity to the variety of observations they made, proved to be intellectually stimulating in itself. Thus was science born.

The Scientific Method

Many believe that science is merely a collection of facts, that it develops by the gathering of these facts and slowly approaches the truth about nature. According to this view, the so-called scientific method is the basis for scientific development. The scientific method is said to have three steps: (1) the observation of relevant facts, (2) the proposal of a hypothesis or theory based on these observations, and (3) the testing of the theory to see if its consequences or predictions are actually observed in practice. This is, however, an oversimplified view of how science is practiced and does not do science justice. Let us examine these steps and see just how valid they are.

Normally scientists study a limited set of phenomena, such as particular aspects of motion or heat or light. The first step—that of careful observation—often involves looking for relationships among different kinds of events. For example, Newton looked for a quantitative relationship between the effort (or force) applied to an object and the object's subsequent motion. If a specific relationship seems to be true in a number of cases, without any exceptions, then it may be hypothesized that the relationship holds in general.

The second step of the scientific method may be the formulation of such a general hypothesis. Or it may be an attempt to explain phenomena by theorizing the existence of a deeper reality. Examples of the latter are the atomic theory of matter and the wave theory of light. No one has ever actually seen an atom with the eye nor seen light travel as a wave; that is why we call them theories or hypotheses rather than facts. The value of such theories is that they can be used to explain many observable phenomena. A good theory, then, results in a more unified view of nature. When a theory brings together many phenomena under a single concept, the hypothesized relationship, or the main axiom (s) of the theory, is called a natural law. Unlike political laws, natural laws are not something imposed on us. Rather they are concise *summaries* of observable phenomena. Examples are Newton's laws of motion, the law of gravitation, and the gas laws.

The third step, the testing of the theory, may involve further experimental testing of hypothesized relationships. Or it may be experimental verification of new phenomena predicted by a new theory. For example, Einstein's theory of relativity predicted that mass could be changed into energy, and energy into mass (this is the meaning of the famous equation $E = mc^2$). That this change might be possible had not previously been thought of, but a few years after Einstein's prediction it was experimentally verified. Incidentally, we see here how science is a unifying process. The concepts of mass and of energy were seen to be one.

Science Is a Creative Activity

The simple view that scientists painstakingly work through the steps of the scientific method is misleading. So is the view that scientists deal only with facts that are waiting to be discovered as if these facts were absolute and not open to interpretation or judgment. Science builds not on facts but on observations and creative ideas. And the observations are open to interpretation and do require judgment.

If we examine the work of great scientists, nowhere will we find the three steps of the scientific method in any simple or sequential way, although ideas resembling each do appear.

Consider the first step: observation of the relevant facts. Scientists can never include everything in a description of what is observed;[1] there is just too much happening (we cannot know, for example, what every atom in the universe is doing at all times). They must judge what is relevant in a given situation, and their judgment is based partly on what they already know and believe. The ancient Greeks, for example, studied the horizontal motion of objects along the ground. To them the relevant observations were that after an initial push an object always slowed down and came to rest. From this they concluded that objects are naturally at rest. Galileo saw it differently, as we shall see in Chapter 2. What was relevant to Galileo was that different objects slowed down at different rates depending on their surfaces. Because he saw something different in the same facts, Galileo was able to establish a new theory of motion.

Thus we see that the practice of science is not an entirely objective activity. The subjective element is very important and, in fact, necessary if new theories are to develop. Being able to see old facts in a new way as Galileo did requires a creative mind.

Creativity in science is most apparent in the formulation of new theories or hypotheses. Consider the idea that matter is made up of atoms; no one had ever *seen* an atom—the idea that they might exist came from someone's mind. Similarly, as we shall see in Chapter 2, Copernicus broke with the traditional view that the earth is at the center of the universe. He hypothesized that the earth and the other planets revolve around the sun. But no one has gone out and recorded the "fact" that the sun is at the center and the planets revolve around it. Certainly Copernicus did not. This hypothesis, this idea, leapt from the mind of a human being. It was a creative act.

The third step in the scientific method, the testing of a scientific hypothesis or theory, is also not a simple straightforward process. First of all, it is often not possible to test a theory directly. As mentioned before, no one has ever seen an atom directly, nor has anyone stood far out in the universe and taken a picture of the solar

---

[1]This is something like the well-known circumstance that five witnesses of an automobile accident will give five different stories of what happened. No one person can grasp all the details, and what each *does* observe depends partly on what that person *expects* to observe; i.e., it depends on one's background.

system to see if Copernicus was right. Often a theory is tested by determining if its predictions agree with experiment in specific situations. If the predictions of a theory are not even close to what is found experimentally, then the theory is discarded or modified. On the other hand, how close must the predictions of the theory be to experimental results in order to say that the theory is verified? Quantitative experimental measurements rarely come out *exactly* as the theory predicts. And no measuring instrument is perfect. Furthermore, it is obviously not possible to test a theory under *all* circumstances.

Clearly, then, a theory cannot be verified absolutely. Therefore, it is perhaps proper to ask not how a theory is confirmed, but rather how it comes to be accepted by scientists. This question, much discussed by philosophers of science,[2] does not have a simple answer. Apparently a number of factors are involved. Normally, a new theory competes with other new theories or with an older theory that it challenges. A particular theory may come to be accepted because of its better agreement with experiment. But this factor alone is usually not sufficient. Equally important is that a newly accepted theory will normally explain a greater variety of phenomena than the competing theory. Such was the case with Copernicus' sun-centered theory of the universe. Copernicus' theory was no better than the old earth-centered theory in predicting the detailed motion of heavenly bodies. But his theory did explain the calendar year and the recurrences of the seasons in a simple way, something the older theory could not do. A theory that gives a more unified view of nature is more useful to a scientist. And it cannot be ignored that the social, cultural, and religious settings of the day also play roles in the acceptance of a theory.

The practice of science is not a mechanical process but rather a creative human activity.

Science and the Humanities

To most people the sciences and the arts and humanities are two very different fields. This split between the arts and sciences was popularized by C. P. Snow in *The Two Cultures* and more recently by Theodore Roszak.[3] There seems to be some truth in this idea of a split. Many scientists and humanists show considerable contempt for one another. Scientists have been known to accuse humanists of being idle dreamers who are out of touch with reality. Humanists,

[2]This question is discussed at length in the stimulating book by Thomas Kuhn, *The Structure of Scientific Revolutions* (Chicago: University of Chicago Press, 1970).
[3]For example, in his *Where the Wasteland Ends* (Garden City, N.Y.: Doubleday, 1972).

on the other hand, often feel that scientists are rigid and out of touch with humanity.

However, a good argument can be made for the sciences and the arts being much more alike than different. For example, J. Bronowski[4] has pointed out that "science is as integral a part of the culture of our age as the arts are." In *Science and Human Values* he carefully notes the similarities between the arts and the sciences. As we saw in a previous section, scientists do not work from a simple collection of facts. They must observe the world and interpret it. Just as artists or poets attempt to represent reality in their work, so too do scientists. Artists or poets look at the world and seek relationships and order. But they put their ideas on canvas, or in marble, or in poetic images. Scientists try to find relationships between different objects and events. To express the order they find, they create hypotheses and theories. Thus the great scientific theories are to be compared to great art and great literature.

To carry the similarity a step further, note for example that artists do not represent exactly everything they see. They emphasize those aspects of reality they feel are most important or relevant. Neither do scientists include everything in the physical world in their theories. They too judge what is relevant.

Of course differences exist between the humanities and sciences. Although scientists are not mere recorders of facts, they are constrained to conform to observations of nature. But artists are not so constrained. Novelty for its own sake is legitimate in art but not in science. Most scientists feel that their goal is to seek a single, accurate account of nature. But two artists can view the same subject and come up with two different creations, and each may be recognized as valid. The sciences and the arts differ in the critique or testing of the finished painting, poem, or scientific theory. Artists do not necessarily seek approval of their work by other artists; but scientists do, and must, seek confirmation of their ideas from other scientists.

Despite these differences, scientists and artists think and work in a basically similar way. Each deals with nature and each creates a vision of reality, be it a painting, a poem, or a scientific theory. But art and science emphasize different aspects of reality. Scientists recreate one side of this world, poets and artists another. Neither has a monopoly on the correct view. But each vision reveals different aspects of reality.

The apparently different ways in which scientists and artists think can sometimes overlap, as pointed out by C. H. Waddington[5] in his book on the relationship between painting and the natural sciences

[4]J. Bronowski, *Science and Human Values* (New York: Harper and Row, 1965).
[5]C. H. Waddington, *Behind Appearance, A Study of the Relations Between Painting and the Natural Sciences in this Century* (Cambridge, Mass.: MIT Press, 1970).

Figure 1–1
La Grande Danseuse d'Avignon, Pablo Picasso (1907). Private
collection, Switzerland.

in the twentieth century. He considers it no accident that Picasso
founded Cubism (Fig. 1–1) in 1905, the same year Einstein proposed his theory of relativity. And Cubism, he points out, has its
roots in Cézanne's struggle to find a way to paint the *structure* of
objects rather than their appearance. Structure, as we shall see in
Chapter 15, has also concerned physicists in the twentieth century.
Around 1910 artists of the Futurist school attempted to represent
speed and movement in their paintings, particularly as they relate
to technology. Later painters absorbed from the new physics of the
quantum theory the idea that matter is less solid and more transparent than previously thought, and that motion cannot really be
frozen into a timeless second.

In our study of physics in the remaining chapters of this book,
we will try to illustrate the humanness of science. We will, where
possible, follow the historical development of ideas. We will see that
the scientists of the past whose theories are now considered out-

moded were not fools. Their vision of reality was different from ours, and perhaps more limited, but they were not necessarily wrong. After all, science, like the arts, develops by building on the ideas of the past. And we will see that scientific development is not the dogmatic process it is popularly thought to be.

2. THE ROLE OF MATHEMATICS

Very little mathematics will be used in this book.[6] Instead, the *ideas* of physics will be emphasized. However, we will need the mathematical concept of proportion.

Direct Proportion

When an increase or a decrease in one quantity[7] results in the same relative change in a second quantity, we say that the second quantity is proportional to the first. For example, when water runs from a faucet at a constant rate into a bathtub, the tub becomes fuller the longer the water is on. We say the amount of water in the tub is proportional to the length of time the faucet was on. Using the symbol $\propto$ to mean *is proportional to,* we can write this relationship as

amount of water in tub $\propto$ length of time faucet is on.

If the faucet is left on for two minutes twice as much water will be in the tub than if the faucet is left on for one minute (assuming, of course, that the rate of flow from the faucet doesn't change). If the faucet is on for only half as long, there will be half as much water in the tub. This kind of proportionality is usually referred to as **direct proportion** to distinguish it from inverse proportion, which we will discuss shortly. In direct proportion, the two quantities increase or decrease together in the same ratio.

The circumference and the diameter of a circle are also related by a direct proportion, Fig. 1–2. If one circle has twice as large a diameter as that of a second circle, the first will also have twice the circumference. If the diameter of one circle is half as large as that of another, the circumference will be half as large, and so on. We can write this proportionality as

circumference $\propto$ diameter.

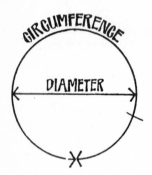

Figure 1–2.
A circle.

---

[6]A brief appendix at the end of the book describes the "powers of ten notation," which we will occasionally use.

[7]By a *quantity* we mean an observable property in nature that can be measured or specified numerically. For example, length, time, weight, temperature, and energy are all quantities.

8

Constants of Proportionality

The ancients found by careful measurement that the ratio of the circumference of any circle to its diameter equals approximately 3.1416, and they called this special number π (pi). The proportionality between the circumference and diameter of a circle can then be written as an equality:

$$\text{circumference} = 3.1416 \times \text{diameter},$$

or

$$\text{circumference} = \pi \times \text{diameter}.$$

You may have seen this relation written as $C = \pi D$, where C stands for circumference and D for diameter. This number, $\pi = 3.14 \ldots$ (keeping only three digits), is called a **constant**, because it does not change. The diameter and circumference can be referred to as **variables**, because they are different for different circles.

We can always make a direct proportionality into an equality by inserting a constant called a **proportionality constant**. The constant of proportionality for the circumference and diameter of a circle is called π and has the approximate value 3.1416. In the example of water running into a tub, we might call the proportionality constant K. Then we can write the proportionality as an equality:

$$\text{amount of water in tub} = K \times \text{time faucet is on.}$$

To find the numerical value of the constant of proportionality K would require a simple measurement (namely, the volume or weight of water that flows from the faucet each second). Of course K will have a different value depending on how far the faucet is turned on; but it will be a constant at any particular setting. This example differs slightly from the situation with the circle because the proportionality constant π is the same for every circle. We therefore say that π is a *universal* constant.

Inverse Proportionality

Sometimes a relationship between two quantities is such that if one quantity is increased the other is decreased. Such a relationship is known as **inverse proportion**. For example, if you have a fixed distance to drive, the time the trip takes is inversely proportional to how fast you drive. Suppose you are driving 100 kilometers between two cities. If you average 60 kilometers per hour (abbreviated km/hr), the trip will take half as long as it would if you averaged 30 kilometers per hour. When the speed is doubled, the time is halved. If the speed is tripled, it takes only a third as much time. We can write an inverse proportionality in the following way:

$$\text{time for a trip} \propto \frac{1}{\text{speed}}.$$

We get *inverse* proportionality by putting the speed in the denominator of a fraction. Notice that when the denominator of a fraction is increased, the fraction as a whole is decreased. For example, one over two (½) is less than one over one ($1/1$); ⅓ is less than ½; and so on. Thus, in the above example, if the speed is doubled, $\frac{1}{\text{speed}}$ is only half as large. This relationship corresponds to the fact that the time required for the trip is only half as much.

An inverse proportionality, like a direct proportionality, can be changed into an equality by introducing a proportionality constant.

3. MEASUREMENT

Figure 1–3

A "standard meter" for the people.

The measurement of distance, time intervals, weight, and other quantities is an important aspect of science. Measurements are made using particular standards or units. For example, length can be measured in the British system in units of feet, inches, and miles, or in the metric system in meters, centimeters, and kilometers. In the past, people have used a variety of units: cubits, leagues, and hands; even the foot varied from place to place. A French person and an English person might measure the same boat as 35 feet and 39 feet in length, for example. (Henry VIII and Louis XIV apparently didn't wear the same size shoe.) It was hard to communicate in the past; but about 200 years ago, universally recognized standards were developed.

The first real international standard was the **meter** established by the French Academy of Sciences in 1791. It was defined as the distance between two fine marks on a particular platinum–irridium bar now kept near Paris.[8] Copies of the standard meter were sent around the world. In fact, a copy remains in the Place Vendôme in Paris (Fig. 1–3), placed there originally so Paris shopkeepers would have access to a standard of measurement.

The British units of length are defined today in terms of the meter: 1 inch = 2.54 centimeters. Because there are 12 inches in a foot,

---

[8]In 1960, a more precise definition of the meter in terms of the wavelength of a particular kind of light was accepted internationally.

Table 1–1
Units of Length and Volume

LENGTH		
Metric		British
1 cm = 10 mm	=	0.394 in.
1 m = 100 cm = 1000 mm	=	39.37 in. = 3.28 ft
1 km = 1000 m	=	3,280 ft = 0.621 mi
British		Metric
1 in.	=	2.54 cm = 25.4 mm
1 ft = 12 in.	=	30.48 cm = 304.8 mm
1 mi = 5,280 ft	=	1.61 km = 1,610 m

VOLUME

1 liter = 1000 cm³ = 1.06 quarts
1 quart = 0.946 liter

there are 12 × 2.54 centimeters = 30.48 centimeters in one foot (ft). Can you figure out how many centimeters there are in one mile (mi)? The relationships between other metric and British units are given in Table 1–1.

In the metric system, larger and smaller units are defined in multiples of 10, 100, and so on, from the standard unit. Thus, a centimeter (abbreviated cm) is 1/100 of a meter (m); a millimeter (mm) is 1/1000 of a meter; and a kilometer (km) is 1000 meters. The prefixes **milli-** (meaning 1/1000), **centi-** (1/100), and **kilo-** (1000) can be attached to other kinds of metric units. For example, a kilogram is one thousand grams. Table 1–2 shows other prefixes that are used, but these three are most common.

The relation between various units in the British system makes calculations difficult (for example 5,280 feet = 1 mile). For this reason, and to make communication and commerce simpler, scientists, and nearly all countries of the world for practical purposes, have adopted the metric system. Congress has decided that the United States will convert to metric units, and the British themselves have already done so.

Sometimes it is necessary to convert from one system to the other. Table 1–1 will help here. For example, suppose you measure something as 20 centimeters long and you want to know how many

Table 1–2 Metric Prefixes				
nano-	=	1/1,000,000,000	=	10^{-9}
micro-	=	1/1,000,000	=	10^{-6}
milli-	=	1/1,000	=	10^{-3}
centi-	=	1/100	=	10^{-2}
deci-	=	1/10	=	10^{-1}
deca-	=	10	=	10
kilo-	=	1000	=	10^{3}
mega-	=	1,000,000	=	10^{6}
giga-	=	1,000,000,000	=	10^{9}

inches it is. From Table 1–1 we see that 1 centimeter is about 0.4 inches. Therefore 20 centimeters equal 20 × 0.4 = 8 inches.

To give you some practice with the metric system, try the following simple exercise:

EXPERIMENT-PROJECT

Use a metric ruler to measure your height in meters or in centimeters. (Do *not* measure in feet and inches and then convert to metric.) Then, take the result of your measurement and use Table 1–1 to convert to British units (feet and inches). Is this how tall you thought you were?

We have so far mentioned only units of length. We will discuss other units when we need to. For now we add that the units of time—hours (hr), minutes (min), and seconds[9] (sec)—are the same in the metric and British systems.

Although learning metric units may seem like a chore, you may be encouraged by the fact that many everyday units are already metric. For example, the units volts and amperes used in electricity are metric and do not require conversion.

4. THE ROLE OF LANGUAGE

Whenever we wish to express an idea, we must be careful about the words we use. Our ideas are affected by the language we use to express them. Much of the difficulty that people have with physics comes from using imprecise language. You may have the correct idea in your

[9]The official abbreviations for hours and seconds are h and s. For clarity, however, we will use hr and sec.

mind, but sloppy language will not convey it to others. Eventually, careless language will confuse even you.

Another barrier to understanding physics is the use of jargon —specialized words or specialized meanings of old words. I feel that even the deepest ideas can be communicated without frequent use of jargon and I have avoided it as much as possible in this book. I have not been able to eliminate new words entirely, however. When a new word or a more specific definition of a common word appears, it is important to pay close attention to this definition and refer to it when necessary.

REVIEW QUESTIONS

1. Scientists follow the three steps of the scientific method:
   (a) strictly
   (b) only vaguely; the three steps of the scientific method do enter one's thinking, however
   (c) not at all

2. Physical laws are
   (a) laws by which nature operates
   (b) summaries of human observations
   (c) laws laid down by a superior being

3. When the increase in one quantity always results in an increase in a second quantity at the same rate, we say the two quantities are
   (a) directly proportional   (c) proportionality constants
   (b) inversely proportional   (d) unrelated

4. When the increase in one quantity is always accompanied by a corresponding decrease in a second quantity, we say the two are
   (a) directly proportional   (c) proportionality constants
   (b) inversely proportional   (d) unrelated

5. How is the total cost of a bag of oranges related to the number of oranges purchased?
   (a) directly proportional   (c) unrelated
   (b) inversely proportional

6. How would you relate the amount of water remaining in a once-full bathtub to the length of time the plug has been out?
   (a) directly proportional   (c) unrelated
   (b) inversely proportional

EXERCISES

1. Do you think that the world, and science itself, would benefit if the general public were more familiar with science and its practice? Discuss.

2. It is sometimes said that science is the new religion complete with high priests and mysteries known only to the select few—the trained scientists. Do you agree? Discuss.

3. Discuss the limitations of science.

4. It has been said that society's ills are the result of science. Scientists often respond that their work is of a pure and intellectual nature and practical applications, i.e., technology, create the problems. Discuss. (It might be useful first to distinguish between science and technology.)

5. Carefully draw a number of circles of different sizes using a compass. Measure both the circumference and the diameter of each. Is the ratio $\frac{\text{circumference}}{\text{diameter}}$ the same for each?

6. Suppose you are driving your car at a steady 50 miles per hour. What kind of proportion is there between how long you drive and the total distance you drive? What do you think is the constant of proportionality in this relationship?

7. What kind of proportion (approximately) relates the amount of water remaining in a once-full bathtub to the amount of time the plug has been out?

8. Approximately what kind of relationship might exist between the steepness of a hill and the speed with which a bicyclist can ride up the hill?

9. Give some simple examples of direct and inverse proportions (or at least approximate ones) that you have observed in everyday life.

10. In the equation $A = r^2$ relating the area of a circle to its radius, what is the constant of proportionality?

11. Discuss the advantages and disadvantages of the metric system compared to those of the British system of units.

12. A frying pan is 250 mm across. How many centimeters is this?

13. How many millimeters are there in one meter?

14. How many miles are there in one inch?

15. Measure the dimensions (length, width, height, diameter, whatever is appropriate) of at least three objects in your room. State what the objects were and the measurements you obtained. (This exercise is intended to give you a feel for the metric system.)

16. It is 8.5 miles from my house to school. How many kilometers is it?

17. A car is 5.2 m long. How long is it in feet?

2

MOTION: IN THE HEAVENS AND ON THE EARTH

Motion of automobiles, animals, people, or the sun is an obvious phenomenon in nature. Motion has fascinated people at least since the time of the ancient civilizations in Asia Minor. But not until the late Renaissance was the problem of motion attacked with vigor and important concepts clarified. Although many contributed to this understanding, Galileo Galilei (1564–1642) and Sir Isaac Newton (1642–1727) stand out above the others.

Motion was perhaps the first aspect of nature to be studied thoroughly and understood. Other phenomena, such as electricity, magnetism, heat, light, and the nature of matter, remained mysteries for a longer time.

Since the time of the ancient Greeks, two distinct types of motion seemed to be central to an understanding of the universe: the motion of celestial bodies and the motion of objects on earth. These two kinds of motion were considered completely separate matters until the work of Galileo and Newton.

A study of the development of ideas about both celestial and terrestrial motion from the ancient Greeks through the great ideas of Galileo is fascinating and forms the subject matter of this chapter.

1. CELESTIAL MOTION: OUR VIEW OF THE UNIVERSE

Have you ever escaped the lights of the city to gaze late at night at the multitude of stars above you? It is a moving experience, which we rarely have today, but it must have deeply impressed the ancients. They saw this sight every cloudless night, for they had neither street lights nor smog to obscure it. The starry heavens have been a source of inspiration for poets and scientists alike for thousands of years. The ancient Greeks and Babylonians were fascinated by the stars and carefully traced their motions.

Observations on the Heavenly Bodies

If you look at the stars on a moonless night, you may notice that they seem to be arranged on a huge inverted bowl. Indeed, the ancient Greeks pictured the stars as fixed on the inside of a giant rotating sphere. Early observers, as they examined this celestial sphere night after night, allowed their imaginations to roam freely. In certain groupings of the stars they saw the outlines of wild animals, great warriors, and beautiful women. These groupings are the **constellations**—the big and little dippers, Virgo, Orion the hunter, Cassiopeia (Fig. 2–1).

Most stars appear to be fixed with respect to each other. The

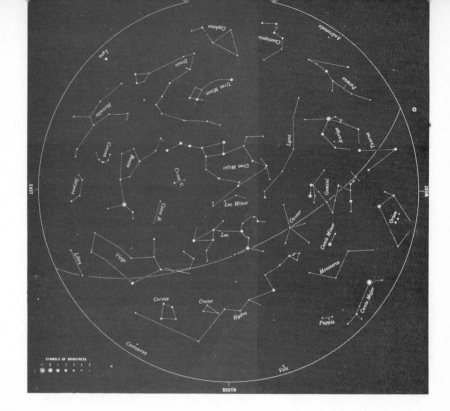

Figure 2–1

A chart of the stars.

Figure 2–2

Time exposure showing movement of stars.

shape of the big dipper, for example, never seems to change. Scientists today, however, tell us that the stars do indeed move with respect to each other. In fact they move at great speed, but they are so far away that movement is imperceptible to the naked eye. Thus the Babylonians and Greeks believed that the stars were fixed in position on the inverted bowl of the heavens. But they did observe a very small number of starlike objects that moved with respect to the "fixed" stars. They discovered five such nomadic objects and called them **planets** (from the Greek *planētēs,* "wanderer"). Although their motion was very slow, the planets clearly changed position over a period of several days or months. The five planets were named for the Roman gods Mercury, Venus, Mars, Jupiter, and Saturn. The sun, the moon, and the planets still serve as the names of the days of the week: *Sun*day, *Mo(o)n*day, Tuesday (from *Tys*, Germanic equivalent of *Mars*), Wednesday (from *Wotan*, meaning Mercury), Thursday (from *Thor*, equivalent of Jove or Jupiter), Friday (*Fria*, the Germanic Venus), and, *Satur(n)*day.

The fixed stars, as well as the sun and moon, appear to rotate from east to west. You can detect this very slow westerly motion by observing the sky at various times during the night. Successive observations will reveal that the stars have indeed moved to the west; on a time exposure photograph of the heavens, Fig. 2–2, each star makes a curved line because of this motion. A complete revolution of the *celestial sphere* takes 24 hours, after which the same stars

Figure 2–3

The moon appears to move slowly among the stars, (A), (B), and (C). After a month (D), it has returned to the same place among the stars.

18

are again directly overhead. The moon also seems to move from east to west, but more slowly; a complete revolution takes about 25 hours. Because the moon moves westwardly more slowly than the stars, its position with respect to the stars seems to move *eastward*. On successive nights the moon's position will be observed to be slightly to the east with respect to the fixed stars, as shown in Fig. 2–3 (A), (B), and (C). About a month later the moon will have returned to approximately the same position among the stars, Fig. 2–3 (D). Careful observations of which stars appear on the horizon just after the sun sets reveal that the sun, too, changes position with respect to the stars. The sun returns to the same position among the stars after one year. In fact, a year could be defined as the time it takes for the sun to pass through the celestial sphere and return to the same position with respect to the fixed stars.

The motion of the planets among the stars is more irregular than that of the sun and moon, but it, too, is generally in an easterly direction with respect to the stars, as shown in Fig. 2–4. The occasional westerly motions, the loops in Fig. 2–4, are referred to as **retrogressions.**

You should not accept these statements as true without examining the night sky yourself. Even without instruments you can make some of the observations described. Throughout the book you will find experiments that you can easily perform. The first one was in Chapter 1. Some may seem trite or trivial, but they are worth doing. You may be surprised to find how much doing a simple experiment can help in understanding the material of this book. Some experiments will take only a minute or two to perform and should be done immediately upon reading them. Others will take longer and are more in the nature of a project, such as this one:

EXPERIMENT-PROJECT

Look at the nighttime sky for five or ten minutes. Note if there is any obvious motion of the stars and moon. It might help to lie on the ground and carefully observe a star above a power line or telephone wire; notice if the star moves, even slowly, past the wire.

Draw a diagram of a fairly large region of the sky—say near the eastern horizon—indicating the position of the brighter stars and any constellations you can see, as well as the moon if it is visible. Be sure to include some reference points such as a fence, a mountain, or a telephone pole.

About three hours later observe the sky again from the same location and draw a *second* diagram showing the same stars (and constellations) and the moon, and note any changes in position.

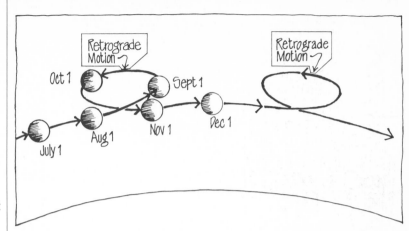

Figure 2–4

The path of the planet Mars among the stars during the course of several months, showing position for different dates.

Finally, on the next night, at the same hour as either of the first two observations, again observe the nighttime sky and draw a *third* diagram of the moon and stars. Compare the three diagrams and note any change in the positions of moon and stars as a whole or with respect to each other. From your observations, briefly describe the motion of the moon and stars.

This section has been a brief description of the motion of the sun, moon, planets, and stars as they appear to an observer on the earth. But we are not satisfied with mere observations. We want to know what is "behind" them. We seek some kind of order or unifying principle for the diverse observations we make. We are thus led to create hypotheses to explain the observations.

The Greeks Conceived of Great Celestial Spheres

Scientific theories about celestial motion can be traced back at least to the time of the ancient Greeks, and more specifically perhaps to the sixth and seventh century B.C. philosophers; the most famous were Thales and Pythagoras. By the fourth century B.C. a detailed theory of the heavenly bodies had been developed. According to this theory the heavens were composed of eight concentric transparent spheres. The sun, the moon, and the five planets were each on one of these spheres, and all the fixed stars were attached to the eighth sphere, Fig. 2–5. Each sphere rotated on a different axis and at a different rate, but the center of each was at the earth. This earth-centered, or **geocentric**, view of the universe was to persist with minor modifications for well over two thousand years. Indeed, references to the "seven celestial bodies and their orbs" (or

19

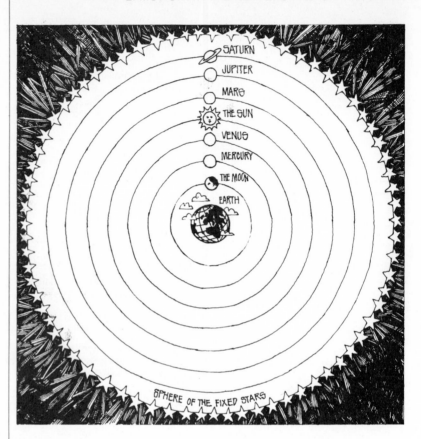

Figure 2–5

The universe, as conceived by the ancients, consisting of eight concentric spheres rotating about the earth.

"spheres") are found in Shakespeare and other great literature over the centuries.

The two greatest philosophers of Greek civilization, Plato and Aristotle, and especially certain of their students, contributed further detail to this concept of the universe. They were very aware that this simple model did not explain adequately the detailed motion of the heavenly bodies, particularly the retrograde motion of the planets. At the same time, Plato and his students raised to the level of dogma the idea that the heavenly bodies move in perfect circles. This view was only natural, since their philosophy was largely concerned with form and perfection. It was inconceivable to them that the orbits of the planets should follow oval paths, for example. To preserve the idea of perfect motion in a circle and to explain the retrograde motion of the planets, Aristotle and his contemporaries hypothesized a complicated set of circular motions. Each planet, the sun, and the moon were attached not to one sphere but to several, each of which rotated about a different axis. The theory required some 50 spheres in all. This combination of motions accounted for many of the ob-

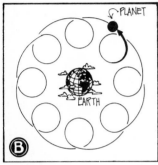

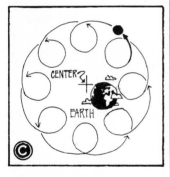

Figure 2-6

(A) Retrograde motion of planets explained with the concept of an "epicycle"—a small circle turning around a center, which itself moves on a larger circle. (B) The net motion of a planet is thus a series of loops. (C) Ptolemy proposed that the earth may not be at the center, but displaced a certain distance.

served features of heavenly motion, but by no means solved all the problems.

Another ancient idea (often attributed to Aristotle and also based on the doctrine of perfection) was the notion that only the "sublunary" (Latin: "below the moon") region of space—the space between the earth and the moon, including the earth—was subject to change and imperfection. The moon and the universe beyond were considered unchangeable and eternal. This idea persisted well into the second millennium A.D.

Ptolemy Uses Circles—the Ultimate Geocentric System

The geocentric view of the universe reached its high point in the work of Ptolemy (A.D. 100–170). By the second century A.D. the cultural center of the West had shifted to Alexandria; there, Ptolemy made extensive astronomical observations and worked out his system of the universe. Ptolemy, too, accepted the notion that the planets must move around the earth in circular orbits. But he did away with the need for spheres to which the planets, the sun, and the moon were attached. Instead he proposed that the planets move in a combination of simple circles, one superimposed on the other. As shown in Fig. 2–6 (A), a planet revolves in a small circle, called an **epicycle**, about a point, which itself describes a larger circle. The combined motion is a series of loops, Fig. 2–6 (B). This model could account for the retrograde motion of the planets, at least in general outline. To bring his model into better agreement with observation, Ptolemy tried other schemes, such as placing the center of each larger circle off to one side of the earth, Fig. 2–6 (C).

Although Ptolemy's system was rather complicated, it was a successful theory. It was based on accurate observations, and from it Ptolemy was able to predict past and future positions of the planets that agreed remarkably well with observations. Although the agreement with experiment was not perfect, it was sufficiently close that almost no one doubted that Ptolemy's theory was basically correct. Indeed, no essential changes were made until Copernicus conceived of a sun-centered, or **heliocentric**, view of the universe 1300 years later.

But we should not leave our discussion of the ancients with the idea that every thinker accepted the geocentric view of the world. A number of ancient philosophers, notably Aristarchus (c. 310–230 B.C.), hypothesized a sun-centered system in which the earth was just another planet orbiting the sun between Venus and Mars. It was a remarkably modern view, but it did not fit in with the prevailing philosophic views of the time. Most thinkers felt it unreasonable to consider the earth anything but the center of the universe. Aristarchus' heliocentric theory was thus not widely accepted and consequently had little impact on the future development of science.

It is sometimes said, or implied, that the Greeks lacked observational power or were somehow blind or ignorant because they believed in an earth-centered universe. Such a harsh judgment lacks historical perspective and overlooks the natural development of ideas. Our contemporary view of the universe did not suddenly occur. It developed, one step at a time, one idea on top of another, with the theories of the ancients as a base. We must also realize that experimental evidence did not indicate the superiority of a heliocentric system, such as Aristarchus', over the geocentric system. Nor was Aristarchus' theory any less complicated than the prevailing system. The ancients were not stupid; they lived in a different age.

The Renaissance: Copernicus Conceives of a Heliocentric System

Nicolaus Copernicus (1473–1543), a Pole, started the revolution that changed our conception of the universe. In his book, *De Revolutionibus Orbium Caelestium* ("On the Revolution of the Heavenly Spheres"), published just before his death, Copernicus revealed his controversial system.

According to Copernicus, the earth is one of the planets, all of which revolve about the sun, Fig. 2–7. It takes the earth exactly one year to make a complete revolution around the sun and return to the same position. Furthermore, the earth rotates about its own axis once a day and this motion gives rise to the apparent daily rotation of the sun, the moon, and the stars. To account for the monthly motion of the moon among the stars, Copernicus proposed that the moon itself revolves around the earth.

Copernicus was not the first to conceive of a heliocentric system, however—there were others before him, as far back as the ancient Greeks, as we have seen. The idea had again begun to flower in Italy during the fifteenth century. The Italy of the late Renaissance was alive with new thought. It was the time of Leonardo da Vinci, Michelangelo, and Columbus. In Italy Copernicus learned of the idea that perhaps the earth moves. Although the heliocentric idea had been discussed by thinkers for a century, it was Copernicus who built it into a comprehensive system.

Many other ideas of Aristotle were questioned at this time. But Copernicus challenged only one, the concept of the earth-centered universe. The fact that Copernicus placed his heliocentric idea amidst medieval Aristotelian ideas does not diminish his greatness. But it does illustrate that the evolution of ideas, like biological ev-

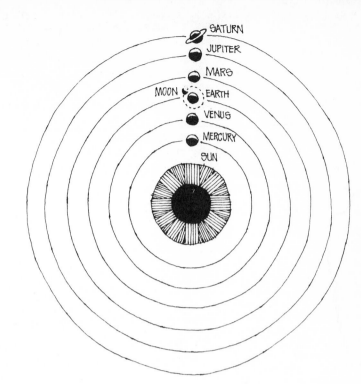

Figure 2–7

Sun-centered universe proposed by Copernicus, which, with the addition of the planets Uranus, Neptune, and Pluto, is how we view the solar system today.

olution, occurs one step at a time. Only later would the other Aristotelian ideas be replaced, one at a time, by the work of Kepler, Galileo, and Newton. Copernicus, however, initiated this revolution. His system remains a landmark in the history of science.

Because he was still predominantly an Aristotelian, Copernicus preserved the requirement that celestial bodies must move in perfect circles. But a single circle was not adequate to describe each planet's orbit around the sun. Consequently, Copernicus' system, like Ptolemy's, used some 50 circles to describe the motion of the planets. Copernicus' system was as complicated as Ptolemy's.

But, did Copernicus' system predict the motion of the planets better than Ptolemy's? By the sixteenth century, serious deviations from Ptolemy's predictions had been observed. Unforrunately, over a period of time, Copernicus' theory by itself did little better. With the increased accuracy of measurement then available, Copernicus' orbit of Mars in particular differed from that observed.

The Copernican theory did not have a great deal to recommend it over the Ptolemaic theory, except for two simplifying elements. First, the calendar year—which refers to the recurrence of the seasons and the return of the sun to the same position among the fixed stars after 365 days—was more simply explained by the Coperni-

23

can theory. In Copernicus' system, the year is simply the time the earth takes to make one revolution around the sun.

Second, the puzzling retrograde motion of the planets disappeared in the Copernican system. The planets only *appeared* to be going backward. This would occur when the earth and the planet in question were passing close to one another, and their relative speeds would give the impression of backward motion. (A similar impression occurs when traveling in a car or train as you pass a row of trees—the trees seem to move backward.)

The Copernican system thus answered one of the most difficult questions of astronomy—why the planets should appear to turn around and go backward. Nonetheless, opposition to the Copernican system was widespread. Objections, particularly in northern Europe, centered on the fact that it violated contemporary philosophic thought and on the "obvious absurdity" that the earth moves about a fixed sun. A century before Galileo was to be confronted by the Catholic church, Copernicus' theory was soundly denounced by John Calvin and Martin Luther.

The controversy over whether the earth moved was not unique in the history of science. Science, like art and poetry or politics, is a human endeavor; and if it is to grow, controversy is inevitable. If science were simply a collection of facts, it would be a dull and static enterprise.

Kepler Uses Tycho's Data to Establish an Accurate Heliocentric System

During the controversy, a great astronomer, Tycho Brahe (1546–1601), was born in Denmark. He was one of the first scientists to realize the power of precise observation. Tycho constructed a large device called a quadrant (Fig. 2–8) with which to view the sky. Although the telescope had not yet been invented, Tycho's instruments were so precise that he could measure the angular position of a planet or star with an accuracy of 1/100 of a degree. For over 20 years he painstakingly plotted the paths of the planets from his observatory near Elsinore Castle, not far from Copenhagen.

But it was an assistant of Tycho's—Johannes Kepler (1571–1630)—who at last brought order out of chaos. Born in southwest Germany near the Black Forest, Kepler was a bit of a mystic. He was deeply interested in the motion of the heavenly bodies and was a strong supporter of Copernicus. He believed that Tycho was the one person whose observations might hold the key to confirming the Copernican hypothesis.

Kepler spent years examining Tycho's data, and from them he was able to improve upon the Copernican system. Kepler wrote a

Figure 2–8

Tycho Brahe and his quadrant.

Figure 2–9

(A) The sum of the distances from a point on an ellipse to one focus plus the distance to the other focus is the same for each point on the ellipse. An ellipse can be made by choosing any two points as foci and tacking two ends of a piece of a loose string at these two points; a pencil held taut against the string and moved under this constraint will outline the shape of an ellipse. (B) A circle is a special kind of ellipse in which the two foci are at the same point.

number of works dealing with his investigations, among which were *A New Astronomy* and *Harmony of the World,* published in 1609 and 1618, respectively. Tucked away among these writings were three findings now famous as Kepler's laws.

Kepler's first law deals with the shape of the planetary orbits. Like other thinkers of his day, Kepler did not doubt that the only proper curve for a planetary orbit was a circle. He tried to fit the orbits of the planets, as measured by Tycho, to a combination of circles and epicycles. He focused his attention on Mars, the planet that deviated most from a perfect circle.

Because all planetary observations are made from the earth and the Copernican theory regards the planets as moving about the sun, a painstaking translation of the observational data was necessary. Kepler had to take each of Tycho's measurements of Mars and determine its position with respect to the sun. For each measurement he had to do a new calculation until he had a complete picture of the planet's orbit around the sun.

Once he had done this, Kepler tried to combine various-sized circles to reproduce on paper the actual orbit of Mars. He tried and tried to fit the data to a combination of circles, but gave up in despair. At this point Kepler reluctantly made a break with the ancient dogma of perfect circles. He decided to see if the actual shape of Mars' orbit might fit some other simple curve. And sure enough, Mars' orbit fit the shape of an ellipse—a particular kind of oval shape, well known to mathematicians, Fig. 2–9. The other planets, he found, fitted elliptical orbits as well. This became the tradition-

25

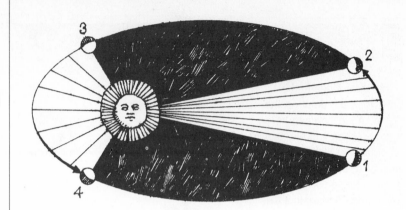

Figure 2–10

Illustration of Kepler's second law. The two unshaded regions have equal areas. It takes the planet the same time to go from point 1 to point 2 as it does to go from point 3 to point 4.

Figure 2–11

The earth moves fastest in its orbit when it is closest to the sun and slowest when farthest from the sun.

breaking hypothesis now known as Kepler's first law: *the paths of planets around the sun are ellipses*.

The ancient rule of the circle had come to an end. Kepler's elliptical orbits were far simpler than the complicated epicycles hypothesized by Copernicus and Ptolemy. The heliocentric theory now possessed a considerable advantage over the geocentric theory.

If a theory is to be viable, one must be able to make accurate predictions based upon it. Kepler's second and third laws dealt with the quantitative aspects of the planetary orbits. In order to predict the past or future motion of a planet in its orbit, it is necessary to find a relationship between the position of the planet and its speed. Kepler sought to determine if the speed was constant on the elliptic orbits, and if not, did a simple relationship exist between speed and position. He did indeed find a simple relationship and stated it quantitatively as follows: *each planet moves so that an imaginary line drawn from the sun to the planet sweeps out equal areas in equal times*. This, the second of Kepler's laws, is illustrated in Fig. 2–10. Simply stated, the law says that when a planet is close to the sun it moves faster than when it is far from the sun. For example, when the earth is at its closest point to the sun, 146,000,000 km (91,000,000 mi) away, its speed is 30.2 km/sec; when it is farthest from the sun, 152,000,000 km (95,000,000 mi), its speed is 29.3 km/sec (Fig. 2–11).

In his search for order in the universe, Kepler brought to light a third law that quantitatively specifies how the period of the different planets increases with their increased distance from the sun. (The period is the time required for a complete revolution about the sun.) Kepler stated his third law very precisely in the following way: *the ratio of the square of the periods of any two planets is equal to the ratio of the cubes of their average distances from the sun*.

Kepler's view of the solar system, as embodied in his three laws, is essentially the view we hold today. Kepler's view is both a simple and an elegant one. The laws that carry his name are significant, too, because they link astronomy to physics rather than theology.

Figure 2–12
Galileo Galilei (1564–1642).

However, Kepler's laws lacked some sort of "reason" or unifying principle. It was soon to be supplied by Newton, as we shall see in Chapter 3. Although Newton was to put the finishing touches on the work begun by Copernicus, Kepler had greatly simplified the heliocentric view of the world. Yet the controversy between the heliocentric and geocentric systems was not over. We should point out, however, that today the controversy is not really an argument about which system is right and which is wrong, although it was probably considered so in the sixteenth and seventeenth centuries.[1] Today we realize that the choice is one of ease in visualization, or of elegance and simplicity. Modern relativity, which we shall discuss in Chapter 6, has shown us that it is as legitimate to say that the sun goes around the earth as it is to say that the earth goes around the sun. It all depends on your point of view. "Everything moves," as Galileo was to say. To describe the orbits of the sun and planets as they go around the *earth* is very complicated. But to describe the motion of the earth and the other planets going around the sun is very simple.

Galileo Supports Copernicus

The story of our view of the universe and how it developed would not be complete without mention of Galileo Galilei (1564–1642) (Fig. 2–12) and his confrontation with the Church. Although Galileo's greatest contributions to scientific thought were not directly related to the motion of celestial bodies, his role in the defense of the Copernican system is important in the history of human thought.

Galileo was the son of a nobleman of low rank who was, however, highly cultured and a composer of merit (his works are still played occasionally). Galileo himself grew up with a considerable contempt for authority and was considered a radical by his colleagues. He particularly scorned the old-fashioned Aristotelian professors who were still common in the universities.

Galileo was among the first to use the telescope and made several important observations. He saw, among other things, that there are many more stars in the sky than are visible to the naked eye, and that the Milky Way itself is made up of a huge number of closely spaced stars. He observed that four moons orbit Jupiter and that the planet Venus exhibits phases just as the moon does. He found that the moon, contrary to prevailing opinion, is not a perfect sphere but is irregular and contains mountains and valleys like the earth. And finally he observed that the sun has spots. These last three observations led to the downfall of the Aristotelian doctrine that the heavenly bodies were perfect spheres, and only the sublunary

[1]Even today it is commonly taught that the heliocentric system is right and the geocentric wrong.

sphere, including the earth, was imperfect. Galileo had shown that the earth and the heavenly bodies were not of a fundamentally different nature. The overthrow of this ancient Aristotelian dogma was a significant achievement.

Galileo felt that his observations of the phases of Venus and the moons of Jupiter—which, he noted, resemble a miniature solar system—were strong evidence for the heliocentric system.

Galileo was a firm supporter of the Copernican system, and it was here that he got into trouble. At first, his teaching of the Copernican theory only aroused the anger of those of his fellow professors who maintained old-fashioned ideas. The Church did not strongly oppose him on the Copernican system as long as he treated it as a hypothesis and not as established fact. In fact, Galileo was encouraged in his studies by a number of cardinals, one of whom later became pope. But in his zeal to discredit the old-fashioned thought of some of his contemporaries, he inevitably stepped on the toes of the clergy. A good case has been made[2] that Galileo's confrontation with the Church was one of personality conflict rather than one of content.

In 1632, Galileo published his great defense of the Copernican system, *Dialogue Concerning the Two Great World Systems*. The dialogue was between three characters, one of whom was portrayed as a simple-minded defender of Ptolemy and Aristotle and was an obvious caricature of the pope himself. Galileo, by now nearly 70, was immediately summoned before the Inquisition and decided to back down. We can be thankful, for soon after the trial he began a new book, *Dialogues Concerning Two New Sciences*, which summarized his detailed findings about the motion of objects on the earth. This book contained Galileo's greatest contributions to science and was important in our present appraisal of Galileo as the "father of modern science." We will discuss these contributions in Section 3 of this chapter.

It is interesting to note that besides his great contributions to science, Galileo also initiated a new style of writing. In his books he used a straightforward style that was much more readable than the long-winded writings of his contemporaries and predecessors.

2. DESCRIPTION OF MOTION

Before looking at Galileo's study of motion, we will first discuss how motion is described.

[2]See Arthur Koestler, *The Sleepwalkers* (New York: Macmillan, 1968).

Speed Relates Distance and Time

When an object is moving from one place to another, we say it is in motion. If it stops moving, we say it is at rest. The *distance* that an object moves is clearly an important thing to know when describing its motion. Equally important is the *time* it takes to move that distance. Distance and time are related to another important property of motion, **speed**. Speed refers to how fast an object is moving; it is the *rate* at which distance is covered. The average speed of an object is defined as the distance the object travels divided by the time it takes to travel this distance:

$$\text{speed} = \frac{\text{distance}}{\text{time}}.$$

If a car is driven a distance of 50 miles in one hour, its average speed is then 50 miles per hour.[3] However, saying that a car is moving at a speed of 50 miles per hour (50 mi/hr) does not imply that the car will necessarily travel for a whole hour. For example, a car that travels 25 miles in a half hour also travels at an average speed of 50 mi/hr because speed = (25 mi)/(½ hr) = 50 mi/hr.

The definition of speed as the distance traveled divided by the time required is the definition of **average speed. Instantaneous speed**, on the other hand, is the speed an object has at any instant of time[4] (this is what the speedometer of a car indicates). The instantaneous speed and the average speed may be quite different. If you were to drive your car 80 kilometers (about 50 miles) in one hour, the average speed would indeed be 80 km/hr. But during that time the instantaneous speed, as registered by the speedometer, may have varied from 70 to 90 km/hr, depending on the traffic.

If a car moves at the same speed for a period of time, with no speeding up and slowing down, we say it has a *constant speed*. In this case the average speed and the instantaneous speed are the same, and either is referred to simply as the speed.

If a car moving with an average speed of 80 km/hr maintains this speed for one hour, it will go a distance of 80 kilometers. If the speed of 80 km/hr is maintained for a half hour, how far will the car have gone? The answer, of course, is 40 kilometers. Thus, if we know the average speed of an object and how long the object traveled at that speed, we can figure out the distance it traveled:

$$\text{distance} = \text{average speed} \times \text{time}.$$

[3]Notice that "per" means "divided by."

[4]Perhaps a better definition of instantaneous speed would be the average speed taken over a very short time interval—an interval so short that the speed can be considered constant for that short time.

This relation is the inverse of the one defining speed. In the example of a car traveling at 80 km/hr for a half hour, this relation tells us that distance = 80 km/hr × ½ hr = 40 km. This result was easily figured out in your head, without using the above relation. We use relations such as the one above to keep the concepts straight in our minds and to help avoid mistakes in more complicated situations.

Relationships Between Quantities Can Be Abbreviated

Often it is convenient to write relationships, such as the ones described above that relate speed, distance, and time, in an abbreviated form. For example, instead of writing out "distance," we simply use its first letter, d; for time we use t; and for speed we use v. (v is short for velocity, which is often taken as a synonym for speed. A distinction between these two words can be made, as we shall see shortly.) Then the definition of speed can be written in abbreviated form as $\bar{v} = \dfrac{d}{t}$. Actually, this relation refers to the average speed, which we shall denote by $\bar{v}$ (the bar over the v means "average"). We can therefore write

$$\bar{v} = \frac{d}{t}.$$

This equation can be used to solve for the speed just as the word equation was earlier. Suppose, for example, that we want to calculate the average speed of a car that travels 300 km in 4 hours. Then the distance d = 300 km, the time t = 4 hr, and $\bar{v} = \dfrac{d}{t} =$ (300 km)/(4 hr) = 75 km/hr. Thus the letters in this equation can be considered symbols for the quantities speed, distance, and time. This equation can be solved (algebraically) for the distance d by multiplying both sides by the same quantity t: $\bar{v} \times t = \dfrac{d}{t} \times t$.

The time t appears in both the numerator and the denominator on the right-hand side and thus cancels out. We get

$$d = \bar{v}\, t,$$

which is the relationship we had before: distance = average speed × time. As an example, let's calculate how far a jet plane goes in 3.5 hours if its average speed is 800 km/hr. The distance $d = \bar{v}t$ = (800 km/hr) (3.5 hr) = 2800 km.[5]

[5]Notice in this example that we must always be consistent in the units we use. That is, if the time is given in hours, then before any calculation is done the speed,

Table 2–1
Equivalent Speeds in Different Units

20 km/hr	=	5.6 m/sec	=	12 mi/hr	=	18 ft/sec
40 km/hr	=	11 m/sec	=	25 mi/hr	=	37 ft/sec
60 km/hr	=	17 m/sec	=	37 mi/hr	=	55 ft/sec
80 km/hr	=	22 m/sec	=	50 mi/hr	=	73 ft/sec
100 km/hr	=	28 m/sec	=	62 mi/hr	=	92 ft/sec

If you are not comfortable using symbols, you can use the words instead. The physics is in the concepts, not in how we write them.

Units of Measurement for Speed

The units most commonly used for speed in everyday life are miles per hour and kilometers per hour, particularly for automobile and train travel. Other common units for speed are feet per second (ft/sec) and meters per second (m/sec), both of which are particularly useful when distances are not great.

Sometimes speed is given in one set of units when another set of units would be more convenient. For example, when figuring out braking distances in automobiles, it is often more useful to know the speed of the automobile in meters per second than in kilometers per hour. Table 2–1 gives comparative speeds in different sets of units. As an example of how these conversions are made, note that a speed of 1 km/hr is equivalent to 1000 m in 1 hr, or 3600 seconds (since there are 60 seconds in a minute and 60 minutes in an hour, and 60 × 60 = 3600). Thus 1 km/hr equals 1000 m/3600 sec equals 0.28 m/sec. A speed of 1 km/hr is therefore equivalent to 0.28 m/sec. Similarly, a speed of 10 km/hr is equivalent to a speed of 2.8 m/sec, and so on. In fact, in Table 2–1 you will see that all speeds in meters per second are 0.28 times the speed in kilometers per hour. The number 0.28 is a "conversion factor." It is used to convert kilometers per hour to meters per second. Conversion factors are also useful in changing miles per hour to kilometers per hour or feet per second.

too, must have hours in it: kilometers per hour or miles per hour, and not, say, feet per second. Also, note how the units of "hour" cancel out. Keeping track of the units in this way helps to avoid mistakes.

EXPERIMENT-PROJECT

Measure the average speed of an ant or another insect or animal. Use an ordinary ruler to measure the distance and a watch with a second hand to measure the time. It will be easier if you find an ant that walks in a straight line.

OR

Check the accuracy of the speedometer of your car. You can do this simply by maintaining a constant speed and measuring the time it takes to go an accurately measured distance. Many freeways and expressways have speedometer check areas in which signs are posted exactly one mile or one kilometer apart. You must use a watch with a second hand so that you can measure the time accurately. To find the speed in miles or kilometers per hour, the time in seconds must be changed to the proper fraction of an hour. Two people are necessary for this experiment: one to drive the car at constant speed, the other to measure the time on the watch. This experiment can be done just as well on a motorcycle or motorbike.

OR

Calculate your maximum average speed while riding a bicycle by measuring the time it takes you to ride a known distance at top speed.

Velocity Has Direction as Well as Magnitude

When specifying the motion of an object, not only the speed but the direction of motion is important. For example, if a friend is leaving from Los Angeles in an airplane traveling at 600 mi/hr, you also want to know where your friend is going, i.e., in what direction. Is he going to San Francisco, the Grand Canyon, Mexico City? North, east, south, or west? Similarly, when you drive your car, sometimes you are driving north, and at other times west, or southeast, or whatever. Thus it is important to specify *direction* as well as magnitude. (Magnitude means the numerical value, such as 50 km/hr.)

We use the term **velocity** to signify both the speed and direction of a moving object. Speed then refers to the magnitude of the velocity; the direction of the velocity is the direction in which the object is moving. When drawing a diagram showing the motion of an object, we often use an arrow to indicate the direction of motion. For example, in Fig. 2–13 a car moving along a curvy road is shown at various positions along the road. At each position an arrow is drawn to show the direction of the car's velocity at that instant.

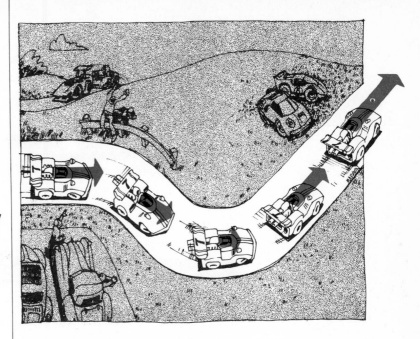

Figure 2–13
Arrows represent the velocity of the automobile at various times as it goes around a curve. Notice that the direction of the velocity changes as the car moves through the turn. The magnitude of the velocity (length of the arrow) also changes.

Notice that constant speed is not the same as constant velocity. For example, an automobile moving at a constant speed of 40 mi/hr as it rounds a curve does not have a constant velocity, because its direction is changing at each instant. An object has a constant velocity only if its speed *and* its direction do not change.

Although the words speed and velocity are often used interchangeably, it is helpful to remember the distinction between them.

Vectors

Quantities such as velocity that have both magnitude and direction are referred to as vector quantities, or **vectors**. Force, as we shall see shortly, is also a vector. Not all quantities are vectors. Time, temperature, and energy have no direction associated with them; they are specified merely by a number and are referred to as **scalars**.

Arrows are often used to represent vectors, as in Fig. 2–13 for the velocity. The arrow is always drawn to point in the direction of the vector it represents. The length of the arrow is proportional to the magnitude of the vector. Thus, in Fig. 2–13 the arrows representing velocity are longer where the speed is greater.

Acceleration

Acceleration is the word used to refer to a rate of change in velocity. Whenever a moving object is changing speed, we say it is accelerating. For example, when an automobile starts from rest and speeds up to, say, 80 km/hr, it is accelerating. If one car can accelerate from rest to 80 km/hr in less time than another, it is said to undergo greater acceleration. Acceleration, then, is defined as the change in velocity divided by the time required to make that change; more specifically, this defines the *average* acceleration. Thus,

$$\text{average acceleration} = \frac{\text{change in velocity}}{\text{time}}$$

Like velocity, acceleration is a rate. Velocity is the rate at which the *position* of an object changes. Acceleration is the rate at which the *velocity* changes. Be careful not to confuse acceleration with velocity. Stepping on the "accelerator" pedal of your car, for example, may or may not cause an acceleration, because maintaining a constant speed (which is zero acceleration) requires you to keep your foot on the pedal.

Let us consider an example. Suppose a car accelerates from 0 to 80 km/hr in 10 sec (Fig. 2–14). Its change in velocity is simply the final velocity minus the initial velocity: 80 km/hr − 0 km/hr = 80 km/hr. Therefore, we can calculate that the average acceleration = $\frac{80\,\text{km/hr} - 0\,\text{km/hr}}{10\,\text{sec}} = \frac{80\,\text{km/hr}}{10\,\text{sec}} = \frac{8\,\text{km/hr}}{\text{sec}}$. The average acceleration is 8 kilometers per hour per second. This means that on the average the velocity changes by 8 km/hr during each second. After one sec-

Figure 2–14

The automobile accelerates from rest up to 80 km/hr in 10 sec.

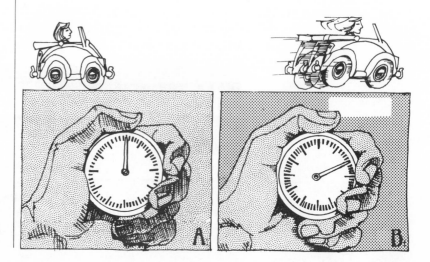

ond the car's velocity will be 8 km/hr, after two seconds it will be 16 km/hr, and so on. This of course will be precisely true only if the acceleration is *uniform*. The *instantaneous* acceleration may be different from the average of (8 km/hr)/(sec) we just calculated. But if the acceleration *is* uniform, the instantaneous acceleration during the 8 seconds will be the same as the average acceleration.

Notice that there are two "pers" and two time units associated with acceleration, because acceleration is the "rate of a rate." If we change kilometers per hour to meters per second, then the acceleration becomes: acceleration = 80 km/hr = 22 m/sec. That is, 2.2 meters per second per second. This result is more simply written 2.2 m/sec², which is read as "2.2 meters per second squared."

The definition of average acceleration can also be written as

$$\text{average acceleration} = \frac{\text{final velocity} - \text{initial velocity}}{\text{time}}.$$

If the initial velocity is zero, as it often is, then this relation reduces to: acceleration $= \dfrac{\text{final velocity}}{\text{time}}$. In symbols, $a = \dfrac{v_f}{t}$, where a stands for average acceleration, t for time, and v_f for the instantaneous final velocity at the end of the time t. This relation can be turned around to yield: final velocity = acceleration × time, or in symbols,

$$v_f = at.$$

For example, if a bicyclist accelerates at 1 m/sec² starting from rest, after 5 seconds the speed will be (1 m/sec²) × (5 sec) = 5 m/sec.

When an object slows down (for example, when the brakes are applied on an automobile), its velocity decreases. This is sometimes called "deceleration" and is merely another example of acceleration. In this case, the final velocity is less than the initial velocity, so the acceleration is negative. Deceleration is thus negative acceleration.

EXPERIMENT-PROJECT

With the help of a friend who has a watch with a second hand, measure the time it takes for your car to accelerate from rest to 50 km/hr or 30 mi/hr, and then calculate the average acceleration.

OR

Calculate your acceleration on your bicycle, in a similar way, as you accelerate from rest to top speed (which you measured in an earlier experiment).

So far we have considered only cases in which the *magnitude* of the velocity changes. If the speed remains constant and the *direction* of the velocity changes, this constitutes an acceleration as well. For example, a child riding on a merry-go-round or a person riding in a car rounding a curve at high speed is aware of an acceleration because the direction of the velocity is changing.

It is an interesting physiological phenomenon that the semicircular canals that lie behind our ears are quite sensitive to accelerations. When these canals are affected, the brain sends messages to the stomach, which then feels tickled or nauseous when excess acceleration occurs. The same strange feeling occurs on a rapidly rotating merry-go-round or barrel-of-fun, or when an elevator starts or slows (accelerates) too rapidly. Similarly, in an automobile we lurch forward or backward when it accelerates or decelerates rapidly, and we lurch to the side when going around a curve at high speed. We lurch because of the acceleration. Clearly, then, acceleration results when either the magnitude or the direction of the velocity, or both, changes. We will discuss acceleration due to change in the direction of velocity in more detail in Chapter 3.

Force

Force is not, strictly speaking, a quantity that describes motion. The motion of a body is completely specified by its *position, velocity,* and *acceleration*. But we know intuitively that force has something to do with motion. It seems to give rise to motion.

What do we mean by "force"? Usually we mean a push or a pull. For example, a person pushing a grocery cart exerts a force on it. When you turn on a water faucet, or a horse pulls a cart, or an automobile pulls a trailer, a force is being exerted. When a hammer hits a nail, the hammer exerts a force on the nail. When the wind blows against a leaf, it is air pushing on the leaf, exerting a force on it. Another example is the "force of gravity," which pulls objects toward the ground. We will later encounter other examples of forces, such as those due to electricity and magnetism.

One way you can tell a force is acting is that a force will change the shape of an object, at least a little. When you exert a force on a balloon, it compresses—it changes shape. The same thing happens if you push on a mattress or a piece of soft wood. Careful measurement shows that even hard materials such as steel are compressed when pushed on. In summary, we can tell that a force is acting on an object if the object is set into motion or its shape is distorted. In Chapter 3 we will discuss the notion of force in more detail and give it a precise definition.

36

Force, like velocity, is a vector quantity. Force has both magnitude and direction. The magnitude is the strength of the force, and the direction is the direction in which the force is exerted. The direction of a force is as important as its magnitude. For example, if you exert a sufficient force sideways on this book, it moves sideways. But if you exert the force upward, the book moves up, Fig. 2–15. When we draw a force vector, we conventionally put the tail of the vector, not the arrowhead, on the object that feels the force.

Figure 2–15

The book moves in the direction of the force applied to it.

3. THE MOTION OF OBJECTS ON EARTH: GALILEO

Among the earliest studies of terrestrial motion were those made by the ancient Greeks, in particular, Aristotle. Although these early studies showed insight, many of the conclusions were based on what was thought to be common sense or at least obvious from everyday experience.

It was Galileo, born in the year that Michelangelo died and Shakespeare was born, the same Galileo who got into trouble with the Church, who clearly established the modern scientific method. Galileo advocated the radical idea that no theory or model of nature was meaningful unless it predicted results in accord with experiments. Galileo felt that one must examine natural phenomena with great care. In fact, many phenomena are so complex that little can be understood from a simple examination of them. Therefore, Galileo reasoned, the simplest situations must be examined first. He sought the simplest examples of natural phenomena and set up simple experimental arrangements to examine them. For analyzing observational results, Galileo introduced an important new tool— that of *idealizing* a situation, such as imagining a perfect, frictionless plane. This theoretical method of idealizing became a cornerstone of science, and its use has allowed otherwise complex phenomena to be analyzed successfully.

To the early investigators there existed two main problems relating to the motion of bodies on earth: the motion of bodies along the ground or on a table top—that is, **horizontal motion**; and the motion of falling bodies—**vertical motion**.

Horizontal Motion

Because physics is the study of natural phenomena—"how nature behaves"—it is usually not worthwhile to simply speculate on how nature behaves. Rather we should examine carefully how nature *does* behave. Therefore, you should perform a few simple experiments as we go along. You have already met other experiments. Here is another that you should perform, *right now*.

37

EXPERIMENT

Take an object, such as a book, and push it across the floor or top of a table. Clearly you must exert a force to make it move. When you stop pushing, you will probably notice that the object either stops immediately or slows down and comes to rest after a short distance. Now try pushing other objects across the floor or table. Use objects with rough surfaces and objects with smooth surfaces. Notice the differences in the amount of force needed to move the objects.

The ancients did such experiments and concluded that the natural state of a body is at rest. According to Aristotle, a body in motion was in a sense "unnatural." A force was always necessary to keep an object in motion. Furthermore, the stronger the force on a body, Aristotle argued, the greater the speed of the body. Some 2000 years later, however, Galileo was skeptical of the Aristotelian view of nature. So he did his own careful experiments. In each of them he made careful measurements of the position, velocity, and acceleration of the bodies in motion. Galileo's greatness rests partly in his ability to separate the relevant from the irrelevant and to get to the heart of a problem. Let us examine the motion of a body along a horizontal surface in a way that Galileo might have.

It should have been evident from the experiment you just did that it requires less force to push a body with a smooth surface across a table than to push an equally heavy rough body. If a lubricant such as oil is between the object and the table, almost no force is required to move the object, and it will continue to move for a considerable distance even though no force is being applied to it. As a continuation of the above experiment, try pushing a moist bar of soap across a wet, level surface. When there is a lubricant, the object eventually comes to rest but only after traveling a considerable distance. By comparing the force necessary to move a rough object across the table, then a smooth object, and finally an object with a lubricant between it and the table, perhaps you will come to the same conclusion as Galileo. He "extrapolated" from his data and theorized that if an object did not rub against the table top at all—or if there were a layer of "perfect lubricant" between them—then, once started, the object would move across the table at constant speed *without any force being applied.* This was a great leap of the imagination on Galileo's part. After all, a perfect lubricant does not exist, so friction between two surfaces cannot be entirely eliminated. For Galileo to imagine what would happen if friction were eliminated required considerable intuition and imagination.

Figure 2–16

F is the force applied by the person, and *F*fr is the force of friction.

In contrast to previous beliefs, Galileo argued that it is just as "natural" for a body to move with constant speed as it is to be at rest. And once a body is moving, it is slowed only if a push or a pull is applied. Galileo interpreted the friction between the moving object and the table as a "resistance," which is similar to ordinary pushes and pulls (the concept of force had yet to be invented—by Newton). Thus, in today's language, friction is interpreted as a force. And in order to push an object across a table, a force is necessary only to balance the friction force that exists between the two surfaces as they pass over each other. When the object is pushed at constant speed, the pushing force exactly equals the friction force and is opposite in direction, Fig. 2–16. The *net* force on the object is, in this case, zero.

To summarize, Galileo concluded that an object naturally either stays at rest or moves with unchanging velocity unless an external force is applied to it. We still accept this important Galilean principle today. It played a crucial role in the further development of mechanics by Newton. It is interesting to note, however, that the full ramifications of this principle were not made clear until Einstein postulated his theory of relativity in the early twentieth century.

Friction

Friction can be a considerable hindrance: it slows down moving objects and causes heating and binding of the moving parts of engines. On the other hand, friction can be very helpful, too. Have you ever tried to walk on slippery ice or seen a car skid on ice?

When friction is a hindrance, lubricants such as oil are used. Even with the best lubricating oils some friction exists. Although an oil lubricant reduces friction to a very small amount, even more effective methods have been developed, but they are not practical for all situations. One of the best ways to reduce friction between two surfaces is to maintain a thickness of air between them. So-called air tracks and air tables use this means of reducing friction to practically zero (Fig. 2–17). Air is forced by a pump through tiny holes so that the moving objects float on a layer of air. Thus friction is reduced practically to zero and Galileo's principle becomes very real.

Falling Bodies

Now for the second problem—vertical motion. We are all familiar with the fact that if we hold an object above the ground and let it go, it falls downward. We say that it falls because of gravity. The gravitational force, or gravitational pull, of the earth pulls the object downward.

Figure 2–17

An air table.

This well-known fact leads to many questions that we can try to answer. What is the nature of the motion of falling bodies? How fast do bodies fall? Do they undergo acceleration? Do some bodies fall faster than others?

It is commonly thought that heavy objects fall faster than light objects. Aristotle held as a basic principle that bodies fall with speeds proportional to their weight. Let us test this principle.

EXPERIMENT

(Do it now!) Take in one hand a reasonably heavy object such as a baseball, a stone, or an eraser; in your other hand take a flat piece of paper, holding it horizontally. Hold the two objects at equal heights above the floor and let them go at the same time, Fig. 2–18 (A). Which reaches the floor first? This experiment seems to be a clear confirmation of Aristotle's principle that lighter objects fall more slowly than heavier objects.

Now modify the experiment. Take the same heavy object and the same piece of paper but this time crumple the sheet of paper into a small wad. Hold the heavy object and the wad of paper at equal heights and let go, Fig. 2–18 (B). What do you observe in this case?

What can you conclude from this experiment? Apparently the speed of a falling body does not depend on its weight. The paper, whose weight is the same whether it is flat or crumpled into a wad, falls faster when it is wadded up; that is, it falls faster when it has a smaller cross-sectional area.

Figure 2–18

(A) A heavy object and a piece of paper are dropped simultaneously. (B) The same heavy object and the same piece of paper, but this time crumpled into a wad, once again dropped simultaneously.

40

Figure 2-19
The Tower of Pisa.

Again it was Galileo who broke with tradition. (Although he performed many experiments, it is interesting that historians are not convinced that he ever used the Leaning Tower of Pisa, Fig. 2-19, as is so often reported, to measure the times of fall for objects of various weight.) Galileo concluded that *in the absence of air, all bodies fall at equal rates*.

It is air resistance, a kind of friction, that slows down the flat piece of paper. If the flat piece of paper had been dropped where there was no air, in a vacuum, it would have dropped as fast as the heavy object, Fig. 2-20. Apparently, air resistance is only significant when a body has a large cross section compared to its weight.

Galileo was able to overturn another of Aristotle's conclusions— that a falling body acquires its speed immediately after it is dropped and maintains that speed throughout its fall. Galileo showed that a falling body is constantly increasing in speed, and the increase occurs at a constant rate. In other words, bodies fall with a *constant acceleration*. Galileo's discoveries on the motion of falling bodies can be summarized in the following sentence:

In the absence of air resistance, all bodies fall with the same constant acceleration.

This conclusion was clearly at variance with previous beliefs. Even today people who have not studied physics might be more likely to accept Aristotle's view. Doubters can be shown the experiment of Fig. 2-18. The fact that objects increase in velocity as they fall was made graphic by Galileo when he pointed out that a block allowed to drop from a height of ten feet will drive a stake much farther into the ground than the same block dropped from a height of one inch. Clearly, the block must be going faster in the former case.

It was stated above that "in the absence of air resistance, all bodies fall with constant acceleration." What about air resistance? For most practical cases, it has a negligible effect. Only in the special case of a very light object with a large cross section (such as

Figure 2-20
A feather and a rock drop at the same rate in a vacuum.

41

A. AIR-FILLED TUBE

B. EVACUATED TUBE

a feather or a piece of paper) or in the case of an object that falls a great distance (such as a skydiver) does air resistance retard an object significantly. Other than those exceptional cases, objects fall with the same acceleration. This acceleration is known as the **acceleration due to gravity** (on earth) and is given the symbol g. The acceleration due to gravity has been measured to be

$$g = 9.8 \text{ m/sec}^2$$
$$= 32 \text{ ft/sec}^2.$$

Some Simple Examples

For those who wish to get a feel for using the results of this chapter in a quantitative way, we now consider some simple numerical examples.

Suppose that an object is dropped from the top of a high tower, Fig. 2–21. Let us calculate the speed of the object after it has been falling for one second and after it has been falling for two seconds. We have learned that when an object starts from rest, the velocity after a given time equals the acceleration times the time, or in symbols, $v_f = at$. In the present case, the acceleration is simply the acceleration due to gravity, 9.8 m/sec² or 32 ft/sec². Let us use the metric system. Then after one second the speed will be $v_f = at =$ 9.8 m/sec² × 1 sec = 9.8 m/sec. After two seconds the speed will

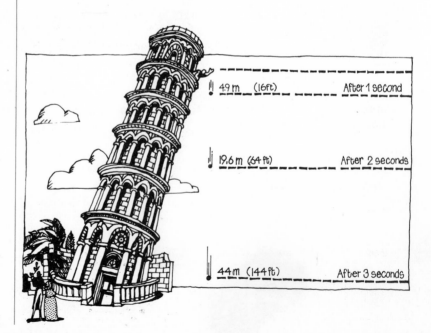

4.9 m (16 ft) After 1 second

19.6 m (64 ft) After 2 seconds

44 m (144 ft) After 3 seconds

Figure 2–21

An object dropped from the top of a tower falls with increasing speed and covers greater distance during each succeeding second.

be $v_f = 9.8$ m/sec² $\times$ 2 sec $= 20$ m/sec (we round off to two places). Thus the object is going 9.8 m/sec after one second and 20 m/sec after two seconds.

Let us now calculate the *distance* the object will have fallen after one second and after two seconds. We have learned that distance equals average velocity times time, or $d = \bar{v}t$. First, we have to figure out the average velocity. Because the velocity increases at a constant rate, the average velocity is simply the average of the initial and final velocities. To find the average value of two numbers we merely add the two numbers and divide by 2. Thus the average velocity *during* the first second is $\dfrac{9.8\,\text{m/sec} + 0\,\text{m/sec}}{2} = 4.9$ m/sec.

Therefore, after one second the object will have fallen a distance ($d = \bar{v}t$) of 4.9 m/sec $\times$ 1 sec $= 4.9$ m. To find out how far it will have fallen after two seconds, we must calculate its average velocity during these two seconds, which is $\dfrac{20\,\text{m/sec} + 0\,\text{m/sec}}{2} = 10$ m/sec.

Therefore, the object will have fallen a distance of 10 m/sec $\times$ 2 sec $= 20$ m; this is 15 additional meters since the end of the first second. After three seconds an object will have fallen 44 m, and so on; see Fig 2–21 and Table 2–2. Notice that the distance covered by a falling object *during each second* increases in time. This occurs because the object's speed is continually increasing.

A quick way to determine how far an object has moved after a given time when accelerating is to use a relationship that relates

Table 2–2
Calculation of Distance Traveled by Dropped Object

Elapsed Time	Acceleration (constant)	Speed After This Time (use $v_f = at$)	Average Speed Since Beginning ($\bar{v} = \frac{0 + v_f}{2}$)	Total Distance Traveled ($d = \bar{v}\,t$)
0	9.8 m/sec²	0	$\bar{v} = 0$	$d = 0$
1 sec	9.8 m/sec²	$v_f = 9.8 \times 1 = 9.8$ m/sec	$\bar{v} = \dfrac{0 + 9.8}{2} = 4.9$ m/sec	$d = 4.9 \times 1 = 4.9$ m
2 sec	9.8 m/sec²	$v_f = 9.8 \times 2 = 20$ m/sec	$\bar{v} = \dfrac{0 + 20}{2} = 10$ m/sec	$d = 10 \times 2 = 20$ m
3 sec	9.8 m/sec²	$v_f = 9.8 \times 3 = 29$ m/sec	$\bar{v} = \dfrac{0 + 29}{2} = 14\frac{1}{2}$ m/sec	$d = 14\frac{1}{2} \times 3 = 44$ m

distance, acceleration, and time. This relation was first derived[6] by Galileo:

distance = ½ × acceleration × time squared.

This is valid whenever the body starts from rest. In symbols this can be written $d = \frac{1}{2}at^2$. For falling objects, the acceleration is that due to gravity, 9.8 m/sec² or 32 ft/sec². Thus $d = \frac{1}{2} \times 9.8 \times t^2$, or $d = 4.9t^2$; this gives the distance in meters, and the time must be given in seconds. Try putting $t = 1$ sec, $t = 2$ sec, and $t = 3$ sec into this formula. You should obtain distances of 4.9 m, 20 m, and 44 m, just as before.

Terminal Velocity

In our discussion of falling bodies, we may have been unfair to Aristotle. He claimed that bodies fall with a constant speed in any particular medium and that this speed depends on the mass of the object. We saw that this is not generally the case for ordinary objects falling in air. The media Aristotle used, however, included water and other liquids. We know now that when a body falls through a fluid it increases in speed. However, especially in liquids, a maximum velocity is reached. This velocity, called **terminal velocity**, is attained when the force of gravity is just balanced by the resistive or frictional force[7] of the medium. At this point, the net force becomes zero, and therefore the velocity will be constant after that. The terminal velocity depends on the size and weight of the object and on how viscous the fluid is; that is, on how much of a resistive force it exerts. In viscous liquids even a heavy object will reach terminal velocity fairly quickly. A feather reaches its terminal velocity in air quite rapidly. But most objects falling in air or other gases reach their terminal velocity only after a tremendous distance of fall. A skydiver without a parachute, for example, reaches a terminal velocity of about 120 mi/hr (200 km/hr) after falling about 1000 ft.

[6]This formula is easily derived, and we use symbols to make it briefer. From the definitions of velocity and acceleration we know that $d = \bar{v}t$ and $v_f = at$, where v_f represents the final velocity *after* any particular time, and $\bar{v}$ is the average velocity during that time period. The average velocity, $\bar{v}$, is simply the average of the initial and final velocities, but since the initial velocity is zero (the object starts from rest) the average velocity is $\bar{v} = v_f/2$. Thus $d = \bar{v}t = (v_f/2)\, t$. If we substitute $v_f = at$ into this relation we get

$$d = \frac{v_f}{2}t \quad = \frac{at}{2}t \quad = \frac{1}{2}at^2$$

[7]The resistive force that fluids exert on an object falling through them is quite complicated and is a field of study in itself. We should note here, however, that this force increases as the velocity of the object increases.

For most practical cases an object hits the ground long before the terminal velocity is reached.

Except for his assumption that terminal velocity is reached immediately, Aristotle's observations were not really erroneous. He just didn't probe as deeply as Galileo did. And because of Galileo we understand motion far better.

Was Aristotle Wrong?

The difference between Galileo's and Aristotle's views of motion is not really one of right and wrong. For example, with regard to horizontal motion, Aristotle might have argued that because friction is always present, at least to some degree, it is a natural part of the environment. It is therefore natural that bodies should come to rest when they are no longer being pushed. For Aristotle, a situation in which friction is absent was unrealistic. And with regard to falling bodies, Aristotle and Galileo also found different aspects relevant, as we just saw.

Perhaps the real difference between Aristotle and Galileo lies in the fact that Aristotle's view was almost a final statement; one could go no farther. But the view established by Galileo could be extended to explain many more phenomena. It was a far more useful view, and it made nature more of a whole. Before Galileo, vertical motion seemed to be governed by different laws from those governing horizontal motion. Because of Galileo we see that the two are linked. Indeed, three-dimensional space is symmetric, and the vertical dimension appears different only because of the gravitational force.

As we shall see in Chapter 3, another great scientist, Sir Isaac Newton, extended Galileo's results. Adding his own ideas to Galileo's, Newton synthesized a beautiful and all-encompassing theory of motion and its causes.

REVIEW QUESTIONS

1. The ancient Greeks viewed the earth as the center of the universe with the planets
   (a) fixed on great spheres
   (b) rotating on ferris-wheel-like orbits
   (c) stationary
   (d) wandering at random

2. According to Copernicus, if the sun is stationary, the earth must
   (a) also be stationary
   (b) revolve on its own axis every 365 days
   (c) move around the sun
   (d) move in a straight line

3. Ptolemy is best known for his
   (a) study of falling bodies
   (b) study of horizontal motion
   (c) comprehensive earth-centered view of the universe

4. Which of the following observations did Galileo *not* make:
   (a) moons of Jupiter
   (b) phases of Venus
   (c) sunspots
   (d) discovery of Pluto

5. Galileo is most noted for his
   (a) astronomical observations
   (b) study of falling bodies
   (c) support of heliocentric system
   (d) all of the above
   (e) none of the above

6. The planets travel in orbits around the sun that are
   (a) circular
   (b) parabolic
   (c) elliptical
   (d) unobservable

7. An object moving with a constant speed
   (a) is not accelerating
   (b) might be accelerating
   (c) is always accelerating
   (d) cannot have forces acting on it

8. When an object is accelerated,
   (a) its velocity necessarily is changing
   (b) its velocity is constant
   (c) its speed is necessarily changing
   (d) its direction must be constant

9. Which of the following is not a vector:
   (a) velocity
   (b) force
   (c) temperature
   (d) none of the above

10. The relation $d = \bar{v}t$ means
    (a) distance equals velocity divided by time
    (b) distance equals time divided by velocity
    (c) distance equals final velocity × time
    (d) distance equals average velocity × time

11. Acceleration is the rate of change of
    (a) position
    (b) energy
    (c) velocity
    (d) force

12. A dog jumps from a tall building. Assuming air resistance can be ignored, on the way down
    (a) its acceleration is constant
    (b) its velocity is constant
    (c) its speed is constant
    (d) all of these
    (e) none of these

13. A 4-lb stone and an 8-lb brick are dropped from the top of a three-story building. To reach the ground, the 4-lb stone compared to the 8-lb brick takes

(a) the same time
(b) twice as long
(c) half as long
(d) 50 percent as long

14. Galileo believed that
    (a) the natural state of an object is at rest
    (b) in the absence of forces, a body continues in motion in a straight line at constant speed
    (c) accelerated motion is the natural state of an object
    (d) constant velocity is unattainable

EXERCISES

1. What does it mean when we say that the moon moves eastward with respect to the stars?

2. On the basis of your own observations of the stars and the moon (see the first experiment in this chapter), do you think the geocentric view of the universe developed by the ancient Greeks was reasonable? Why or why not?

3. Have you observed anything that would lead you to believe that the earth travels around the sun?

4. On the basis of Kepler's second law, determine whether the earth is moving around the sun faster in the summer or in the winter. Note: the earth is farther from the sun in summer than in winter.

5. Write the days of the week in one of the Latin lanuguages (Spanish, Italian, French). Do you recognize the names of the planets in them? Explain.

6. When a track star runs the mile in four minutes, what is his average speed in miles per hour?

7. What is the speed of a sprinter who runs the hundred-yard dash (300 ft) in 9.2 seconds? What would be the time for the mile at this pace?

8. At an average speed of 15 km/hr, how far will a bicyclist travel in 4½ hours?

9. What must be the average speed of a sprinter if the 100-m dash is to be done in 9.0 seconds?

10. How far will an automobile travel in 15 minutes if its average speed is 90 km/hr?

11. The distance from New York to Chicago is 1400 km. How long will this trip take if you can average 90 km/hr?

12. The distance from San Francisco to Los Angeles is 380 miles. How fast must you drive to make the trip in 6 hours?

13. What is the speed of the earth around the sun in kilometers per hour and miles per hour? Use the average of the values given in Fig. 2–11.

14. Does a greater speed necessarily imply a greater acceleration? Give examples.

15. Can you assume that a car is not accelerating if its speedometer shows a steady speed of 40 km/hr? Explain.

47

Figure 2–22

16. A car travels around a curve with a speed of 30 km/hr. If it rounds the same curve at 40 km/hr, is its acceleration greater? Why or why not?

17. Will acceleration be the same when a car rounds a *sharp* curve at 50 km/hr as it is when it rounds a *gentle* curve at the same speed? Why or why not?

18. A ball is thrown straight up in the air. What is its velocity when it reaches its highest point? What is its acceleration at this point?

19. Seasickness and airsickness are often referred to as motion sickness. Is it motion per se that makes people sick, or is it a particular aspect of motion? What might be a more appropriate term?

20. Compare the acceleration of a car that speeds up from 60 to 70 km/hr in 5 seconds with a cyclist who accelerates from rest to 10 km/hr also in 5 seconds.

21. What is the acceleration of a car that goes on a straight road from rest to 80 km/hr in 8 seconds?

22. What is the acceleration of a bicycle that starts from rest and reaches 20 ft/sec after 5 seconds?

23. What is the acceleration of a car that increases its speed from 60 km/hr to 90 km/hr in 6 seconds?

24. A horse accelerates at a rate of 3.0 m/sec^2. What will be its speed after 2.5 seconds if it starts from rest? How far will it have traveled?

25. What can you infer is happening in Fig. 2–22 (B) that is not happening in Fig. 2–22 (A)?

26. Can an object keep moving without being pushed?

27. Is it possible to test directly Galileo's assertion that a body in motion will continue in motion with unchanging speed as long as no external force is acting on it?

28. How does your answer in the above question affect the validity of Galileo's assertion?

29. An object is dropped from a high tower. Calculate (a) its speed in feet per second and (b) how far it has fallen (in feet) after 1, 2, 3, and 4 seconds. Use $g = 32$ ft/sec^2.

30. How many meters will an object fall in the first 10 seconds after being dropped from an airplane?

31. For how long does an object dropped from a high tower have to fall before it reaches a speed of 90 km/hr (25 m/sec)?

3

MOTION: NEWTON'S SYNTHESIS

Figure 3–1

Isaac Newton (1642–1727).

Sir Isaac Newton (Fig. 3–1) is generally recognized as having been one of the great thinkers of the Western world. Born on Christmas Day, 1642, he entered Trinity College at Cambridge in 1661 and there became interested in the problem of motion. The college was closed in 1665, however, because of an outbreak of the Plague. Newton returned to the farm on which he had been raised and there, between the ages of 23 and 24, he worked out the ideas that would bring him lasting fame. He formulated the laws now known as Newton's three laws of motion, which deal with motion and force in general. It is believed that during this time he developed his theory of gravitation; but it was not published until 20 years later in his celebrated treatise, the *Principia*.[1]

Unlike Galileo, who thrived on controversy, Newton was shy and retiring. He was aware of the inevitable criticism that falls on new and creative ideas and sought to avoid any kind of controversy. Perhaps this is why he waited 20 years before presenting his unifying theory of gravitation to the world—and only then at the vigorous urging of his friends. In spite of his fears, Newton's work was accepted comparatively rapidly.

The revolution begun by Copernicus and Galileo was brought to completion by Newton. Not only did he further the idea of the unity of horizontal and vertical motion of objects on the earth, but through his three laws of motion and the law of gravitation, Newton made it clear that terrestrial and celestial motion could be considered basically the same. The word **synthesis** in the title of this chapter means "bringing together." All motion, whether on earth or in the heavens, was seen to be described by the same natural laws.

Figure 3–2

An object with a large inertia is (A) hard to start moving, (B) hard to stop moving once it is in motion, and (C) hard to shift from a straight-line path by pushing sideways on it.

1. NEWTON'S LAWS OF MOTION

Newton's First Law: The Law of Inertia

Newton's synthesis of the phenomena of motion began with the formulation of his three laws of motion. The first of these laws is in part a restatement of Galileo's principle, though in broader terms:

Every body continues in its state of rest or of uniform motion in a straight line unless it is compelled to change that state by forces impressed upon it.

[1]The *Philosophiae Naturalis Principia Mathematica* (Mathematical Principles of Natural Philosophy), published in 1687, contains nearly all of Newton's work on motion.

Figure 3–3

As we saw in Chapter 2, it took a great leap of the imagination for Galileo to see that the state of uniform motion—by which is meant motion with constant velocity—was just as natural as the state of rest. This was a startling innovation, for it was no longer necessary to find an explanation or cause for uniform motion.

The tendency of objects to maintain their state of rest or their state of uniform motion in a straight line was given the name **inertia** by Galileo. Consequently, Newton's first law is sometimes called the **law of inertia**.

Inertia and Mass

The harder it is to change the state of motion of a body, the more inertia the body is said to have. An object such as a refrigerator or a heavy truck has a large amount of inertia, whereas a pencil or a safety pin has very little inertia. An object with a large inertia, such as an automobile, is hard to start moving, even if there is no friction. It is also hard to stop it moving once it is in motion. (Have you ever tried to stop a coasting automobile singlehandedly?) Furthermore, if an object with a large inertia is moving with constant speed in a straight line, it is difficult to shift the object from its straight-line path by pushing sideways on it (Fig. 3–2). Thus *inertia is the property or tendency of a body to resist any change in its state of motion, be it starting, stopping, or changing it from a straight-line path.*

The term **mass** is used to refer to the amount of inertia an object has. The more inertia a body has, the greater is its mass. The term mass can also be said to refer to the intuitive concept of "quantity of matter" in an object. Mass is a very important property of any body, and it plays a prominent role in Newton's theory of motion.[2]

EXPERIMENT

Place a sheet of paper under a book on the edge of a table. Pull the paper: (a) gradually; (b) all the way out with a sudden jerk. Are your observations consistent with the law of inertia?

OR

Suspend a block of wood or a rock by a piece of thread as shown in Fig. 3–3. Connect a second piece of thread to the bottom of the object. If you pull this thread with a slowly increasing force, will the thread break above or below the block? If instead you exert a sharp powerful tug on the thread, where will the thread break? Now try this experiment both ways. Explain your observations. Were your predictions fulfilled?

[2]Another important property of an object is its **volume,** which is the amount of space it takes up. We tend to associate mass and volume because objects with a large volume usually have a large mass. But this is not always true. A balloon, for example, may have a greater volume than an apple, but the apple has the greater mass.

Weight versus Mass

A body with a large mass is not only hard to start, hard to stop, and resistant to any change in its motion. It is also hard to lift. The property of a body that makes it hard to lift is called its **weight**. To be more specific, weight is defined as *the gravitational force* (or pull) *acting on an object*. The greater the pull of gravity on a body, the harder it is to lift. Weight and mass are often confused, but they are different concepts. They are related in that the weight of an object is directly proportional to its mass. The greater the mass of a body the greater is its weight and the harder it is to lift. But a careful distinction must be made between weight and mass. Weight is a force, the force of gravity acting on a body, whereas mass is the amount of inertia (or quantity of matter) in the object.

To see more clearly the difference between mass and weight, consider what happens when an object is taken to the moon. It will have lost none of its matter, so its mass will be unchanged. In the absence of friction, it will be just as hard to start or stop moving as on the earth. But it will be much easier to lift, because the force of gravity on the moon is only one-sixth as great as on the earth. Thus the object will weigh only one-sixth as much on the moon as it did on earth, even though its mass is the same.

Newton's Second Law: $F = ma$

Using Galileo's ideas as a basis, Newton found it relatively easy to state his first law. By building on the foundation laid by his predecessor, Newton was able to proceed more deeply into the nature of motion. Galileo had sought a correct *description* of motion and, succeeding in this, he was satisfied. But Newton, starting with Galileo's findings behind him, was able to ask a deeper question: what is the *cause* of a change in motion? What is it that makes a body at rest start moving, or once moving change its speed or direction? To find an answer to this question Newton considered the role of force.

Newton's first law tells us that if no force is exerted on an object, it will continue to stay at rest or continue to move with constant velocity in a straight line. If the velocity is zero at one instant, it stays zero; if it is 14 m/sec in a northeasterly direction, it stays that way, as long as no forces act on the object. But what happens when a force is exerted on the body? Newton saw that the velocity will change. A force exerted on an object may make it speed up or, if the force opposes the motion, slow down. In either case, the velocity changes. If the force acts sideways on a moving object, the *direction* of its velocity can be changed. Thus, when a force is exerted on an object, the object experiences an *acceleration*.

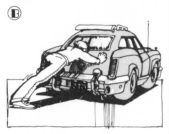

Figure 3–4

The same force applied to a child's wagon gives rise to a greater acceleration than when applied to an automobile.

Our own experience tells us that forces give rise to acceleration and even suggests that a simple relationship exists between force and acceleration. For example, if you push a child's wagon very gently in a straight line for a few seconds, it will accelerate from rest to some velocity, maybe 3 mi/hr. If you push twice as hard, the acceleration is twice as great—it only takes half the time to accelerate the wagon from zero to 3 mi/hr. Newton observed that the acceleration of a body invariably is in direct proportion to the net applied force. If the net force is doubled, the acceleration is doubled; if the net force is tripled, the acceleration is tripled, and so on. We use the term **net force** in case there is more than one force acting. For example, friction is usually present, so the net force on the wagon would be the force applied minus the friction force. More on this later.

The acceleration of a body also depends on its mass. Suppose you push a stalled automobile and a child's wagon with the same force (Fig. 3–4). The automobile will undergo a much smaller acceleration than the wagon. The car's velocity will change much more slowly than the wagon's. The more mass a body has, the more slowly it accelerates. Newton found that the acceleration of a body is always inversely proportional to its mass. With these observations, Newton was able to state his second law of motion:

The acceleration of a body is directly proportional to the net force acting on it and inversely proportional to the mass. The direction of the acceleration is in the direction of the applied force.

Newton's second law can be written in symbols as

$$a \propto \frac{F}{m},$$

where a stands for acceleration, F for force, and m for mass; the symbol $\propto$ means "is proportional to." Notice in this proportionality that if the force F is doubled, the acceleration will be doubled. But because the mass is in the denominator, if it is doubled the force will be halved—this is what is meant by inverse proportion. If the appropriate units are chosen for force and mass, the above proportionality[3] can be written as an equality: $a = \frac{F}{m}$. By multiplying both

[3]Ordinarily when a relationship is discovered, such as the proportionality $a \propto \frac{F}{m}$, a constant of proportionality is needed when the equality is written; for example, $a = k\frac{F}{m}$, where k is a constant. However, in this situation it is possible to choose the unit of force so that $k = 1$.

sides of this equation by the mass *m,* the familiar form of Newton's second law is obtained:

$$F = ma.$$

In words, *the mass of an object multiplied by its acceleration is equal to the applied force.* This is an alternate but exactly equivalent way of stating Newton's second law.

Newton's second law basically relates the description of motion —velocity and acceleration—to the cause of motion, force.

In Chapter 2 we discussed force in a rather loose and intuitive way. Now we can state the modern definition of force Newton himself invented: *force is any action capable of accelerating an object.*

Units of Force and Mass

Mass, in the metric system, is measured in grams or kilograms (a kilogram is 1000 grams). It is of course necessary to define exactly what is meant by one gram or one kilogram. We use as a standard a particular block of platinum and iridium alloy kept at the International Bureau of Weights and Measures near Paris. By definition its mass is one kilogram.

Force, in the metric system, is measured in *newtons* (abbreviated N). In accordance with Newton's second law, $F = ma,$ one newton is the force required to impart an acceleration of one meter per second per second to a one-kilogram mass. That is, 1 newton = 1 kg • m/sec².

In the English system, the common unit of force is the pound. We are most familiar with using the pound to specify weight, but it is used for any other kind of force as well. The unit of mass is called the *slug,* but it is rarely used. One slug is defined as that quantity of mass such that a force of one pound will impart to it an acceleration of one foot per second per second. Thus 1 pound = 1 slug • ft/sec².

As an example, the net force needed to accelerate a 20-kg grocery cart to 2 m/sec² is $F = (20 \text{ kg})(2 \text{ m/sec}^2) = 40$ N.

Gravitational Force

Galileo discovered that the acceleration of any freely falling object at the surface of the earth, in the absence of air resistance, is the same regardless of the mass of the object. Applying Newton's second law $(F = ma)$ to the gravitational force, for *a* we can use *g,* the acceleration due to gravity. Then the force of gravity on an object, which is its weight, can be written as the mass of the body times the acceleration due to gravity, or in symbols, weight = mg. Thus the weight of a body, which is defined as the gravitational force acting on it, is directly proportional to its mass.

Since g is 9.8 m/sec² in the metric system, Newton's second law tells us that a mass of 1 kilogram will be pulled earthward by a force $F = mg = 1$ kg $\times$ 9.8 m/sec² = 9.8 newtons. This is just the weight. One kilogram weighs 9.8 newtons. Similarly the weight of a 2-kilogram mass will be 19.6 newtons and so on. In the English system, Newton's second law tells us that the gravitational force on a mass of 1 slug will be $F = mg = 1$ slug $\times$ 32 ft/sec² = 32 lb; a mass of 1 slug thus has a weight of 32 pounds. If you know the weight of an object in pounds, you can find its mass by dividing by 32. For example, a 3200-pound automobile has a mass of 100 slugs.

It is the custom when using British units to specify the *weight* of an object rather than its mass. Consequently the unit of mass, the slug, is not well known. In the metric system it is customary to specify the mass of an object in grams or kilograms, and rarely do you specify the weight (in newtons). Thus you may give your mass, say 50 kilograms, or your weight, 110 pounds. Because mass and weight are not the same quantity, we cannot state a conversion factor between them as we can for centimeters and inches, for example. However, we can state an equivalence: 1 kilogram is equivalent to 2.2 pounds; but this is valid only on the surface of the earth. On the moon, for example, 1 kilogram mass will weigh only 0.4 pounds, because the force of gravity is less on the moon.

Newton's Third Law: Action and Reaction

With the second law of motion Newton was able to describe quantitatively how force causes motion. But he was still not satisfied. He next asked: where do forces come from? He concluded that a force applied to any object is always applied *by another object*. A farmer pushes a wheelbarrow; a horse pulls a cart; a person lifts a bag of groceries; a hammer pushes on a nail; the wind (air) blows the leaves on a tree; a magnet attracts (pulls) a paper clip.

In each example one body exerts the force and another body feels it. The farmer exerts the force and the wheelbarrow feels it; a hammer exerts the force and the nail feels it. But is it really so simple? Is one body the pusher and the other body the pushed? Newton did not believe that nature would be so unsymmetrical. Observations and intuition led him to believe that the two bodies must be treated equally. For example, the hammer does exert a force on the nail. But the hammer is brought to rest in the process. Thus there

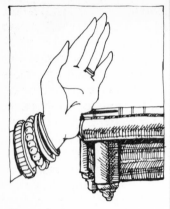

Figure 3–5

When your hand pushes on a table, the table pushes back. Notice how squashed the hand looks.

Figure 3–6

Ice skater pushes on a wall and moves backward because the wall exerts a force on her.

56

must also have been a force on the hammer to decelerate it, and this force must have been exerted by the nail. Observations such as this one led Newton to his third law:

> **Whenever one object exerts a force on a second object, the second object exerts an equal and opposite force on the first.**

Newton's third law is often stated as: "to every action there is an equal and opposite reaction." However, as the earlier statement emphasizes, the "action" and "reaction" forces act on *different* bodies. It is easy to become confused and incorrectly assume that the action and reaction forces act on the same body. They don't.

We experience countless examples of Newton's third law every day. In fact, *every* time a force is applied, a reaction force is present. Sometimes, however, it is hard to see where it comes from. Let's look at some examples.

Try exerting a force on the edge of a table by pushing your hand against it hard. Your hand exerts a force on the table, and the table exerts an equal and opposite force on your hand (Fig. 3–5). You know you are pushing on the table, but you may find it hard to imagine that the table actually pushes back. But look at your hand—it is squashed. This is unmistakable evidence, as we saw when we discussed the fact that a force distorts the shape of an object, that the table indeed is pushing on your hand. Furthermore, you can feel the table pushing on your hand—it hurts.

Action and reaction can also be observed readily by standing on an almost frictionless surface such as ice. The ice skater in Fig. 3–6 pushes against the wall of the ice rink. She exerts a force on the wall and then she starts moving backward. The force she exerts on the wall could not make her move. To make her start moving there has to be a force exerted *on her*. It is the wall that exerts this force on her. This force is the reaction to the force she exerted on the wall.

How Can Inanimate Objects Exert Forces?

It is difficult for most people to accept the fact that an inanimate object such as a wall can exert a force. We generally tend to associate force with active bodies. Humans, animals, and engines can exert forces; so can an object in motion, such as a flying rock or a falling wall. But can a wall or table at rest exert a force? The squashed hand in Fig. 3–5 is clear evidence that it can. And so is the skater moving backward in Fig. 3–6. But how does the wall or table do it?

An object such as a wall can exert a force because it is elastic, although admittedly not very elastic. Pushing on it is like pushing on a stiff piece of stretched rubber. No one can deny that a stretched piece of rubber can exert a force. A slingshot, or merely the shooting of a wad of paper with a rubber band, testifies to this. But all materials are elastic, at least to some extent—rubber is only more so. Because of its elasticity, a solid material can exert a force on an object. When you pull a wad of paper against a rubber band, the rubber exerts a force on the wad and your hand. (If you release the wad, it is accelerated and flies across the room.) Similarly, the force your hand feels when it pushes on a wall is the elastic force due to the "stretched" wall. The stretching is often too small to be noticeable. But you may notice the stretch when you push on a thin board (Fig. 3–7), the side of a refrigerator, a soft piece of wood, or when you bend a paper clip slightly.

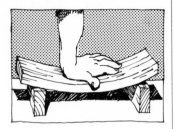

Figure 3–7
A hand bends a board.

More Examples of Newton's Third Law

If you are in a rowboat or a canoe at rest in the water and throw a package from the boat onto the shore as in Fig. 3–8, you will suddenly realize that your boat is moving away from the shore. This is another illustration of Newton's third law. You exerted a force on the package in the process of throwing it. This force gave the package the acceleration it needed to reach the shore. But the package exerted an equal and opposite force on you and this gave you and your boat an acceleration outward; that is, you started moving away from shore.

Figure 3–8
Throwing a package onto shore from a boat that was previously at rest causes the boat to move outward from shore (Newton's third law).

Figure 3–9

Lunar module detachment, landing, and takeoff, as well as maneuvering in empty space, require acceleration obtainable by firing rockets.

Figure 3–10

You move forward when walking because the ground exerts a forward force on you.

The motion of rockets, Fig. 3–9, depends on action and reaction in a similar way. It is a common misconception that a rocket accelerates because the expelled gases push against the ground, or against the atmosphere. But this is not what gives rise to rocket motion at all. The rocket expels gases because it exerts a strong force on them. The gases, then, exert an equal and opposite force on the rocket (Newton's third law). It is this force *on the rocket* that propels it forward. Thus, the motion of the rocket is analogous to the motion of the boat in Fig. 3–8, except that hot gases, instead of a package, are thrust out the back. A space vehicle can maneuver, accelerate, and change direction far out in space where there is little or no atmosphere, just by firing rockets in the proper direction. If the motion depended on the gases pushing against the atmosphere, these maneuvers would be impossible.

Life would be very different if for some reason Newton's third law did not hold. For example, let us analyze how a person walks. Suppose you are standing still; your velocity is zero. In order to start walking, you push your foot against the ground. You exert a force on the ground. By Newton's third law, the ground exerts an equal and opposite force on you, Fig. 3–10. This latter force, on *you,* causes you to move forward.

The fact that it is the force the ground exerts on you that makes you move forward is set forth by Newton's second law. In order to accelerate an object—in this case you—a force has to be exerted on that object. The only force exerted on you is the one exerted by the ground.

Similarly, when an automobile engine causes the wheels to turn, the tires exert a force against the ground. The ground, in turn, exerts an equal and opposite force on the car, and this latter force causes it to move forward. In these last two examples we also see the importance of friction. Without friction, we would be unable to exert a horizontal force on the ground and so the ground could not exert an equal and opposite force on us to make us move.

**Action and Reaction
Forces Act on
Different Bodies**

As mentioned earlier, Newton's third law is sometimes used incorrectly. A famous example is that of the horse and cart, Fig. 3–11(A). This particular horse is apparently well read and thinks to itself, "When I exert a forward force on the cart, the cart exerts an equal and opposite force backward. So, how can I ever move? No matter how hard I pull, the backward reaction force always equals my forward force, so the net force must be zero. I'll never be able to pull

Figure 3–11

The horse and wagon argument. In (B) the forces acting *on the horse* are shown as solid arrows; the forces acting *on the wagon* are shown as dotted arrows; the forces acting on the ground are not shown.

this cart." But alas, the horse cannot get away with such lazy excuses. Although it is true that the action and reaction forces are equal, it has forgotten that they are exerted on different bodies! The forward action force is exerted by the horse *on the cart,* whereas the backward reaction force is exerted by the cart *on the horse*. To determine whether the *horse* moves or not, we must consider only the forces *on the horse* and then apply $F = ma$, where F is the net force on the horse, a is the acceleration of the horse, and m is the mass of the horse. The forces on the horse that will affect its forward motion are the force of the ground pushing forward on it (the reaction to its pushing backward on the ground) and the cart pulling backward, Fig. 3–11(B). When the ground pushes on the horse (because it pushed on the ground) harder than the cart pulls back, the horse will move forward. The cart, on the other hand, moves forward because the pulling force exerted by the horse is greater than the frictional force pulling backward.

This example illustrates that whenever you want to determine how an object will move, you must consider *only* the forces that act

59

on that one object. The horse was wrong because it was confused about which forces acted on which object. It is necessary, therefore, to specify *on* what object the force is acting and *by* what object the force is being exerted. For example, we should say "the force exerted *by* the horse *on* the cart," or "the force exerted *by* the earth *on* the horse." The prepositions *on* and *by* are very important in this situation. Use them whenever you specify a force.

2. NET FORCE AND ADDITION OF VECTORS

Net Force

Newton's second law states that the acceleration of a body is directly proportional to the net force acting on the body and inversely proportional to the mass of the body. What is meant by the term **net force**? If only one force is acting on the object, that force is the net force. If two or more forces are acting, the *net force is the sum of all those forces, taking into account the direction of each.* Because force, like any vector, has both magnitude and direction, adding forces is not as simple as adding ordinary numbers. We must now investigate how to add vectors. We are interested in force vectors at the moment, but the rules we will learn also apply to other types of vectors, such as velocity.

Addition of Vectors

As mentioned in Chapter 2, we can draw an arrow to represent a vector such as a force. The arrow is drawn so that (1) the direction of the arrow is the direction of the vector, and (2) the length of the arrow is proportional to the magnitude of the vector. For example, a force of 10 pounds might be represented by an arrow one inch long; then a 30-pound force would be represented by an arrow 3 inches long.

To see how two vectors can be added, we first consider an object being pulled by two men, Fig. 3–12(A). They each exert a force of 50 pounds in the same direction; the total force on the object in this case is 100 pounds. That is, the two 50-pound forces have the same effect on the object as a single 100-pound force. Thus the net force, which is the sum of the two individual forces, is 100 pounds. If the two men pull in opposite directions, each with a force of 50 pounds, Fig. 3–12(B), the object won't move at all; the net force is zero. In Fig. 3–12(C) the two men are pulling in opposite directions, but the one on the left is pulling with a force of only 35 pounds. The net force in this situation is 15 pounds to the right. These examples suggest that two vectors can be added by the following set of operations, which are illustrated in Fig. 3–13:

Figure 3–12
Two men pulling on a heavy
load.

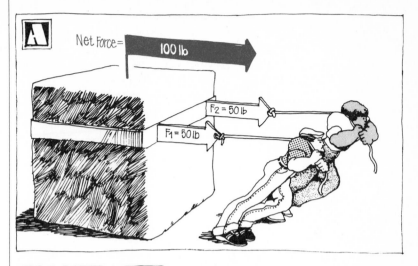

61

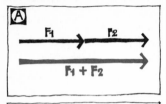

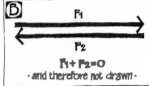

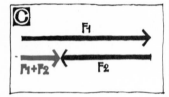

Figure 3–13

"Tail-to-tip" method of adding vectors.

1. Draw the two force vectors to scale; call one of them F_1, the other F_2. (It doesn't matter which is called F_1 and which F_2.)
2. Carefully move F_2 so that its tail is at the tip of F_1. Be sure you do not change the length or direction of F_2.
3. The sum of the two force vectors is the arrow drawn from the tail of F_1 to the tip of F_2. The sum vector is labeled $F_1 + F_2$. (We use boldface type to remind us that we are adding vectors.)

This set of rules gives the expected result for each example in Fig. 3–12. These rules also apply when the two forces do not act along the same line. For example, suppose two 50-pound forces act at right angles (90°) to one another as in Fig. 3–14(A). In this case, it is clear that the object will move along a line at 45°. Therefore the sum of the two forces must be along the line at 45°. When our rules for adding two vectors are used, Fig. 3–14(B) results. The sum of the two forces $F_1 + F_2$ is indeed at 45° according to our rules. The magnitude of the total force, $F_1 + F_2$, can be measured with a ruler directly on Fig. 3–14(B). If the scale is one centimeter = 15 lb, F_1 and F_2 are each 3.3 cm long. The sum $F_1 + F_2$ measures about 4.7 cm, which corresponds to a total force of about 70 pounds.[4] This result can be checked experimentally, and indeed, experiment shows that a single force of 70 pounds at an angle of 45° has exactly the same effect as two 50-pound forces at right angles to one another. In fact, experiment confirms the validity of our vector addition rules in general.

[4]This same result can be obtained using arithmetic and Pythagoras' theorem that the sum of the squares of the sides of a right triangle equals the square of the hypotenuse, $c^2 = a^2 + b^2$. Since F_1, F_2, and $F_1 + F_2$ in Fig. 3–14(B) make up a right triangle with $F_1 + F_2$ being the hypotenuse, the length or magnitude of $F_1 + F_2$ equals the square root of $50^2 + 50^2 = 2500 + 2500 = 5000$; and the square root of 5000 is 70.7 lb.

Figure 3–14

Adding two nonparallel forces to get the net, or total, force using the "tail-to-tip" method.

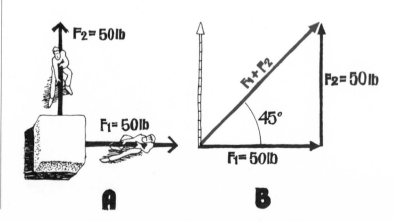

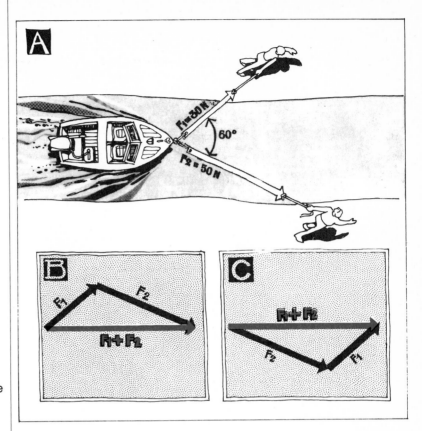

Figure 3–15

(A) Two unequal forces act on a boat. (B) and (C) Adding the two forces to get the net force using the tail-to-tip method. Notice that the same result is obtained no matter which force is drawn first.

The fact that two 50-pound forces should add up to a total force of 70 pounds may at first seem strange. But remember that forces are vectors and their direction is important. Two men who each pull with a force of 50 pounds in the same direction do exert a total force of 100 pounds, as in Fig. 3–12(A). If they pull at an angle to one another as in Fig. 3–14, some of their cooperative effect is lost and we would expect the total, or net force, to be less than 100 pounds. If the angle between the two forces is 180°—that is, if they pull in opposite directions—the net force is zero, as in Fig. 3–12(B). Thus, when the two forces are at an angle of 90° to one another, we expect the force to be less than 100 pounds but more than zero. Only when the two forces are in the same direction will they add as in ordinary arithmetic.

Figure 3–15 shows a more general case of two unequal vectors acting at a particular angle to one another, in this case 60°. These two forces can be added in the usual way to obtain the net force, $F_1 + F_2$, Fig. 3–15(B). When adding two vectors, it doesn't really matter which one is moved, F_1 or F_2, as long as the tail of the moved vector is placed at the tip of the other. Figure 3–15(C) shows that

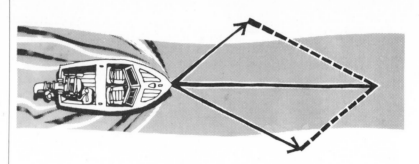

Figure 3–16

The same net force as in Fig. 3–15 is obtained here using the parallelogram method of adding vectors.

the same result is obtained if the vector **F₁** is moved. Use a ruler to measure the length of **F₁** + **F₂**; you should find its magnitude is 70 newtons.

This technique for adding vectors is called the **tail-to-tip** method. Another way to determine the sum of two vectors is the **parallelogram method**. A parallelogram is a four-sided plane figure in which each pair of opposite sides is parallel and equal; for example, ⎰⎰. In the parallelogram method, the two vectors are left just as they are (Fig. 3–16). Then a parallelogram is drawn with the two vectors as two of its sides. The sum vector is then the diagonal as drawn from the common origin to the opposite corner of the parallelogram, as shown. Because the opposite sides of a parallelogram are of equal length and parallel (i.e., at the same angle), the construction of Fig. 3–16 is equivalent to the tail-to-tip method of Fig. 3–15, and the same net force is obtained.

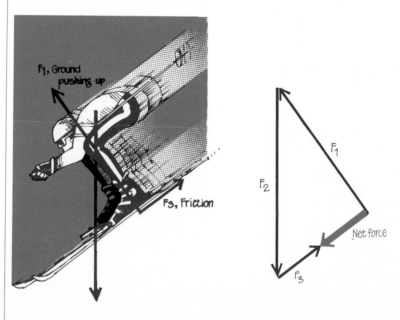

Figure 3–17

(A) Forces acting on a skier. (B) Net force on skier found using tail-to-tip method of adding vectors.

64

Figure 3–18
(A) A person pushes a lawn mower with a force **F** directed along the handle. (B) The force is resolved into vertical and horizontal components. (C) The horizontal component can be increased by reducing the angle without increasing the total force on the mower.

When three or more vectors must be added, the tail-to-tip method is the easiest to use. Figure 3–17(A) shows three forces acting on a skier. Figure 3–17(B) shows how to obtain the sum, which is the net force on the skier.

Resolution of Vectors

We have seen that two (or more) vectors can be added to give a single vector, the sum, whose effect is exactly the same as the original two vectors. It is often useful to do the opposite; that is, to consider any single vector as the sum of two other vectors. These latter vectors are called the **components** of the original vector. Most often the components of a vector are chosen so that they are perpendicular to each other, with one component horizontal and the other vertical. For example, look at Fig. 3–18(A), where a man pushes his lawn mower with a force **F** directed along the handle. We want to find the components of **F** in the horizontal and vertical directions. The process of determining these components is called the **resolution** of the vector, or *resolving the vector into its components.* Keep in mind that the two components must add up to the original vector. Here is how to resolve a vector: draw a rectangle around the given vector, with that vector as a diagonal, Fig. 3–18(B). The components are then the two sides of the rectangle, with origin at the same point as the original vector. It should be clear that the horizontal and vertical components, labeled F_h and F_v, respectively, add up to give the original vector **F** by the parallelogram method of adding vectors.

This example also illustrates the usefulness of resolving vectors. The force exerted on the handle of the lawn mower had horizontal and vertical components, as indicated by F_h and F_v. That is, the man pushes on the lawn mower in the horizontal and vertical directions at the same time. However, only the horizontal component of the force, F_h, propels the lawn mower forward. The vertical component, F_v, merely pushes the lawn mower against the ground and in a sense is wasted. If the man exerted the same force at a smaller angle [Fig. 3–18(C)], more of his force would go into horizontal motion; thus he could exert a smaller total force **F** along the handle and still obtain the same F_h and therefore the same horizontal motion. However, there are practical disadvantages to the smaller angle. His back would probably get tired from bending over; furthermore, the vertical component of force ensures good contact of the lawn mower with the ground and thus a more even cutting job. In this case a proper balance between the horizontal and vertical components of the force is advantageous.

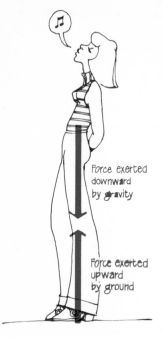

Figure 3–19

The two forces acting *on* a person standing still.

Equilibrium Means the *Net* Force Is Zero

When the net force on an object is zero, the object is said to be in **equilibrium.** An example of an object in equilibrium was shown in Fig. 3–12(B). Newton's second law tells us that if the net force on an object is zero, the acceleration must be zero. Any body at rest is in equilibrium.

What about an object at rest on the earth? The gravitational force does not suddenly disappear, as we all know when we step on a scale to read our weight. Even though we are not accelerating, the same gravitational force pulls downward. There must therefore be an upward force that balances the downward force of gravity. This force is applied by the ground, Fig. 3–19. The ground pushes *up* on the person with a force that has the same magnitude as the downward gravitational force. We say the object is in *equilibrium* ("equal forces" in Latin; we discuss this case in detail in Chapter 4). People sometimes confuse this situation with the action-reaction pairs of Newton's third law. Be careful. In Fig. 3–19 the two forces are *not* the action-reaction pairs required by Newton's third law, because they are both acting on the same body. The "reaction" forces are not shown in Fig. 3–19.

When dealing with objects moving *horizontally* on the earth, we often don't mention the gravitational force downward. Now we see that this is legitimate because it is balanced exactly by the ground's upward force. There is therefore no net force in the vertical direction.

3. MOTION IN A CURVE

An object moves in a straight line whenever the net force is either zero or along the direction of motion. When the net force on an object acts *sideways* to the direction of motion, it begins to move in a curved path.

Projectile Motion

If for some reason gravity were not acting when you threw a baseball, the baseball would continue outward in a straight line indefinitely. But of course the force of gravity does act on the baseball and brings it down to the earth. A thrown baseball, a kicked football, a speeding bullet are all examples of **projectiles** acted on by the vertical force of gravity. We will consider now the motion of projectiles whose path does not take them far from the earth's surface.

Consider first a ball rolling off the edge of a table, Fig. 3–20. Galileo first realized that this kind of motion is most simply dealt with by considering the horizontal and vertical motions separately and then combining them to get the complete motion. The velocity vector always points in the direction in which the object is moving at any given point, and hence it is always tangent to the path of the object. Following Galileo, we consider the vertical and horizontal components of the velocity, v_v and v_h, separately at each point. The velocity vector and its components are shown in Fig. 3–20 at several points along the path; these points are separated by equal time intervals. The force of gravity is acting vertically downward at every point and therefore gives rise to a vertically downward acceleration. Because acceleration means a changing velocity, the vertical component of velocity is continually increasing. Thus the arrows representing v_v in Fig. 3–20 are longer for each successive point on the path. On the other hand, if we ignore air resistance, there are no forces in the horizontal direction and v_h does not change at all; it has the same magnitude at each point shown in Fig. 3–20.

One result of this analysis is the prediction that if a second ball is dropped straight down from the edge of the table just as the rolling ball passes the edge, the two balls will reach the ground at the same time, Fig. 3–21. Galileo first predicted this because he realized that the vertical motion of the ball in Fig. 3–20 is exactly the same as that of a vertically falling object. To convince yourself that this is true, do one of the following experiments.

Figure 3–20

The velocity vector and its horizontal and vertical components are shown at several equally spaced time intervals.

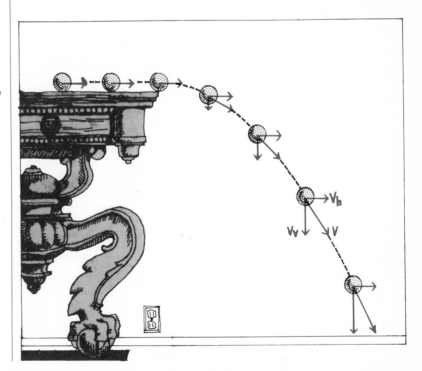

67

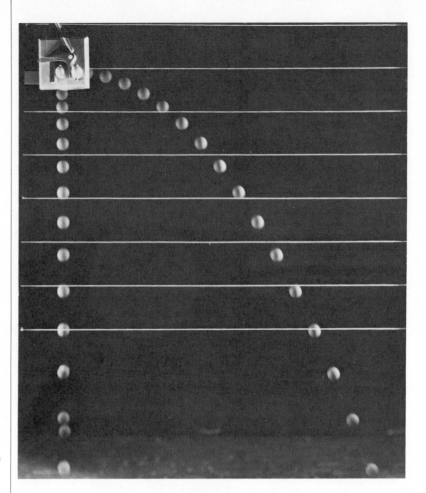

Figure 3–21

A ball is projected horizontally at the same time that a second ball is dropped straight down. Both reach the ground at the same time.

EXPERIMENT

Place a small pencil on the edge of a table or top of a bookshelf (the higher the better), so that it is almost falling off, Fig. 3–22. Now roll a marble (the bigger the better) swiftly so that it just barely bumps the end of the pencil, which then falls vertically. Listen for the sounds they make when they hit the ground. To assure a simultaneous start, you may find it helpful to roll the marble down a ruler to guide it as shown and to balance the pencil on the edge of a sheet of paper.

OR

Stand on top of your house or another building and throw a tennis ball out horizontally at the same time you drop one vertically.

Figure 3–22

When an object is projected upward at an angle, such as when a football is kicked, the analysis is essentially the same. Figure 3–23 shows this situation. The horizontal component of velocity remains constant as before. The vertical component of velocity is initially quite large. But because the force of gravity acts downward, this component decreases in time until the football reaches a maximum height, at which point the vertical component of velocity is zero. The vertical component then begins to increase in the downward direction, becoming larger and larger as time goes on.

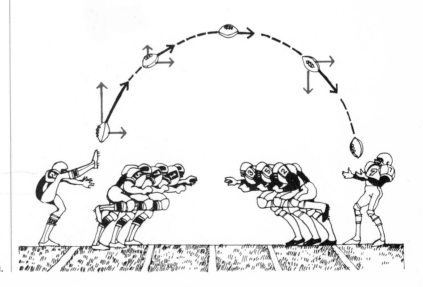

Figure 3–23
Path taken by a kicked football, showing the velocity and its components at several points along the path.

Figure 3–24

Air resistance reduces the distance a projectile will travel.

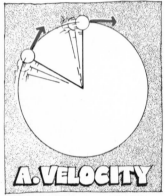

Figure 3–25

A ball on the end of a string is swung in a circle. In (A) the velocity vectors are shown at two different places; in (B) the force vectors that keep the ball moving in the circle are shown.

Air resistance is often so small that it can be ignored, as we have done above. It becomes important, however, if the projectile is not too massive compared to its volume, or if the projectile is to go a long distance—for example, an artillery shell. Air resistance acts as a retarding force and thus decreases both the horizontal and vertical components of velocity. The path of a projectile is modified by air resistance in the manner shown in Fig. 3–24.

Circular Motion

Many objects move in a circular or nearly circular path. The moon revolving around the earth follows a nearly circular path as does the earth in its orbit around the sun. A ball attached to the end of a string can be swung in a circle about a person's head; and a child on a merry-go-round moves in a circle.

The simplest circular motion occurs when the object moves with constant speed. Even though the velocity is not changing in magnitude, it is continuously changing in direction, as shown in Fig. 3–25. Therefore, the object is accelerating and there must be a force acting on it. If there were no force acting on the object, it would go in a straight line. But for the object to move in a circle, a force must pull it out of a straight-line path. This force is directed toward the center of the circle, as shown in Fig. 3–25(B). At each point along the path, this force (always pulling toward the circle's center) pulls the object away from its natural tendency to go in a straight line.

The magnitude of the force required to keep an object moving in a circle depends on how fast the velocity vector is changing; that is, on how large the acceleration is. By using geometry, it is possible to show that the acceleration of an object moving in a circle is given by the formula $a = \dfrac{v^2}{r}$, where v is the speed of the object and r is the radius of its circular path. Although we will not prove this formula, we can indicate that it does make sense. The greater the speed v, the more rapidly the velocity vector is changing direction and therefore the greater the acceleration. (Why the acceleration is proportional to the speed *squared* is not quite so obvious.) Similarly, for a given speed v, the larger the radius r the less rapidly the velocity vector changes and the less the acceleration; this accounts for the inverse proportionality to r. To cause an object moving in a circle to accelerate requires a force $F = ma = \dfrac{mv^2}{r}$. Because the acceleration points toward the center of the circle, it is sometimes called a "center-seeking" or **centripetal** acceleration. This situation is very different from the projectile situation in which the acceleration, and the force that causes it, always points in the same downward direction.

There is a common misconception that some kind of mysterious *centrifugal force* ("center-fleeing" force) is pulling outward on the object rotating in a circle. To see why this is a misconception, let us consider a simple example. Figure 3–26 shows a person revolving a ball on the end of a string in a circle above his head. If you have ever done this, you know that you feel a force pulling outward on your hand. The misconception arises when this pull is interpreted as an outward "centrifugal" force pulling on the ball, which is transmitted along the string to the hand. But this is not what is happening at all. To keep the ball moving in a circle, the person continually pulls *inwardly* on the ball. Because of Newton's third law, the ball exerts an equal and opposite force on the hand. This is the force your hand feels—the reaction to your pulling in on the ball. The only force on the ball (except for gravity) is the one you exert inwardly on it. There are no others. In particular, there is no outward "centrifugal force" acting on the ball.

EXPERIMENT

Tie an object such as a rock or a tin can to the end of a string and swing it horizontally around your head, as in Fig. 3–26. As the object is swinging, suddenly let go of the string. Note very carefully the position of the object at the instant you let go, and observe in which direction it moves.

Figure 3–26

Ball swung in a circle, showing force acting on the ball and the "reaction" force acting on the hand.

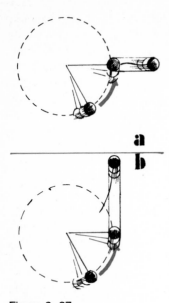

Figure 3-27

When you let go of the string, does the ball fly off as in (A) or as in (B)?

If a centrifugal force were pulling outward on the object, you would expect it to fly outward when you released it, as shown in Fig. 3-27(A). But if our discussion is correct, the object will fly off tangentially in a straight line, because there is no longer a force pulling it inward, Fig. 3-27(B). Which figure, 3-27(A) or 3-27(B), represents what you observed in your experiment?

Another example of circular motion occurs when you are riding in a car as it rounds a curve. If the curve is sharp, it feels as if you are being pulled outward. But again this is not a centrifugal force. What you feel is the result of the natural tendency of an object to go in a straight line. See Fig. 3-28. When the car turns into the curve, your body, because of its inertia, has a tendency to continue going straight. So you slide, or are thrust, toward the driver. In a sense, the car is actually curving in front of you and you tend to keep going straight. For you to move in the curved path and stay inside the car, a force is necessary. Either the driver of the car pushes inward on you, or the friction of the car seat on your back is sufficient to keep you in place.

Of course, the automobile also naturally moves in a straight line and can go around a curve only if it is accelerated by a force acting toward the center of the circular path. The road exerts this force on the tires through friction.

If the friction force between the road and the tires is not great enough to provide the required acceleration, the car will skid. A lower velocity means less force is required and therefore there is less likelihood of skidding.

Figure 3-28

A force must be exerted on the passenger if he or she is to move in a curve.

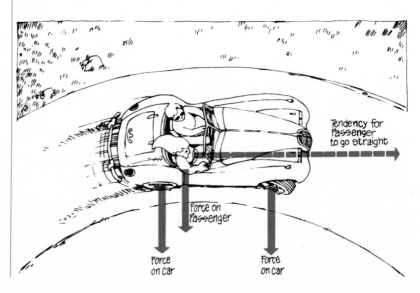

Figure 3–29

The force of gravity. The earth pulls on the apple; the apple pulls on the earth.

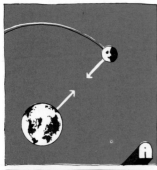

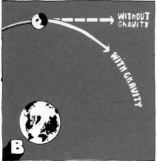

WITHOUT GRAVITY

WITH GRAVITY

Figure 3–30

(A) The earth pulls on the moon; the moon pulls on the earth. (B) The moon is pulled out of a straight-line path by the force of gravity.

4. GRAVITY

Circular motion per se did not start Newton thinking about the nature of gravity, but it did play a role. Newton was concerned, among other things, with understanding the motion of the planets. Some thinkers still held that the orbits of the planets needed no explanation because they merely represented a *natural* motion. But many thinkers in Newton's time recognized that a force was needed to keep the planets in their nearly circular orbits. Newton felt that the origin of this force was the sun. And on the basis of Kepler's laws Newton showed that the strength of this force must be inversely proportional to the square of the distance between a planet and the sun.

Newton was also concerned with the apparently unrelated problem of gravity. He had already concluded that since falling bodies accelerate, they must have a force on them, a force we call the gravitational force. As we have seen, he had also observed that whenever a body has a force exerted on it, some other body exerts that force. That is, every force is really an interaction between two bodies. Every object on the earth feels the downward force of gravity. But what object, Newton asked, *exerts* this downward force? Since the direction of the gravitational force is always toward the center of the earth, no matter where the object is, the most obvious candidate is the earth itself. It is the *earth,* Newton said, that exerts the gravitational force on objects at its surface (Fig. 3–29).

The next step was a crucial one. Newton began to wonder if the gravitational force had something to do with the force that holds the planets in their orbits. An early biographer reports that Newton himself said that the idea came to him when he was sitting in his garden and he noticed an apple drop from an apple tree. At the time, Newton was trying to understand what kind of force could hold the planets in orbit around the sun and even more importantly what kind of force held the moon in its nearly circular orbit around the earth. Observing the falling apple, Newton was struck by a sudden inspiration: if the force of gravity acts all the way to the top of trees and even at the tops of mountains, perhaps it acts all the way up to the moon!

Here we see Newton in the creative act: the bringing together of two separate ideas and finding a unity in them. Newton saw the moon as falling toward the earth just like any other body. This statement may seem strange, but remember that an object tends to go in a straight line if no forces act on it. The moon would move in a straight line if no forces were acting on it. Instead, it keeps *falling out* of a straight line path because of the gravitational attraction of the earth, Fig. 3–30. Newton thus came to the remarkable conclusion that *the gravitational force of the earth holds the moon in its nearly circular orbit.*

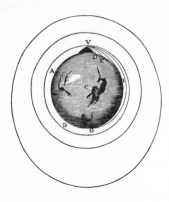

Figure 3–31

Drawing from Newton's *Principia* showing the path of a projectile projected horizontally at various high speeds.

The Moon as a Projectile

To substantiate his argument that the moon is held in its orbit by the gravitational pull of the earth, Newton used the drawing shown in Fig. 3–31. It shows that an object projected with a large horizontal velocity will travel part way around the earth before striking the ground (lines VD and VE in the figure); the larger the velocity of the projectile, the farther it travels. If its velocity is great enough, the projectile will continue all the way around the earth without ever hitting the ground. It has then taken a circular or elliptical path, Fig. 3–31. Thus, Newton was the first to recognize the possibility of artificial satellites circling the earth, but the high velocity necessary to attain an orbit was not achieved for almost three centuries. However, what Newton was trying to make plausible was not the possibility of artificial satellites, but the idea that the moon is merely a high-speed projectile that circles the earth under the influence of the gravitational force.

Force of Gravity Decreases with Distance (Squared)

At the surface of the earth, the force of gravity accelerates objects at the rate of 32 ft/sec² or 9.8 m/sec². But what is the acceleration of the moon, Newton asked. Using the formula for the acceleration of an object moving in a circle, $a = \frac{v^2}{r}$, Newton calculated the acceleration of the moon. He found that it was 3600 times smaller than the acceleration of objects at the surface of the earth, 0.009 ft/sec² versus 32 ft/sec². Newton explained this much smaller acceleration by proposing that the gravitational pull of the earth is weaker on more distant objects.

Since the radius of the earth is 4000 miles (6400 km), and the moon is 240,000 miles away, the moon is 60 times farther from the earth's center than are objects at the surface of the earth, Fig. 3–32(A). Now $60 \times 60 = 60^2 = 3600$, again that number 3600. This suggested to Newton that the gravitational attraction of the earth for any object must decrease inversely with the square of its distance from the earth's center.

$$\text{Force of gravity} \propto \frac{1}{(\text{distance})^2}.$$

This is illustrated in Fig. 3–32(B). The moon, which is 60 earth radii away, will feel a force only $\frac{1}{60^2} = \frac{1}{3600}$ times as strong as it would if it were at the surface of the earth. This theoretical result is what is found experimentally.

Figure 3–32

(A) Objects at the earth's surface are 4000 miles (6400 km) from the center of the earth; the moon is 240,000 miles (380,000 km) away.
(B) Force of gravity due to the earth decreases inversely as the square of the distance. For example, 4000 miles from the earth's surface (8000 miles from its center), the force of gravity is ¼ what it is at the earth's surface.

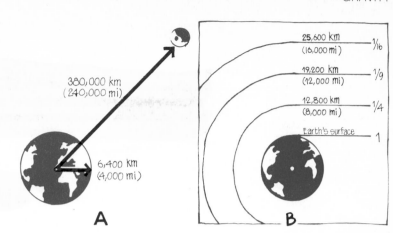

The Law of Universal Gravitation

In the same way that all objects on the surface of the earth have the same acceleration due to gravity of 32 ft/sec², so all objects 240,000 miles from the earth will have the same acceleration due to the earth's gravity, 0.009 ft/sec². The actual force on the object, as we saw earlier, depends also on its mass. It is, in fact, proportional to the mass.

When the earth exerts its gravitational force on an object, such as the moon, then according to Newton's third law the other object must exert an equal and opposite force on the earth, Fig. 3–30(A). Because of this symmetry, Newton argued that the force of gravity must be proportional to *both* the masses involved. Thus, the force of gravity is proportional to the product of the earth's mass and the mass of the object, and inversely proportional to the square of the distance between their centers:

$$F \propto \frac{\text{mass of earth} \times \text{mass of object}}{(\text{distance})^2}$$

The distance used here is the distance from the center of the earth to the center of the object. This is not to say that the force of gravity originates at the center of the earth. Rather, as Newton was able to show with the use of calculus, all parts of the earth exert a gravitational pull on other objects, but the net effect is the same as if the total force acted from the center.

As mentioned earlier, Newton had determined that the force that holds the planets in their orbits decreases by the inverse square of the distance of the planets from the sun. This led him to believe that the same kind of gravitational force was acting between the sun and each planet. It also reinforced the idea that the gravitational force

75

decreases with the square of the distance. With all these arguments in hand, Newton proposed his famous **Law of Universal Gravitation**:

Every body in the universe is attracted to every other body with a force that is proportional to the product of their masses and inversely proportional to the square of the distance between them. This force acts along the line joining the two bodies.

In symbols, this law can be written as

$$F \propto \frac{mm'}{d^2},$$

where m and m' are the masses of the two bodies and d is the distance between them. To make this proportionality into an equality, we need only introduce a constant of proportionality, which we call G. Thus, $F = G\frac{mm'}{d^2}$. The constant G is a universal constant; that is, it has the same value no matter what bodies are involved. According to Newton, then, the force of gravity is not unique to the earth. It exists between all objects.

The law of universal gravitation might seem at first a rather outlandish claim. *Every* body attracts every other body? That means a pencil attracts a paper clip; and a rock attracts other rocks and all other objects as well. But we are not aware of any such attraction in our daily lives. The only attraction that we notice is the attraction of the earth for other objects. This is because the constant G is very small. So the gravitational force between two objects will be very small unless the mass of one of them is extremely large, as the earth's is. That is why we only notice the gravitational force due to the earth and not that between other objects.

The force between two ordinary-sized objects was first measured experimentally in 1798 by Henry Cavendish (1731–1810), who thus confirmed Newton's hypothesis, see Fig. 3–33. At the same time, Cavendish was able to measure the value of the constant G.[5] Once he had measured G, Cavendish was able to determine the mass of the earth, because the force F exerted on any body of mass m at the surface of the earth and d (radius of the earth) were known. Only m', the mass of the earth, was not known, and it could be calculated by the above formula. The result is 6×10^{24} kg. Thus, Cavendish was the first to "weigh" the earth.

The law of universal gravitation should not be confused with Newton's second law of motion, $F = ma$. The former describes a particular force, gravity, and how its strength varies with the distance and the masses involved. Newton's second law, on the other hand, does not describe a particular kind of force. Instead, it relates

[5]The accepted value today is $G = 3.4 \times 10^{-8}$ lb•ft²/slug², or in metric units, 6.7×10^{-11} N•m²/kg².

Figure 3–33

A schematic diagram of Cavendish's apparatus. A light horizontal rod, to which two spheres are attached at the ends, is suspended at its center by a thin vertical fiber. When another sphere (A) is brought close to one of the suspended spheres, the attraction causes the latter to move, which in turn, twists the fiber slightly. Cavendish had earlier determined the force required to twist the fiber a given amount. He was thus able to measure the magnitude of the gravitational force between two objects of known mass.

the force applied to a body—be it a pushing force, a magnetic force, or a gravitational force—to the mass of the body and its acceleration. It is a more general law; it relates any force whatever to the motion of the body on which it acts.

Gravity Is a Force Acting at a Distance

One aspect of this gravitational theory bothered even Newton. Other common forces acted by contact—a push or a pull. Or at least there was something material to transmit the force; for example, a person can tie a rope to a rock and then pull on the rock by pulling on the rope. The rope transmits the force from the person to the rock. But with gravity, the force can be exerted over a distance without any matter to transmit it. A falling apple feels the pull of the earth even though it isn't touching the earth. The moon, 240,000 miles from the earth, feels the gravitational attraction of the earth. The fact that the gravitational force should be a "force acting at a distance," is difficult to understand or accept. Partly because of this difficulty, Newton did not publish his results immediately.

Earth Satellites

Newton was the first to indicate the possibility of an earth satellite, Fig. 3–31. To launch a satellite, a rocket carrying the satellite is fired off vertically. When the rocket reaches a height of a few miles, where the frictional drag of the earth's atmosphere is very small, it gives a horizontal thrust to the satellite, Fig. 3–34. If its velocity is large enough, the satellite will go into a circular orbit. The velocity required for a circular orbit not far above the earth's atmosphere is about 18,000 mi/hr (29,000 km/hr). If its velocity is less than this, the satellite returns to the earth like an ordinary projectile. If its velocity is larger, the satellite will go into an elliptical orbit around the earth, just as the planets revolve in elliptical orbits around the sun. However, if its velocity is greater than 25,000 mi/hr (40,000 km/hr), the satellite will completely escape the earth, never to return. The gravitational force of the earth will not be strong enough to hold it back. This minimum velocity that an object must have to escape is known as the **escape velocity**.

The Planets

Kepler had discovered experimentally that the planets trace elliptical orbits around the sun. Newton showed theoretically that a force

Figure 3–34

A space vehicle can come back to earth, become an earth satellite, or escape from the earth, depending on its initial velocity. At any velocity between 18,000 and 25,000 mi/hr (29,000 to 40,000 km/hr), the satellite will move in an elliptical orbit.

law that depends on the inverse square of the distance necessarily implies that the orbits are ellipses. This he used as evidence for his law of universal gravitation. However, because each body in the universe attracts every other body, each planet must exert a force on the other planets as well. Since the mass of the sun is much greater than that of any of the planets, the force on one planet due to any other planet will be very small by comparison. But because of this small force, each planet should show some departure from a perfectly elliptical orbit, especially when a second planet is fairly close to it, Fig. 3–35. Indeed, careful measurements show that departures from perfect ellipses, or **perturbations** as they are called, do exist.

Such perturbations led the way to the discovery of the planets Neptune and Pluto. The orbit of Uranus, for example, did not follow a perfect ellipse and its deviation could not be accounted for by perturbations due to the other known planets. In the mid-nineteenth century, careful calculations using Newton's law of gravitation indicated that perturbations of Uranus's orbit could be accounted for if there were an unknown planet beyond it. The position of this new planet was calculated, and telescopes focused on that region of the sky immediately showed the new planet. It was given the name Neptune. Similar but much smaller perturbations of Neptune's orbit led to the discovery of Pluto in 1930.

In addition to the nine known planets, a large number of tiny objects known as **asteroids** orbit the sun between Mars and Jupiter.

Figure 3–35

Gravitational force due to one planet can cause perturbation of another planet's orbit.

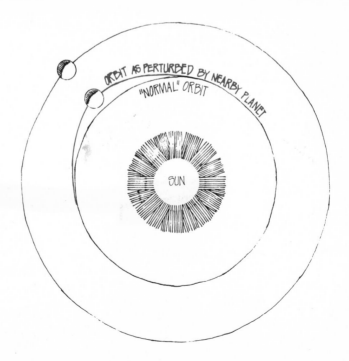

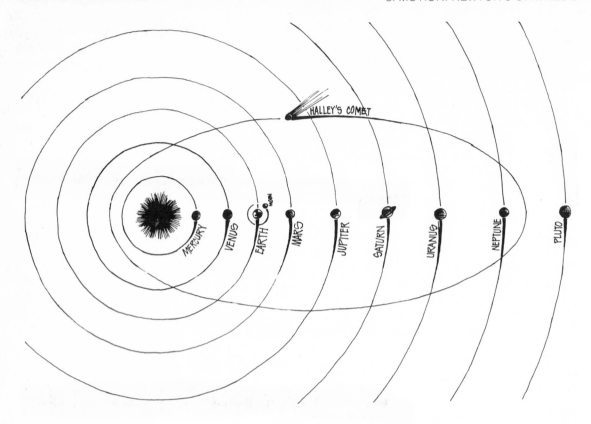

Figure 3–36

The solar system, including the orbit of Halley's comet.

A number of comets, which are also relatively small objects and often have very elongated orbits, also orbit the sun. The most famous is Halley's comet, which passes the earth every 75 years, Fig. 3–36.

The Ocean Tides Are Due to the Gravitational Attraction of the Moon

Thousands of years ago, it was noticed that a correlation exists between the tides and the moon. But this connection remained a mystery until the seventeenth century. Newton reasoned that the two high tides (Fig. 3–37) per day are due to the gravitational attraction of the moon. Because the gravitational force decreases with the square of the distance, the moon pulls harder on that part of the earth closest to it. This pull has only a slight effect on the land masses, but the great masses of water that make up the oceans are free to move. The water in the oceans is attracted toward the moon, resulting in a high tide on the side of the earth facing the moon, Fig. 3–38. Simultaneously a high tide occurs on the opposite side of the earth. The latter high tide is because the pull of the moon on the far side of the earth, since it is a greater distance from the moon, is less than the pull on the central part of the earth. In a sense, the water on the far side of the earth is *left behind,* because its acceleration is less than that of the earth as a whole.

Thus, the ocean water congregates on the sides closest and farthest from the moon and is drawn away from the regions between. Since the earth revolves on its axis once a day beneath the moon, each place on the earth's surface experiences two high tides and two low tides per day.[6] Because the high tides stay almost in line with the moon, it is as if the solid earth moved beneath the tidal bulges.

Because there is friction between the solid earth and the moving water, and because the sun also exerts a force on the ocean's water, the high tides do not necessarily correspond precisely to the time at which the moon is directly overhead. These factors also affect the height of the tides. For example, the highest high tides and the lowest low tides occur when the sun and moon are lined up so that both are pulling in the same direction. These are the so-called **spring tides**. When the sun and moon are at right angles, the tides are smallest (**neap tides**). See Fig. 3–39.

5. THE IMPACT OF THE NEWTONIAN SYNTHESIS

Newton's work has had a considerable influence on the world of thought. With the three laws of motion and the law of universal gravitation, Newton was able to describe the motion of objects on earth and how this motion is caused. He was able to mathematically deduce Kepler's laws of planetary motion and to explain the motion of the planets around the sun and the motion of the moon around the earth. Galileo's contention that the heavens and the earth are not two separate realms of activity was further strengthened by Newton's work. The motions

[6]Actually there are not two high tides every day at each point on the earth's surface, but two high tides every 25 hours or so. This is because the moon returns to its same position overhead about every 25 hours.

Figure 3–37

A pier or dock reaching out into the ocean showing water level (A) at high tide and (B) at low tide.

Figure 3–38

The tides on the earth are due to the movement of water attracted to the moon by gravity.

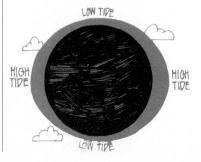

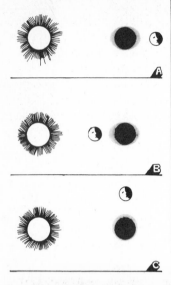

Figure 3–39

The tides are greatest when the sun, moon, and earth are in line (A) and (B), and least when they make a right angle (C).

of celestial and earthly bodies were seen to follow the same laws. Nature made no distinction between the heavens and the earth. The universe is one.

Causality

Suppose two events occur sequentially. If upon repetition of the first event, the second is repeatedly found to occur without exception, such as a rock striking a window and then the window breaking, we often say that the first event *caused* the second—the ball striking the window caused the window to break. Similarly, when we let go of a pencil, experience tells us that it always falls to the floor; we say that the force of gravity caused it to fall. This notion of **causality**, or "cause and effect" as it is sometimes called, was well known long before the time of Newton. It was used, for example, by Thomas Aquinas in the thirteenth century in his arguments for the existence of God. But with Newton's laws, causality took on a more forceful meaning. The motion, or rather the acceleration, of any body was seen to be *caused* by the force applied to it. Newton's laws are thus often referred to as **causal laws**.

Newton's View of the Universe Influenced Other Fields

Newton's brilliant work, which is so encompassing as to be a theory of the universe, dominated science for well over two centuries. The influence of Newtonian mechanics extended far outside the realm of physics, however. By the eighteenth century a philosophy had developed based largely on the new mechanics. In this philosophy nature was considered a collection of particles moving in predictable paths according to natural laws. The universe was pictured as a big machine whose parts move in a predictable and predetermined way. In fact, the great French mathematician Pierre Simon de Laplace (1749–1827) claimed that if the positions and velocities of all the particles or atoms in the universe—and the forces between them—were known at one time, the future course of the universe, including human behavior, could in principle be calculated. Thus, according to Laplace, the unfolding of events in the universe is already determined.

The mechanistic and deterministic view of the natural world was slowly absorbed by the general culture, and the Age of Reason was, at least in part, a direct outcome. The existence of exact laws in physics suggested to the minds of the eighteenth and nineteenth centuries that similar laws might be found in other fields of human thought. Thinkers in such diverse fields as history, politics, ethics, and psychology attempted to construct theories based on precise natural laws. For example, Woodrow Wilson pointed out that the political philosophy of the British Whigs, which influenced the American Revolution, was a copy of Newtonian physics.

Even in the twentieth century the mechanistic view is a potent force. Behavioral psychology, for example, is recognizably based on the Newtonian conception of the world. According to the developers of this school, a person's behavior is "caused" by the variety of experiences he or she has had from childhood on. Human freedom finds little room here.

The famous defense attorney Clarence Darrow carried the mechanistic view of humanity to the extreme. In the 1920s he would, on occasion, defend a client whose guilt was beyond doubt by arguing that the person was a victim of heredity and environment. The guilty person's actions were determined by the fixed laws of nature, he argued, and therefore how could society, which helped to form his client's character, presume to punish him or her?

Whether such extreme views are tenable cannot be answered by science alone. As we saw in Chapter 1, science is a human endeavor. The laws "discovered" by Newton were not preexisting entities waiting to be uncovered. They were, rather, magnificent summaries of observations created within someone's mind.

Thus, questions such as those concerning human freedom can probably not be answered by a single aspect of human endeavor, such as science. On the other hand, science may be able to shed some light on these questions. In Chapter 15 we shall see, for example, that modern discoveries in physics indicate that the physical world does not obey strict deterministic laws, and that in the tiny world of the atom the idea of causality fails.

REVIEW QUESTIONS

1. The fact that an object resists a change in its motion is called
   (a) inertia
   (b) weight
   (c) gravity
   (d) centrifugal force

2. The mass of a body is
   (a) how heavy it is
   (b) a measure of its inertia
   (c) dependent on gravitational pull
   (d) different on the moon than on earth

3. A horse pulls on a cart with a force of 100 lb. The cart pulls back on the horse with a force
   (a) dependent on the acceleration of the cart
   (b) of less than 100 lb so the horse can move it
   (c) of more than 100 lb if the horse cannot move it
   (d) of 100 lb
   (e) none of these

4. If we know the magnitude and direction of the net force acting on a body of given mass, Newton's second law allows us to calculate its
   (a) position
   (b) speed
   (c) velocity
   (d) acceleration

5. When a boy pulls a wagon, the force that makes him move forward is
   (a) the force the wagon exerts on him
   (b) the force he exerts on the wagon
   (c) the force he exerts on the ground with his feet
   (d) the force the ground exerts on his feet

6. The same force applied continuously to a moving body gives it
   (a) accelerated motion
   (b) centripetal acceleration
   (c) uniform motion
   (d) constant velocity

7. Newton sought to discover what gives rise to forces. His answer was that a force applied to an object
   (a) is applied by another object
   (b) is due to energy
   (c) is due to motion of the object
   (d) none of these

8. The action and reaction forces of Newton's third law
   (a) act on different bodies
   (b) act on the same body
   (c) act along the same line but are not necessarily equal in magnitude
   (d) are always equal in magnitude but may not act along the same line

9. A rocket accelerates because
   (a) the gases it expels push against the ground
   (b) the gases push against the rocket
   (c) the ground pushes up on the rocket
   (d) the rocket exerts a force on itself

10. You are flung sideways when your car travels around a sharp curve because
    (a) you tend to continue moving in a straight line
    (b) there is a centrifugal force acting on you
    (c) the car exerts an outward force on you
    (d) of gravity

11. When a football is kicked, the vertical component of its velocity is greatest
    (a) just after it leaves the kicker's foot
    (b) at the highest point of its flight
    (c) just before it hits the ground
    (d) both (a) and (c)

12. How would earth's gravity affect a rifle bullet fired horizontally?
    (a) the bullet would begin to fall when it had lost most of its speed
    (b) the rate of fall would depend on the horizontal speed
    (c) the bullet would fall just as fast as if it had no horizontal speed
    (d) none of these

13. When you observe an object moving in a circle, what can you infer?
    (a) there is a centrifugal force on the object
    (b) something is pushing the object around in the circle
    (c) a force directed toward the center of the circle is acting on the object
    (d) there are no forces acting on the object.

14. According to Newton, the moon is held in its orbit around the earth by
    (a) gravity   (c) the tides
    (b) centrifugal force   (d) the sun

15. Newton explained the ocean tides as being due to
    (a) his second law
    (b) his third law
    (c) the gravitational pull of the earth
    (d) the gravitational pull of the moon

EXERCISES

1. Why does a child in a wagon fall backward when another child suddenly pulls the wagon forward?

2. (a) Is it proper to say that "bodies at rest tend to stay at rest and bodies in motion tend to stay in motion as long as no force is acting" *because of* inertia? Or should we say that we use the word inertia to *describe* this property of bodies? (b) Comment on the statement "We often think that by giving a name to some concept we then understand it."

3. A rear-end collision between a soft-drink truck and a car occurs. A lawsuit develops over who is at fault. The truck driver claims the car backed into him. The auto driver claims the truck hit him from behind. The only evidence is that quite a number of soda bottles fell forward into the truck driver's seat. From the evidence, can you tell who was at fault?

4. What is the principle behind a magician's ability to pull a tablecloth from beneath china and glassware without breaking them?

5. Why do automobiles use more gasoline for city driving than for highway driving?

6. Why do you exert more force on the pedals of a bicycle when you first start out than when you have reached a constant speed?

7. When a tennis ball is dropped to the floor, it bounces back up. Was a force required to make it come back up? If so, what exerted the force, and how?

8. Analyze the motion of your leg during one stride while walking, using Newton's first and second laws.

9. Why does a person wearing a cast on an arm or leg feel more tired than usual at the end of the day?

10. Whiplash is a common result of an auto accident when the victim's car is struck from the rear. Why does the victim's head get thrown backward in this kind of accident?

11. What is the net force being exerted on a 1400-kg car accelerating at 3 m/sec²?

12. What is the acceleration of a 6000-kg rocket if the net force on it is 30,000 N?

13. When you are running and you want to come to a quick stop, you must decelerate rapidly. Analyze, in the light of Newton's second and third laws, the origin of the required force.

14. When you lift a bag of groceries, you exert an upward force on the bag. Newton's third law says that there is a "reaction" force to your upward force on the bag. What object exerts this reaction force and what object feels it? In which direction does this force act?

15. Explain, using Newton's third law, why when you walk on a log floating in water, the log moves backward as you move forward. Draw a diagram of the log and a person walking on it, and show the motion of each.

16. Why does it hurt your toe when you kick a rock?

17. Can two vectors, neither of which is zero, add up to a vector that is zero? Show how on a diagram.

18. Can two vectors of equal length add up to give a sum vector whose length is the same as each of the original vectors? Show on a diagram.

19. A force of 8 pounds is exerted on an object in one direction and a force of 6 pounds is exerted on the same object in a perpendicular direction. What is the net force on the object? Use a diagram to show the direction of the net force.

20. You push on a small car with a force of 600 N at a 45° angle to the horizontal. How much force actually goes into moving the car along a level road?

21. A child is pulling a sled in the snow. Draw a diagram showing the force the child exerts on the sled, and the horizontal and vertical components of this force.

22. How much would you weigh on the moon? Compare to your weight on earth.

23. Calculate your mass in kilograms. What would be your mass on the moon?

24. When you are standing on the ground, how large is the upward force exerted on you by the ground? Why doesn't this force make you fly up in the air?

25. A rock weighs 10 pounds. What is the net force on it when you hold it? What is the force on it just after it is dropped? What is the force on it just before it hits the floor?

26. The pilot of an airplane traveling at about 200 km/hr wishes to drop supplies to flood victims stranded on a small island. Draw a diagram showing the airplane and the island; show the approximate point where the plane must release the package of supplies and the subsequent path of the package through the air.

27. At what point in its path is the speed of a projectile the least?

28. You are hiding in a tree from a sniper who is quite far away. You see the sniper aim directly at you and fire. Should you remain where you are or drop from the tree as the rifle is fired?

29. A diver running 5 m/sec reaches the edge of a vertical cliff and dives toward the water below. She reaches the water after 2 sec. How high was the cliff? How far from the base of the cliff did she hit the water?

30. If a bucket of water is swung in a vertical circle at a high enough speed, the water won't spill at the top of the circle when the bucket is upside down. Explain.

31. A racing car travels around a curve of radius 30 meters at a speed of 30 m/sec. What is its centripetal acceleration? Express the answer in "g's"; that is, how many times larger than the acceleration due to gravity, $g = 9.8$ m/sec², is this?

32. Why are curves on a highway banked?

33. Why does a car tend to skid on an icy curve?

34. Explain why a car is less likely to skid when traveling around an icy level curve if it does so at a low speed.

35. Sometimes it is said that water is removed from clothes in a spin dryer by centrifugal force. Why is this incorrect?

36. If you were asked "What keeps a satellite in its orbit around the earth," what would you say?

37. Astronauts spending considerable time in outer space are affected by weightlessness. Gravity can be simulated, however, by having the space vehicle rotate. This is conveniently done if the space vehicle is doughnut shaped and rotates slowly like a wheel. Explain how this could simulate gravity.

38. The farther you are from the earth, the weaker is the force of gravity acting on you due to the earth. What would your weight be if you were in a slowly moving space vehicle 4000 miles above the earth's surface?

39. Distinguish between the gravitational force exerted on the moon by the earth and the gravitational force exerted by the moon on objects at its surface.

40. Would you expect the escape velocity from the moon to be greater than, less than, or equal to the escape velocity from the earth? Why?

41. Is the acceleration due to gravity the same everywhere?

42. Does an egg exert a gravitational force on the earth? How large a force?

43. Would your weight on top of Mt. Everest be more or less than your weight at sea level?

44. Suppose both the mass of the earth and the mass of the moon were double their present values but that the distance between them remained the same. By what factor would the force of gravity between them change? Would the speed of the moon around the earth have to increase, decrease, or remain the same?

45. If the moon rises shortly after the sun sets, do you think the tides will be large or small? What will the tides be like if the moon is directly overhead when the sun sets?

4

ROTATIONAL
MOTION
AND
BODIES
IN
EQUILIBRIUM

A. Initially

B. A short time later

Figure 4-1

A rotating wheel.

Torque, Rotational Inertia, and Newton's Laws

1. ROTATIONAL MOTION

Up to now we have been discussing what is called **translational motion**—that is, the motion of bodies as a whole without regard to any rotation they might undergo. We now consider **rotational motion**—the motion of a body rotating about an axis. Rotating wheels, a swinging door, and the earth itself exhibit rotational motion.

Rotational motion is normally described using angular quantities. We can specify the position of a rotating wheel by giving the angle θ that a line from the center of the wheel makes with its original position, Fig. 4–1. The angular velocity is the rate at which the angle θ changes. It is commonly specified by giving the angular frequency in revolutions per minute (rpm) or in revolutions per second (rev/sec). If the angular velocity changes, then there is said to be an angular acceleration (defined as the rate at which the angular velocity changes).

The analysis of translational motion using Newton's laws can be extended to rotating bodies. Instead of ordinary velocity and acceleration, we use the corresponding angular quantities. And in place of force we use the concept of **torque.** Torque is closely related to force, but before we define it, try the following experiment.

EXPERIMENT

Open a door slightly. Now exert a force on the door near its end. Be sure to push at right angles to the door, Fig. 4–2(A). Next, exert the same force on the middle of the door as shown in (B). Third, exert a force of the same magnitude at a 45° angle near the end of the door as shown in (C). Finally, exert a force on the end of the door (D). In each case, note how quickly the door opens.

You undoubtedly found that the door rotated more quickly in (A) than in (B) and (C), and that in (D) the door didn't rotate at all. Indeed, careful observations show that the angular acceleration of the door is not only proportional to the magnitude of the force. It is also proportional to the **lever arm**, which is designated ℓ in Fig. 4–2; the lever arm is defined as the perpendicular distance from the axis of rotation to the line along which the force acts. Note that in (D), $\ell = 0$; so regardless of how hard you push it; the door will not

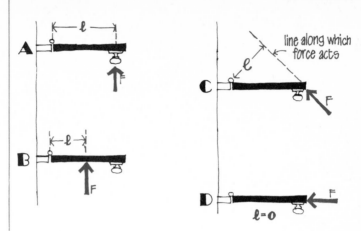

Figure 4-2
Applying a torque to a door.

rotate. The product of the magnitude of the force and the lever arm is defined as the **torque**:

$$\text{torque} = \text{lever arm} \times \text{force}.$$

In symbols, $\tau = \ell \times F$, where τ stands for the torque. If in Fig. 4–2(A) you push on a 2-foot-wide door with a force of 5 pounds, the torque would be 10 lb•ft. In (D) the torque is always zero, because $\ell = 0$.

From our discussion and the experiment you tried, it should be clear that *torque* (and not simply force) gives rise to rotational motion. That is, torque plays the same role in rotational motion that force does in ordinary translational motion. Indeed, the rotational equivalent of Newton's second law states that *the angular acceleration is proportional to the applied torque.* Now we must examine what plays the role of mass (or inertia) for a rotating body.

You have probably noticed that rotating objects, such as a spinning bicycle wheel or top or the earth itself, tend to continue spinning. This tendency is the rotational equivalent of inertia and we call it **rotational inertia.** Indeed, the rotational equivalent of Newton's first law states that *a rotating body continues to rotate uniformly as long as no net torque acts to change this motion.*

An object of large mass normally has a large amount of rotational inertia. However, the mass is not the only thing that determines how much rotational inertia an object has. How this mass is distributed is also important. If more of the mass is concentrated a large distance from the axis of rotation, the rotational inertia will be greater. For example, a large-diameter cylinder will have much greater rotational inertia than a small-diameter cylinder of the same mass (Fig. 4–3). The former will be much harder to start rotating and harder to stop once it is rotating. In other words, a greater torque will be

Figure 4-3

Two cylinders of equal mass. The one with the larger diameter has the greater rotational inertia.

Figure 4–4
A seesaw.

needed to give the same angular acceleration to a larger-diameter wheel than to a smaller-diameter one, or, for the same net torque, the angular acceleration will be less if the rotational inertia is greater. Thus we can say that *the angular acceleration of a body is directly proportional to the net torque applied to it and inversely proportional to the rotational inertia.* This is the rotational equivalent of Newton's second law, $a = F/m$ (see Chapter 3).

As a simple example of torque, let us consider a child's seesaw. It is well known that if one child on the seesaw is heavier than the other, the pivot support must be closer to the heavy child. Exactly where the pivot should be can be found by the torque equation, since the torques produced by the weight of each child must balance one another. In Fig. 4–4, child A tends to rotate the seesaw counterclockwise and child B tends to rotate it clockwise. There will be a balance when these two opposite torques have the same magnitude (so the sum of them, or the net torque, is zero). Thus, if ℓ_A = 3 ft, the torque produced by child A is 50 lb × 3 ft = 150 lb•ft. To balance this, ℓ_B must be 5 ft so that the weight of child B will also produce a torque of 150 lb•ft, that is,

$$\ell_B = 150 \text{ lb•ft}/30 \text{ lb} = 5 \text{ ft.}$$

Center of Gravity and Center of Mass

If you throw an object through the air so that it rotates—say a football going end-over-end or the bowling pin shown in Fig. 4–5—its motion may appear wobbly, but one point in the object follows the smooth parabola curve of a projectile. This point is called the **center of gravity** or **center of mass** of the body.[1] This point is the average position of all the particles in the body. To determine the translational motion of a body, we can assume that *the force of gravity on the body acts at the center of gravity.* Thus, the motion of a body

Figure 4–5

The motion of a body, in this case a bowling pin moving as a projectile, can be considered as the translational motion of its c.g. plus rotational motion about its c.g.

[1]Technically, these terms differ, but for most practical purposes they are the same.

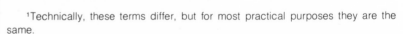

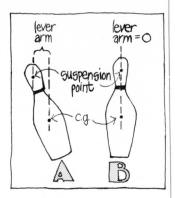

Figure 4–6

(A) A torque acts on a body if its c.g. is not directly below the point of support. The body can remain at rest (B) only if the c.g. is below the suspension point.

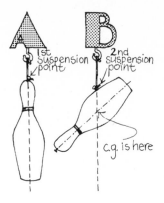

Figure 4–7

Determining an object's center of gravity by suspending it from two different points and drawing vertical lines.

such as the bowling pin in Fig. 4–5 can be considered the translational motion of its center of gravity (c.g.) plus rotational motion about its center of gravity.

The center of gravity of regularly shaped objects that are uniform in composition, such as cylinders, spheres, and rectangular solids, is at their geometric center, as you would expect. To determine the c.g. of an irregular body, we make use of the following fact. When an object suspended from a point lies at rest, its center of gravity lies directly below (or at) the point of suspension. If this were not true, a torque would act to rotate the object as shown in Fig. 4–6(A). Only when the c.g. is below the suspension point does the lever arm, and therefore the torque, equal zero, Fig. 4–6(B). The center of gravity of an irregularly shaped body can thus be found by suspending it from two different points and drawing in the vertical line for each case as shown in Fig. 4–7. The intersection of these lines is then the center of gravity.

Figure 4–8

The center of gravity of a high jumper or pole vaulter can actually pass beneath the bar.

93

> ### EXPERIMENT–PROJECT
> Cut an arbitrary shape, such as that shown in Fig. 4–6 or 4–7, from a piece of cardboard and then determine its center of gravity.

The center of gravity of an object may not lie within the body itself. This is clearly true for a doughnut—its c.g. lies at the center of the hole.

Another interesting example is that of a person in a bent position as shown in Fig. 4–8. This diagram shows a high jumper. Note that the c.g. actually passes *beneath* the bar. By contorting his or her body in such a way, a jumper can clear a higher bar for a given take-off speed.

2. BODIES IN EQUILIBRIUM

Conditions for Equilibrium

We have been concerned mainly with bodies in motion. Now we consider a body at rest. Because the velocity and acceleration are both zero in this case, Newton's second law tells us the net force on the object must also be zero. But that does not mean there are no forces acting at all. In fact, it is virtually impossible to find an object on which there are no forces. In everyday situations, gravity acts on every body. And if a body is at rest, some other force must be acting on it to balance the force of gravity. For example, we saw in Fig. 3–19 that the ground pushes up on a body with a force exactly equal to the force of gravity (the person's weight) pulling down.

A body at rest is said to be in equilibrium. The study of the forces acting on a body at rest is useful in many situations ranging from the forces on muscles in the human body to the internal forces within a building or other structure. After we know one or more of the forces acting on a body, we can often determine what other forces are acting by using the fact that the sum of all the forces is zero.

One reason the subject of bodies at rest is so important is that materials change shape under the action of forces. If the forces are great enough, the object may break or fracture. Thus, architects and engineers use the ideas of equilibrium, as do physical therapists and others in the medical field.

For a body to be in equilibrium, it is clear that *the sum of all the forces acting on it must be zero*. This is called the "first condition for equilibrium," but it is not sufficient. Consider, for example, the log in Fig. 4–9. There are equal forces acting in opposite directions, so the net force on the log is zero. But there is a net *torque* on the log and so it will rotate. Thus, for a body to be at rest the *sum of all the torques acting on it must be zero* also. This is the second (and final) condition for equilibrium.

We have already seen an application of this principle in Fig. 4–4. The seesaw was in equilibrium when the two torques were balanced —that is, they were equal in magnitude but acted in opposite directions.

Forces in Our Muscles and Joints

The principle of equilibrium, or **statics** (meaning "at rest") as it is sometimes called, can be used to find the strength of forces in the human body. Let us calculate the force exerted by the biceps muscle in your upper arm when you hold a 15-lb object in your hand. (We use pounds rather than newtons since they are so much more

Figure 4–9
The net force is zero, but the log rotates because there is a net torque.

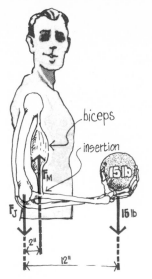

Figure 4–10

Forces exerted on the lower arm when a 15-lb weight is in the hand.

familiar.) Fig. 4–10 shows the bones and biceps muscle in an average person's arm. The arm is held in the position shown because the biceps muscle contracts and exerts an upward force F_M on the lower arm. This force acts at the point at which the muscle is attached to the forearm; this "insertion" point is normally about 2 inches (1/6 ft) from the elbow joint, which acts as a pivot point. Two other forces act on the arm (we ignore the weight of the arm itself, as it is relatively small): the force of the 15-lb object pushing down on the hand, which acts about 12 inches (1 ft) from the elbow as shown and the force F_J that the upper arm exerts on the lower arm at the joint. The latter gives rise to no torque, since its lever arm is zero. The torque due to the 15-lb object tends to rotate the arm downward (or clockwise about the elbow) and the torque due to F_M tends to rotate the arm upward (or counterclockwise). Because the arm is in equilibrium, these two opposite torques must be equal in magnitude:

$$(15 \text{ lb}) \times (1 \text{ ft}) = (F_M) \times (1/6 \text{ ft}).$$

So F_M equals (15 lb)(6) = 90 lb.

The force exerted by the muscle is much larger than the weight of the object lifted, because the muscle is connected to the lower arm at a point close to the elbow. A person whose muscle is inserted slightly farther out, say 2½ inches instead of 2 inches, would have a considerable advantage in lifting and throwing. Indeed, successful athletes often have muscle insertions farther from the joint than those of the average person.

The net force on the arm in Fig. 4–10 must be zero. Since F_M = 90 lb acts upward and the 15-lb weight acts downward, there clearly must be another force acting downward to balance them. This force must be the one exerted by the upper arm at the elbow joint F_J. Because the sum of the downward forces, F_J + 15 lb, must equal the upward force F_M = 90 lb, F_J must be 75 lb. Thus, the force exerted at the joint F_J and that exerted by the muscle F_M are much larger than the weight of the object lifted. Indeed, throughout the body, the muscles and joints are subjected to forces that are often as great as 500 lb or more.

As another example, let us examine the forces involved when a person bends over to pick up something, as shown in Fig. 4–11(A). The upper body is supported by back muscles that act at an angle of about 15° to the body axis. The force exerted by these muscles is represented by F_M in Fig. 4–11(B). This force acts about the hip joint with a lever arm of approximately 4 inches in a person 6 ft tall; its torque balances the torque due to the weight W of the upper body acting at its center of gravity with a lever arm of approximately 16 inches. The upper body contains about two-thirds

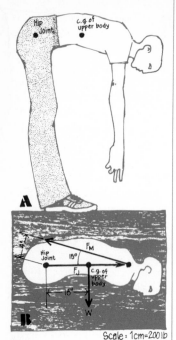

Scale: 1cm=200lb

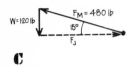

Figure 4–11

Forces exerted on the back when bending over.

Figure 4–12

Objects in (A) stable equilibrium and (B) unstable equilibrium.

of the total body weight. In a 180-lb person, then, the weight of the upper body W is about 120 lb. Thus,

$$(120 \text{ lb})(16 \text{ in.}) = (F_M)(4 \text{ in.}),$$

so $F_M = 4 \times (120 \text{ lb}) = 480 \text{ lb}$! On a vector diagram, Fig. 4–11(C), we can draw the vectors W and F_M to scale, and the vector that completes the triangle (so the sum of all forces is zero) is the force F_J that must be exerted at the base of the spine. Using a ruler, we measure F_J to be about 460 lb. This incredibly large force is exerted on the lower part of the spine, which is composed of small bones called vertebrae, which are separated from one another by flexible fluid-filled disks. These disks are thus compressed under tremendous force when a person bends over. It is no wonder that backaches are common and that occasionally a disk will rupture. You can protect your back by squatting with your knees bent to pick up an object and then letting your legs lift you instead of bending over.

3. STABILITY AND BALANCE

If a body in equilibrium at rest is displaced slightly, three things can happen: (1) it may return to its original position, in which case we say it is in **stable equilibrium**; (2) it may move even farther from its equilibrium position of its own accord, in which case we say it is in **unstable equilibrium**; or (3) it may remain in the new position, in which case we say it is in **neutral equilibrium**.

To make this clear, let us look at examples of each possibility. Remember that the force of gravity on an object can be considered to act at its center of gravity. A ball suspended from a string as shown in Fig. 4–12(A) is in stable equilibrium, for if it is displaced to one side as shown, it returns to its original position. A pencil standing on its tip is in unstable equilibrium [Fig. 4–12(B)], for if it is displaced even slightly from the exact balance point, there will

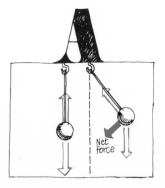

Figure 4-13
Object in (A) and (B) is stable but in (C) is unstable.

Figure 4-14
With your nose touching the end of the door, put your feet astride the door and try to rise up on your toes.

be a torque that acts to make it fall. Finally, a ball on a flat table is in neutral equilibrium, for if it is displaced to one side, it remains in its new position.

In most situations, such as those that occur in the human body or in buildings, we are interested mainly in maintaining stable equilibrium, or **balance**. In general, when a body's center of gravity is below its point of support, it will be in stable equilibrium. The ball on the string in Fig. 4-12(A) is an example. If its c.g. is above the point of support, the situation is more complicated. Consider a milk carton standing on its end as shown in Fig. 4-13(A). If it is tipped slightly (B), it will return to its original position due to the torque that acts on it. But if it is tipped too far (C), it falls over. The critical point is reached when the c.g. of the object is beyond the base of support. This is because the upward force on the base of the object can act only within the area of contact; and if the force of gravity acting at the c.g. is beyond the base, there is a net torque that causes the object to fall. Thus, *a body will be stable if its c.g. lies within the base of support.*

When the c.g. is above the base of support, we see that stability is relative. A milk carton lying on its side is much more stable than one standing on its end, for it will take a greater effort to tip it over. The pencil in Fig. 4-12(B) is an extreme case, because its base of support (the pencil tip) is so very tiny. In general, the larger the base and the lower the center of gravity, the greater is the stability of the object.

In this sense, humans are much less stable than four-legged animals, which have a broader base of support because of their four legs and a lower center of gravity. Humans have only their two feet as a base of support. A person who is walking or performing other sorts of movement must continually shift his or her body so that its c.g. is over its base of support. We are able to do this without conscious thought, although our brains had to learn how. What babies are learning when they "learn" to walk is how to keep their c.g. over their feet.

EXPERIMENT

Stand with your heels and back against a wall. Now try to bend forward. Why can't you do it without falling over? What do you normally do when you bend over that keeps you from falling?

AND

Stand facing the edge of an open door with your nose touching the door and your feet on either side of it as shown in Fig. 4-14. Now try to rise up on your toes. Why can't you do it?

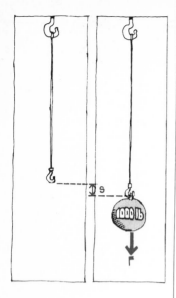

Figure 4–15

A wire is stretched an amount s proportional to the magnitude of the force F.

4. ELASTICITY AND FRACTURE

An ordinary object in equilibrium has forces exerted on it. These forces can change the shape of the body and if strong enough can cause it to break, or **fracture**.

Elasticity and Hooke's Law

Robert Hooke (1635–1703) found that the amount of deformation of a body is proportional to the amount of force being applied. For example, suppose a long wire is suspended as shown in Fig. 4–15. When a weight is hung on it, the wire stretches. The amount of elongation or stretch s is proportional to the force F exerted on it:

$$s \propto F.$$

This relationship is usually written as an equation,

$$F = ks,$$

and is known as Hooke's law. Thus, if a force of 2000 newtons stretches a wire 1 mm, a force of 4000 newtons will stretch it 2 mm, and so on.

Hooke's law is found to hold for most materials but only up to a point. If the force is too large, the object stretches excessively and eventually breaks. Fig. 4–16 shows a graph of force versus elongation s for a typical material. Up to a point called the **elastic limit**, the object follows Hooke's law quite well; and when the force is released, the object returns to its original shape. This ability is called **elasticity**. If the force exceeds the elastic limit, the object becomes permanently deformed. If the force is increased much more than the elastic limit, the object breaks. The force at the breaking point is called the **ultimate strength** of the material.

Tension, Compression, and Shear

Let us consider the wire in Fig. 4–15 again. Not only is there a force pulling down on the wire, there is also a force pulling up at the top (it's in equilibrium, remember). This force is shown in Fig. 4–17(A). The same force also exists within the material itself. Consider the lower part of the wire shown in (B). This lower section is also in equilibrium (it is at rest), which means that the upper part of the wire exerts a force on the lower part. This relationship must be true at any point in the wire. Thus, external forces applied to an object give rise to forces (or **stress**[2]) within the material itself.

The wire in Fig. 4–17 is said to be under **tension** or **tensile**

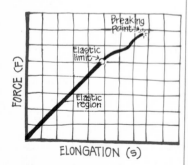

ELONGATION (s)

Figure 4–16

Graph of the force applied to a material versus its elongation.

[2]*Stress* usually refers to the force per unit area of cross section.

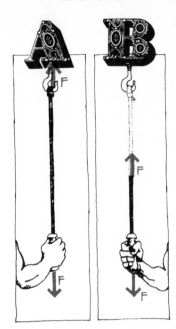

Figure 4–17

(A) When a force is exerted downward on the wire, the support at the top pulls upward. (B) The force exists within the wire itself, since any section of the wire must be in equilibrium.

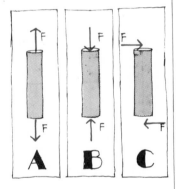

Figure 4–18

The three types of stress.

stress. Two other common kinds of stress are **compression** and **shear**. All three types are illustrated in Fig. 4–18. When an object is under tension (A), it is being stretched. But when the two forces act *toward* each other, as in (B), they act to shorten or compress the object. This stress is called *compression,* and Hooke's law works in this case, too. Anything that supports a weight, such as the columns of a Greek temple (Fig. 4–19), is subjected to compression. *Shear,* on the other hand, occurs when equal and opposite forces act along different lines, as in Fig. 4–18(C); shear tends to change the shape of the object from, say, a rectangle to a parallelogram. If you lay this book on a table and push across its top with your hand, the book will be undergoing shear.

Fracture

If the stress in an object is too great, the object will break. This can happen as a result of tension, compression, or shear, as shown in Fig. 4–20. The point at which fracture occurs depends on the cross-sectional area of the material—the greater the area, the greater the force it can withstand. Hence, the **ultimate strength** of a material is specified as the force per unit area (N/m^2 or $lb/in.^2$, usually) at which that material will normally break under a given kind of stress. Table 4–1 gives the ultimate strengths of a number of materials under tension, compression, and shear. It can be seen that some materials are weak under tension and shear compared to their

Figure 4–19

The columns of a Greek temple (this one is in Sicily) are under compression.

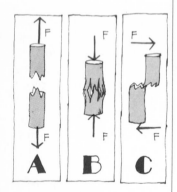

Figure 4–20

If the stress in an object is too great, the object fractures.

strength under compression. Because materials vary somewhat in composition, the numbers in the table are only a guide. When designing a building, architects generally use a safety factor of 5 to 10. That is, they assume a material can withstand only 1/5 to 1/10 of its ultimate strength.

As an example, suppose a marble column in a Greek temple (Fig. 4–19) supports 600,000 kg. The total weight is then $mg = (600,000 \text{ kg}) (9.8 \text{ m/sec}^2) = 6 \times 10^6$ N. Marble has an ultimate compressive strength of 80×10^6 N/m² according to Table 4–1. If we use a safety factor of 10, we say the column can support 8×10^6 N/m². Since it only needs to support 6×10^6 N, the cross-sectional area needs to be $(6 \times 10^6 \text{ N}) (8 \times 10^6 \text{ N/m}^2) = 0.75$ m².

Table 4–1
Ultimate Strengths of Materials (Force/Area)

MATERIAL	TENSILE STRENGTH		COMPRESSIVE STRENGTH		SHEAR STRENGTH	
	N/m²	lb/in²	N/m²	lb/in²	N/m²	lb/in²
Steel	500×10^6	70×10^3	500×10^6	70×10^3	250×10^6	40×10^3
Concrete	2×10^6	0.3×10^3	20×10^6	3×10^3	2×10^6	0.3×10^3
Marble	very weak		80×10^6	12×10^3	very weak	
Granite	very weak		170×10^6	25×10^3	very weak	
Wood (pine)	40×10^6	6×10^3	35×10^6	5×10^3	5×10^6	0.8×10^3
Bone (limb)	130×10^6	20×10^3	175×10^6	25×10^3		

Since $A = \pi D^2/4$ for a circle, the minimum diameter D is $D = \sqrt{4\ (0.75\ \text{m}^2)/(3.14)} = 1.0$ m.

5. APPLICATIONS TO ARCHITECTURE: ARCHES AND DOMES

The arts and humanities often overlap the sciences, and this is particularly obvious in architecture. Many of the architectural features we admire have practical purposes as well as decorative effects. Consider, for example, the methods used to span a space from the simple beam to arches and domes.

In the so-called post and beam, or post and lintel, construction, two upright posts support a horizontal beam. It is still used a great deal today; but until the introduction of steel in the nineteenth century, the span of a beam was very limited, because the strongest building materials were stone and brick. The size of a span was therefore limited by the size of available stones. But of equal importance is the fact that all these materials, though strong under compression, are exceptionally weak under tension and shear. And as shown in Fig. 4–21, all these stresses occur within a beam, because it sags (slightly) under its own weight. If the weight is too great, the beam will fracture on its bottom surface where the tension is greatest. That only a minimal space could be spanned using stone is evident in the closely spaced columns of the great Greek temples (Fig. 4–19).

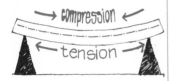

Figure 4–21

A beam sags under its own weight and thus undergoes compression (the upper half is shortened), tension (the lower half is elongated), and also shear.

The semicircular arch, introduced by the Romans (Fig. 4–22), had aesthetic appeal and was also a great technological innovation for it could span a wider space. The virtue of a well-designed semicircular arch is that its wedge-shaped stones undergo mainly compressive stress even when they support a considerable load such as the wall and roof of a cathedral. The stones are forced to squeeze against each other and are thus put under compression. (See Fig. 4–23.) By using many well-shaped stones, an arch could be made to span a considerable space. However, it required much buttressing on the sides to support the horizontal components of the force. This was often supplied by walls leading off laterally, and in many of the great basilicas of the Renaissance these were conveniently used to enclose side chapels.

Figure 4–22

A roman arch.

The pointed arch was first used in Europe about A.D. 1100 and was an essential part of all the great Gothic cathedrals. It was apparently introduced for constructional purposes rather than for aesthetics, because it was used where heavy loads had to be supported—such as beneath the tower of a cathedral and as the central arch across the nave; the lesser arches in these early churches remained round. The builders realized that because of the steepness of the pointed arch, the forces due to the weight above

101

Figure 4–23

In a well-designed circular arch, the stones are mainly under compression, for which they are strong.

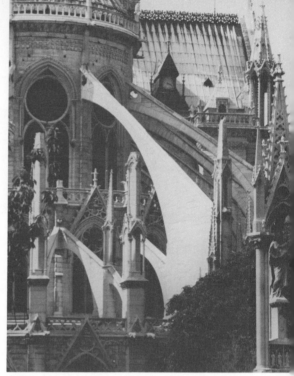

Figure 4–24

Flying buttresses that support the walls of Notre Dame Cathedral in Paris. They are lighter than the heavy buttressing needed for churches built with round arches.

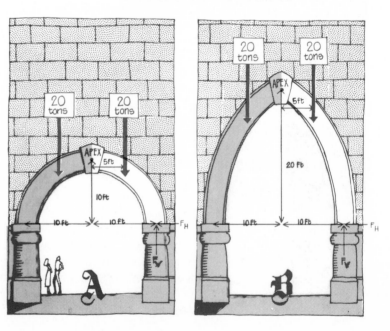

Figure 4–25

Comparison of round (A) and pointed (B) arches.

would be brought down more nearly vertical. Thus, less buttressing was required. The pointed arch took the load off the walls, allowing more openness and light and the glorious use of stained glass windows. The limited amount of buttressing still required was supplied on the exterior by graceful flying buttresses (Fig. 4–24), which were themselves half arches.

Let us now see why horizontal forces occur at the sides of an arch and why they are smaller for a pointed arch than for a round one. In Fig. 4–25 we compare a round arch and a pointed arch that span the same space, 20 feet. The round arch is thus 10 ft high and the pointed arch has been chosen to be 20 ft high. Each supports the same 40-ton weight above, which for simplicity is assumed to act as shown on the two halves of each arch. Let us focus our attention only on the right half of each arch. Since there is a downward focus of 20 tons due to the weight above, the supports must exert an upward force at the base of each arch of $F_v = 20$ tons. Notice that these two forces do not act along the same line, so there seems to be a net torque. Thus the arch would rotate (and crumble) if there were not an additional horizontal force F_H at the base of the arch. If we calculate all the torques about the apex (top) of the arch as if this were an axis, the sum of all torques must be zero. The counterclockwise torque due to F_v is greater than the clockwise torque due to the 20-ton weight, because the lever arm is greater (10 ft versus 5 ft). Hence F_H must point inward to give an additional clockwise torque to achieve balance. It is clear from the diagram that the lever arm for F_H in (B) is 20 ft whereas that in (A) is only 10 ft. To provide the same torque, F_H need be only half as great in (B) as in (A). Thus we see that the horizontal buttressing can be less for a pointed arch because it is higher.

An arch spans a more or less two-dimensional space, whereas a dome—which is essentially an arch rotated on its axis—spans a three-dimensional space. The Romans were the first to build large domes; they were hemispherical in shape and some of them still stand, such as that of the Pantheon in Rome (Fig. 4–26). The technique for building a large dome seems to have been lost by the time of the Renaissance. The problem became critical with the building in Florence of a cathedral designed to have a dome 140 feet in diameter to rival that of the Pantheon. By the early fifteenth century the cathedral was finished except for the dome. A dome, like an arch, is not stable until all the stones are in place. Previously, as each stone was added to a dome, the partially completed structure was supported by a wooden framework, called "centering." But there were no trees big enough to span the 140-ft space required for the Florence cathedral nor strong enough to support such a massive dome. Indeed, that the Romans could have built the Pantheon was a source of wonder. (Even today we do not know pre-

Figure 4–26
The Pantheon in Rome.

Figure 4–27

The skyline of Florence, showing Brunelleschi's famous dome.

cisely how it was done.) A public competition for the design of the Florence dome held in 1418 was won by Brunelleschi. Besides the problem of "centering," Brunelleschi had to contend with the fact that the drum on which the dome was to rest had been completed without external buttressing, nor was there a place to put any. Thus the dome had to exert the minimum horizontal force.

Brunelleschi solved this by constructing a pointed dome (see Fig. 4–27), because only a pointed dome would exert a sufficiently small side thrust on the supporting drum. And he did away with centering altogether. Instead, Brunelleschi built the dome in horizontal layers, each of which was bonded to the previous one, which held it in place until the last stone of the circle was added. The closed ring was then strong enough to support the next layer, and so on. It was an amazing feat. Not until the present century were domes larger than Brunelleschi's built, and these use newly developed building materials such as steel and prestressed concrete.

REVIEW QUESTIONS

1. The rotational equivalent of force is
   (a) rotational inertia
   (b) lever arm
   (c) angular acceleration
   (d) torque

2. The rotational equivalent of mass for translational motion is
   (a) rotational inertia
   (b) lever arm
   (c) angular acceleration
   (d) torque

3. The rotational inertia of a body does *not* depend on
   - (a) its mass
   - (b) its angular velocity
   - (c) how the mass is distributed
   - (d) any of these

4. The following bodies all have the same mass. Which has the greatest rotational inertia?
   - (a) A sphere of radius R
   - (b) A cylinder of radius R
   - (c) A thin hoop of radius R
   - (d) A cube whose side is length R

5. The center of gravity of a body
   - (a) may be outside the body
   - (b) is always inside the body
   - (c) is arbitrary
   - (d) must be at its geometric center

6. A body in equilibrium has
   - (a) no forces acting on it
   - (b) no torques acting on it
   - (c) no net force acting on it
   - (d) acceleration

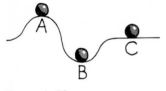

A

C

B

Figure 4–28

7. The ball at point A in Fig. 4–28 is in
   - (a) stable equilibrium
   - (b) unstable equilibrium
   - (c) neutral equilibrium
   - (d) none of these

8. The ball at B in Fig. 4–28 is in
   - (a) stable equilibrium
   - (b) unstable equilibrium
   - (c) neutral equilibrium
   - (d) none of these

9. An object is most stable if its center of gravity is
   - (a) high and the base is narrow
   - (b) low and the base is wide
   - (c) low and the base is narrow
   - (d) below the point of support

10. A body is said to be elastic when a given force is applied if
    - (a) it fractures
    - (b) it returns to its original shape when the force is removed
    - (c) it remains in the new shape given it by the applied force
    - (d) the force exceeds its elastic limit

11. According to Hooke's law, the force needed to stretch a body by an amount s is
    - (a) proportional to s
    - (b) inversely proportional to s
    - (c) proportional to $1/s \, s$
    - (d) independent on s

12. An object fractures when
    - (a) its ultimate strength is exceeded
    - (b) it is elastic
    - (c) the elastic limit is reached
    - (d) it reaches unstable equilibrium

13. When a rope is used to pull a car, the rope is under
    (a) compression   (c) elasticity
    (b) tension   (d) shear

14. A shearing stress on a body tends to
    (a) increase its length   (c) make it wider
    (b) shorten its length   (d) change its shape

15. The pointed arch has an advantage over a round arch because to span the same space
    (a) it is less likely to break
    (b) there is less horizontal force required
    (c) it requires fewer stones
    (d) it does not have to be in equilibrium

EXERCISES

1. Describe the motion of the divers in Fig. 4–29(A) and (B) in terms of translational and rotational motion.
2. A long round log can be rotated about various axes—for example, about the "symmetry" axis through its center, about an axis perpendicular to this (like spinning a baton), or about its end. In which situation would the rotational inertia be the least? The most?
3. A heavy metal cylinder known as a flywheel is attached to the crankshaft of a car and turns with it. What purpose does it serve?
4. Why do high-quality record turntables have a heavy platter?
5. Racing bicycles have fairly large-diameter wheels. Can you see a reason why cyclists can attain high speeds faster if they use very lightweight tires?

Figure 4–29

6. Why does a tightrope walker carry a long, saggy, horizontal pole?

7. Animals that depend on being able to run fast have slender lower legs with flesh and muscle concentrated higher up near the shoulder. Why is this distribution of mass advantageous?

8. A cylindrical hoop and a solid cylinder of the same radius and mass roll down an incline. Which do you think will reach the bottom first?

9. The center of gravity of a meter stick is at its midpoint. Why is this not true of your leg?

10. Show on a diagram how your center of gravity shifts when you change from a standing to a bending position.

11. When two children are balanced on a seesaw, why can one get the advantage by leaning backward?

12. How would you describe the motion of a football kicked end over end?

13. Using the ideas of center of gravity and torque, explain why a ball rolls down a hill.

14. Why is it easier to do sit-ups with your arms extended in front of you?

15. Where should the pivot point be placed under a 3.0-m-long board if it is to be balanced when a 20-kg child and a 40-kg child are on either end?

16. Give an example of a situation in which the net torque on a body is zero but the net force is not. Would such a body be in equilibrium?

17. If the following forces are exerted on a body, can the body be in equilibrium: 100 lb, 200 lb, 50 lb? Explain.

18. If forces of 200 N, 300 N, and 400 N are applied to a body, can it be in equilibrium?

19. If the biceps muscle in Fig. 4–10 were attached 2½ inches from the elbow instead of 2 inches, what force would be required of the biceps muscle?

20. Why is there less strain on your back if you touch your toes while seated on the floor with outstretched legs than if you touch them from a standing position?

21. The ball at C in Fig. 4–28 is in what kind of equilibrium?

22. Do you think the daredevil rider in Fig. 4–30 is really in danger? Explain. What if the people below were removed?

23. In what state of equilibrium is a square box when it is (a) on its edge, (b) on its face?

24. Examine how a pair of scissors or shears cuts through a piece of cloth or cardboard. Is the name "shears" reasonable?

25. Why do you lean backward when carrying a heavy bag of groceries?

26. Why is it harder to overturn a body with a low c.g. and a broad base than it is to overturn a body with a high c.g. and a narrow base?

27. If you double the weight on a column of a Greek temple, how much more does it shorten?

28. A piano string stretches 2 mm when a force of 400 N is applied. How much will it stretch if the force is increased to 1000 N?

29. Why do you bend forward when carrying a backpack?

30. One marble column has twice the diameter of a second one. How

Figure 4–30

107

much more force can the first one withstand than the second without breaking?

31. Which is more likely to fracture under compression, a marble column or a wood column of the same diameter? What about under tension? (Hint: See Table 4–1.)

32. Why must there be a horizontal force exerted at the base of an arch?

33. A round arch and a pointed arch have the same height. Which spans the greater space? Which requires greater horizontal buttressing? Explain.

34. A round arch and a pointed arch require the same horizontal force at their base. Which spans the greater space? Explain.

35. Show that the horizontal force required at the base of the arches in Fig. 4–25 is $F_H = 10$ tons for the round arch (A) and $F_H = 5$ tons for the pointed arch (B). (Hint: Calculate each torque about the apex of the arch and set the sum of them equal to zero.)

5

ENERGY
AND
MOMENTUM

The concept of energy is basic to every one of the sciences. And yet it is difficult to define exactly what energy is. We usually think of energy as distinct from matter. Yet matter can *contain* energy. For example, we say that a child who runs around a great deal is "very energetic." Or we say that it "takes lots of energy" to pedal a bicycle uphill. Energy is thus associated with motion or activity. But we also say that food contains energy; this is the energy our bodies use when we do work. And we say that gasoline contains energy; when the engine of a car burns gasoline, this energy is released and makes the car go. Thus energy can also be associated with objects at rest. In this chapter we will develop the concept of energy and discover how useful a concept it is.

1. WORK

Work Is Force Acting Through a Distance

Before we discuss energy itself, let's look for a moment at the closely related concept of work. In everyday language the word work has various meanings. In physics, however, work is defined in a very specific way: work is what is accomplished by the action of a force. Whenever a force is applied to an object, the work done on the object by that force is defined as *the product of the magnitude of the applied force times the distance through which the force acts*:

$$\text{Work} = \text{force} \times \text{distance.}$$

The man pushing the refrigerator in Fig. 5–1(A) does an amount of work equal to the magnitude of the force he applies times the distance he moves the refrigerator. If he pushes the refrigerator 10 feet with a force of 20 pounds, he does 20 lb × 10 ft = 200 ft•lbs of work.

In the British system of units, which we used in this example, work is measured in foot•pounds. In the metric system, work is measured in newton•meters. A special name is given to this unit, the joule: one joule equals one newton•meter.

Work is done on an automobile to accelerate it. Work is done to push a grocery cart or pedal a bicycle. A person who lifts a package vertically does work on the package, Fig. 5–1(B). However, a person may exert a force and yet do no work, in the specific sense of the word, if the force does not act through a distance. For example, a man pushing against a brick wall does no work on the wall if the wall does not move. The product of force and distance is zero because the distance is zero. Similarly, when you hold an object, as in Fig. 5–1(C), you do no work on it even though you may get tired holding it. Only a force that gives rise to motion is doing work.

Figure 5–1

Work is done on the object in (A) and (B). No work is done on the object in (C) and (D).

Another example of a force that does no work is shown in Fig. 5–1(D). A person carrying a package is walking at constant speed. Thus no horizontal force is needed (Newton's first law). But the person does exert a force on it that is vertically upward, perpendicular to the direction of motion. This upward force has nothing to do with the horizontal motion of the package. Thus we say that the upward force is doing no work. Whenever the force is perpendicular to the motion, no work is done. That is what we meant when we defined work as the product of the applied force times the *distance through which the force acts.*

The force does not necessarily have to be parallel to the direction of motion in order for work to be done by that force. Work will be done even if the force is at an angle to the direction of motion as long as the angle is not 90°. For example, a person pushing a lawn mower, as in Fig. 3–18, does work. To calculate the work done in such cases, we multiply the distance times the *component of the force that is parallel to the direction of motion.*

Simple Machines: The Lever

When two children of the same weight sit on either end of a seesaw, the board balances if it is supported in the middle. But if one child

111

is heavier than the other, the support must be moved closer to the heavier child to keep the board balanced, as we saw in Chapter 4. If one child is twice as heavy as the other, then the lighter child must be twice as far from the balance point as the heavier child. This same principle is used in the simple lever, which can be used to lift very heavy loads, Fig. 5–2. Suppose we want to lift a 500-pound rock. An ordinary person would have difficulty exerting the 500-pound upward force necessary to lift the rock directly. But by using a lever with a long "lever arm," as shown in Fig. 5–2(B), that same person can lift the 500-pound rock by exerting a much smaller force. For example, a force of only 100 pounds is needed to lift the 500-pound rock if the distance of the applied force from the support is five times the distance of the rock from the support.

Simple machines like the lever are said to offer a "mechanical advantage." Other simple machines are the wheel and axle, pulley systems, and the inclined plane. The mechanical advantage of a simple machine is defined as the output force divided by the input force. With the lever in Fig. 5–2, the output force is the force that lifts the rock, namely 500 pounds; and the input force is the force applied by the man, 100 pounds. The mechanical advantage is $\frac{500}{100} = 5$.

Do Simple Machines Save Us Work?

By using a lever we can multiply a force (as in the above example where the applied force is multiplied by a factor of 5). It may look as if we are getting something for nothing. But are we? Look at Fig. 5–2(C). The man pushes down with a force of 100 pounds over a distance of 5 feet while the 500-pound rock goes up only 1 foot. We don't really get something for nothing. What is gained in force is lost in the distance the object can be moved. This does not lessen the usefulness of the lever—indeed, the rock couldn't be lifted at all without it.

Although the force has been multiplied, the work done has not been increased. The work done by the man, called the work input,

Figure 5–2

A heavy rock (A) can be lifted by using a lever (B), which in this case offers a mechanical advantage of 5. The distance the rock is raised (C) is only 1/5 of the distance through which the man pushes.

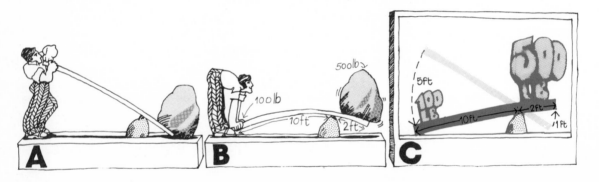

is the force he exerts multiplied by the distance through which he exerts the force: 100 lb × 5 ft = 500 ft•lb. The work done on the rock, the work output, is also 500 ft•lb (500 lb × 1 ft). So, for this simple machine, we can say

work input = work output.

We can make the same statement about other simple machines as well.[1]

Why is it that this quantity we call work should be the same for the input as it is for the output? Is there some larger rule or law operating here? Yes. As we shall see presently, what we have here is an example of the "conservation of energy."

2. ENERGY

Energy can be defined as *the ability to do work*. Energy takes many forms. We begin with the simplest: kinetic energy and potential energy.

Kinetic Energy: Energy of Motion

When a moving object strikes another object, it can do work on that object. A moving hammer, for example, can do work on a nail, Fig. 5–3. When the hammer strikes the nail, it exerts a force on the nail and moves it through a distance. Similarly, when a flying cannonball strikes a brick wall, it exerts a force on the wall and knocks it down. Thus, an object in motion has the ability to do work and, by definition, it therefore has energy. This kind of energy, energy of motion, is called **kinetic energy** ("kinetic" comes from the Greek word for "motion"). Mathematically, the kinetic energy of an object is defined as

$$\text{Kinetic energy} = \frac{1}{2}mv^2,$$

where m is the object's mass and v is its velocity. Notice that because the velocity term is squared, when the velocity of an object is doubled, its kinetic energy is quadrupled ($2^2 = 4$); and, as we will see below, the object is able to do four times as much work.

It was a Dutch scientist, Christian Huygens (1629–1695), who first suggested that this particular combination of mass and velocity, which we call kinetic energy, was somehow special. He found that during the collision of two hard elastic balls, such as billiard balls, the total kinetic energy (he called it "vis viva," or "living force") re-

Figure 5–3

When the moving hammer strikes the nail it exerts a force on the nail, moving it through a distance. The kinetic energy of the hammer goes into work.

113

---

[1]In our example, we have assumed that there is essentially no friction. If friction is present, the work output will be less than the work input.

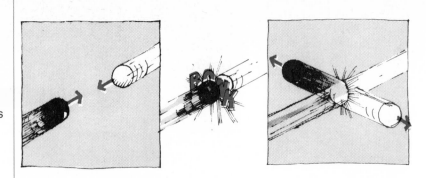

Figure 5–4

When two hard elastic balls collide, the total kinetic energy of the two balls does not change. What one ball loses in kinetic energy, the other ball gains.

mains constant. That is, the sum of the kinetic energies of the two balls—the kinetic energy of ball 1 plus the kinetic energy of ball 2—before the collision is equal to the sum of their kinetic energies after the collision, even though the kinetic energy of each of the balls may change as a result of the collision, Fig. 5–4. When a physical quantity (such as total kinetic energy in this example) remains unchanged during a process, that quantity is said to be **conserved**. That something should be conserved, even under the special circumstances of two colliding billiard balls, is remarkable. That's why Huygens thought "vis viva," or kinetic energy, must somehow be a special quantity in nature.

The Connection Between Work and Kinetic Energy

If a net force F acts on an object over a distance d, work is done on the object and it is accelerated from rest to some velocity v. It can be shown[2] that the kinetic energy acquired by the body as a result of the net force F acting on it is exactly equal to the net work that is done on it: $\frac{1}{2}mv^2 = F\,d$. (By the *net* work we mean the work done by the *sum* of all forces acting on the body, rather than, say, the work done by only one of the forces.) If the body is already moving when the force is applied, the work done goes into changing its kinetic energy. Thus,

Net work done on a body = change in kinetic energy of the body.

As an example, consider what happens when you throw a baseball, Fig. 5–5(A). Your hand exerts a force on the baseball, and this force

---

[2]The net work done on the object is $W = F\,d$. The net force F on the body gives rise to an acceleration because of Newton's second law, $F = ma$; the acceleration is given by $a = \frac{v}{t}$, where v is the final velocity after the force has acted, and the initial velocity is assumed to be zero. The distance d is given by $d = \bar{v}t = \frac{v}{2}t$. Substituting these into $W = Fd$, we find $W = Fd = mad = m\frac{v}{t}d = m\frac{vv}{t\,2}t = \frac{1}{2}mv^2$. That is, the kinetic energy of the body equals the work done on it.

acts through a distance. Because of the work you do on the ball, the ball leaves your hand with a certain velocity and with a kinetic energy, $\frac{1}{2}mv^2$; this is just equal to the work done on it by your hand, Fig. 5–5(B). The reverse process occurs when the ball strikes a fielder's glove; the ball exerts a force on the glove, and the glove moves backward a short distance. The ball, in the process of stopping, does work on the glove.

Thus the connection between work and kinetic energy operates in two directions. First, if work is done on an object, the kinetic energy of the object increases. Second, if an object has kinetic energy, it can do work on something else.

A similar example is the motion of the hammer shown in Fig. 5–3. The carpenter does work on the hammer, which thereby gains in kinetic energy. The hammer, because it has kinetic energy, can then do work on the nail.

The connection between work and kinetic energy has a direct bearing on automobile safety. A car traveling 80 km/hr has four times the kinetic energy of a car traveling 40 km/hr. So it will take four times as much work to stop a car going 80 km/hr as to stop one going 40 km/hr. Since, for a given car, the force that can be applied by the brakes is essentially the same at all speeds, the car traveling 80 km/hr will go four times as far as the car going 40 km/hr before it comes to a stop.

Because of the direct connection between work and energy, energy must be measured in the same units as work: joules in the metric system and foot•pounds in the English system. However, we must make a distinction between work and energy. Energy is something a body *has*. Work is something a body *does* (to some other body). And a body can *do* work only if it *has* energy. Note, too, that both work and energy are scalars (they do not have a direction).

Potential Energy: Energy of Position

An object can have energy not only by virtue of its motion (kinetic energy), but also by virtue of its position or shape. This form of energy is called **potential energy**.

A wound-up watch spring, for example, has potential energy because of its shape. As the spring unwinds, it releases this energy and does work in moving the watch hands around. The spring acquired its potential energy because work was done on it by the person who wound the watch. Just as in the case of kinetic energy, an object can acquire potential energy when work is done on it; and an object that has potential energy can do work on something else.

An ordinary coil spring has potential energy when it is compressed. A ball placed against a compressed spring, and then released, acquires kinetic energy, as shown in Fig. 5–6. The

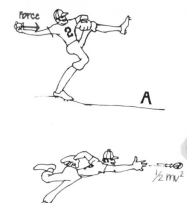

Figure 5–5

(A) When you throw a baseball you exert a force on it through a distance; that is, you do work on it. (B) As a result, the ball acquires kinetic energy in an amount equal to the work done on it.

115

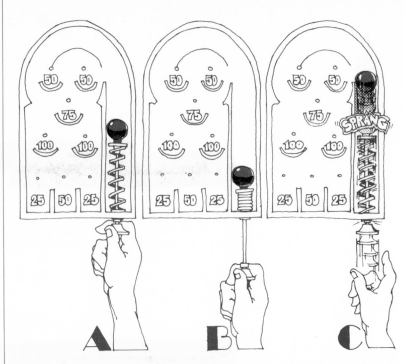

Figure 5–6

A compressed spring has potential energy; it can do work on an object, giving it kinetic energy.

compressed spring has potential energy; when the spring is released, it does work on the ball, which acquires kinetic energy.

An object held high in the air has potential energy because of its position. The object is said to have **gravitational potential energy**, because it has potential energy by virtue of the gravitational attraction of the earth. When the object is released, it falls with increasing speed toward the ground. The object's potential energy is changed into kinetic energy, and the object can do work. A pile driver has an enormous amount of potential energy when it is lifted into the air, and when it is released, it can do a great deal of work.

EXPERIMENT

Find three empty soft-drink cans (the ones that are rather weak and "spongy"). Lift a large stone and a small stone from the floor and place them on a table. Put a can on the floor below each stone. Then push each stone off the table so that it falls on the can below. Which does more work? Now place the third can on the floor and drop one of the stones on it from twice the height as before. Does it do more work this time? On what two factors does the potential energy of a stone depend? Remember that energy—in this case, potential energy—is defined as the ability to do work.

When you lift an object a height *h* from the floor, you are exerting a force equal to its weight. So the work you do is the product of the object's weight times the height *h*. Similarly, if you let the object fall, it can do an amount of work equal to the product of its weight times the height it has fallen. Thus we can say that the gravitational potential energy of an object is equal to the product of its weight times its height above the ground. Since its weight is the product of its mass times the acceleration of gravity, *mg,* we can say

$$\text{Gravitational potential energy} = \text{weight} \times \text{height}$$
$$= mgh.$$

A 3000-lb pile driver lifted 50 feet into the air has 3000 lb $\times$ 50 ft = 150,000 ft•lb of potential energy. A 5-kg stone 20 meters above the ground has (5 kg)(9.8 m/sec^2)(20 m) = 980 joules of potential energy.

Gravitational potential energy plays a role in all sorts of practical situations. One example is the action of a pile driver. Another is the potential energy of water at the top of a waterfall or in a reservoir behind a dam; when the water falls, it can do work. If the water falls into turbines at the bottom of a dam, the work done can be transformed into electric energy, as we shall see in Chapter 14.

In each of these examples an object has the *capacity* to do work even though it may not actually be doing work and isn't even in motion. That is why energy of position is called "potential" energy. A coiled-up watch spring isn't doing any work when it is coiled up, nor is the compressed spring shown in Fig. 5–6. But each has the *potential* to do work. Its potential energy *could* be transformed into kinetic energy. Similarly, a rock or a pile driver high above the ground may just be resting there, but each has the potential for doing a lot of work.

These examples show that energy can be *stored,* for later use, in the form of potential energy.

Other Forms of Energy

There are other forms of energy besides the kinetic and potential energy of ordinary-sized objects. These include thermal energy, electrical energy, nuclear energy, and the chemical energy stored in food and in fuels such as gasoline, oil, and natural gas. With the rise of the atomic theory of matter in the last century, scientists have come to the view that these other forms of energy are simply kinetic energy or potential energy at the atomic or molecular level. For example, according to the atomic theory, thermal energy is merely the kinetic energy of molecules. When an object is heated, the molecules that make up the object move faster and the object feels hot.

The energy stored in food and gasoline, on the other hand, is potential energy—energy stored by virtue of the relative positions of the atoms within a molecule. In order for this potential energy to be used to do work, it must be released, usually through a chemical reaction. This is much like a compressed spring which can do work when it is released. In our own bodies, enzymes allow the release of energy stored in food molecules. In an automobile, the violent spark of the spark plug allows the mixture of gas and air to release its energy and do work against the piston to propel the car forward.

Electric, magnetic, and nuclear energy are other examples of kinetic and potential energy. In later chapters we will have more to say about those forms of energy.

Energy Can Be Transformed

In the last few pages we discussed several examples of the change, or transformation, of one type of energy into another. A rock held high above the ground has potential energy. As it falls it loses potential energy, because its height above the ground decreases; at the same time it gains kinetic energy, because its velocity increases. Its potential energy is being transformed into kinetic energy.

Often the transformation of energy involves a transfer of energy from one body to another. The potential energy of the compressed spring in Fig. 5–6 is transformed into the kinetic energy of the ball. Water plunging to the bottom of a dam turns the blades of a turbine; the kinetic energy of the water is transformed into the kinetic energy of the rotating turbine blades. The potential energy stored in a bent bow can be transformed into the kinetic energy of an arrow.

In each of these examples, the transfer of energy is accompanied by the performance of work. The spring does work on the ball; water does work on the turbine blade; and the bow does work on the arrow. This observation gives us a new insight into the nature of work: *Work is done whenever energy is transferred from one object to another.* This is true when a person throws a baseball, Fig. 5–5, or pushes a heavy object across the floor, Fig. 5–1(A). In both cases, chemical energy within the person, obtained from food, is transformed into kinetic energy of the object that is pushed. The work done on the object is a manifestation of the transfer of energy from the person to the object.

**3.
CONSERVATION
OF ENERGY**

Energy Is Conserved

Whenever energy is transformed from one form to another, or transferred from one body to another, we find that no energy is lost in the process. As

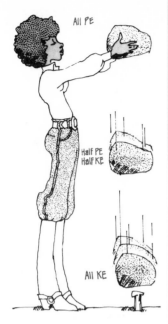

an example, let us look at what happens when a rock is dropped from a height h above the ground, Fig. 5–7. Before it is dropped, it has a potential energy equal to its weight times height. As it falls, it loses potential energy because the height is decreasing; meanwhile its kinetic energy is increasing, and just before it strikes the ground, it has only kinetic energy. The kinetic energy it has at the bottom is exactly equal to the amount of potential energy it had originally at the top. The potential energy was completely transformed into kinetic energy. No energy was lost.

When the object has fallen halfway to the ground, it has both potential and kinetic energy. Since its height is only half what it was originally, it has only half as much potential energy; the other half has been transformed into kinetic energy. At this point it has equal amounts of kinetic and potential energy. The sum of its kinetic energy plus its potential energy at this halfway point is equal to the amount of potential energy it had originally. In fact, the sum of its kinetic and potential energy, which we call its *total* energy, remains the same for all points along its path. Thus, although the kinetic energy and the potential energy each change throughout the motion, the sum of the two remains constant.[3]

This example illustrates one of the most important principles of physics, the law of the **conservation of energy**:

Energy is never created or destroyed. Energy can be transformed from one kind into another, but the total amount of energy remains constant.

Like all laws of physics, this one is based on experimental data. It has been found to hold true in every experimental situation in which it has been tested.

Other Examples of the Conservation of Energy

When the billiard balls in Fig. 5–4 collide, the sum of the kinetic energies of the two balls was found by Huygens to remain essentially constant. Since there is no potential energy involved, the law of conservation of energy applies to their kinetic energies alone. If, as a result of the collision, one ball loses kinetic energy and slows down, the other ball will have its kinetic energy increased by exactly that amount. Any energy lost by one ball is gained by the other.

Another example is illustrated in Fig. 5–8, which shows a car on the top of a hill, on the left. It could be a real car, with its engine off, or it could be a toy car. In either case we will assume that friction

Figure 5–7

As an object falls it loses potential energy and gains kinetic energy. This energy can be used to do work—for example, to drive a stake into the ground.

[3]If you are wondering what happens to the energy when the object strikes the ground, you will find the answer in just a few pages.

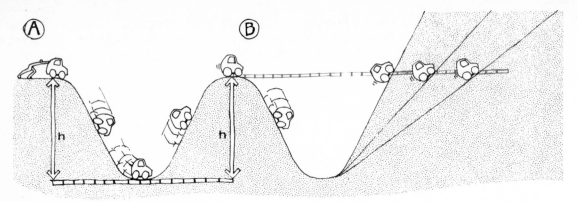

Figure 5–8

(A) The car loses potential energy and gains kinetic energy as it goes down the hill; as it rises up the hill on the right, it loses kinetic energy and gains in potential energy, arriving at the top with the same potential energy it had earlier. (B) Starting from rest on the hill at the left, the car will reach the same *height* on the hill at the right no matter what the angle of that hill.

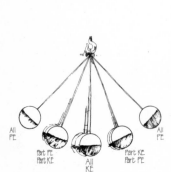

Figure 5–9

The simple pendulum illustrates conservation of energy.

is so small that it can be ignored. The car begins to coast downhill, increasing in speed and in kinetic energy as it loses potential energy. It reaches its maximum kinetic energy at the bottom of the valley. As the car climbs up the hill on the other side, its kinetic energy is gradually changed back to potential energy. When the car comes to rest, its kinetic energy is zero and all of its energy is again potential energy. Since energy is always conserved, the car will have the same amount of potential energy as it had initially on the hill. And, because potential energy is proportional to height, it will come to rest at a height above the ground equal to its original height. If the second hill is the same height as the first, the car will just reach the top as it stops. If the second hill is lower, not all of the car's kinetic energy will be transformed into potential energy and the car will continue over the top and down the other side. If the second hill is higher, the car will reach a vertical height equal to its original height on the first hill (and then roll back down). The length and angle of the second hill do not affect how high the car will go vertically, as shown in Fig. 5–8(B). This is because the potential energy depends only on the vertical height *h*.

Another example, and one that more closely approximates a friction-free system, is that of a simple pendulum, which is merely an object hung from a support (Fig. 5–9). If the pendulum is pulled to one side and then released, it swings to the other side and then back, reaching an equal height on each side (as long as friction is negligible). At the lowest point of its swing it has its maximum kinetic energy; at the greatest height of its swing on either side it has no kinetic energy and a maximum potential energy. Its kinetic energy at the bottom is equal to its potential energy at the top. At intermediate points it has both kinetic and potential energy, the sum of which remains constant throughout the swing.

Friction Gives Rise to Thermal Energy

Up to now we have neglected friction; but it is always present in real situations, although sometimes it is small enough to ignore.

Since friction affects the swing of a pendulum, the pendulum bob does not reach the same height on each swing and eventually will come to rest. The car moving between the two hills in Fig. 5–8(A) will not reach quite the same height on the second hill, because of friction.

Do these observations throw doubt on the concept of the conservation of energy? After all, aren't we saying that the sum of the kinetic and potential energies of an object decreases instead of staying constant?

It was questions of this kind that prevented a comprehensive law of conservation of energy to emerge until the mid-nineteenth century. Not until then was heat, which is an inevitable product of friction, recognized as another form of energy. That friction produces heat is readily observed when you rub your hands together rapidly. Whenever friction is present, heat is produced. Quantitative studies by nineteenth-century scientists, particularly the Englishman James Prescott Joule (1818–1889), demonstrated that energy is indeed conserved in all natural processes when heat is considered a form of energy. This form of energy we call thermal energy.

An object moving on a level surface slows down because some of its kinetic energy becomes thermal energy. In Fig. 5–8, as the car travels between the two hills, some of its kinetic energy is transformed into thermal energy because of friction. Both the ground and the car's tires heat up (the heating up of the tires is easily noticeable when a car is driven a considerable distance). Thus, at the bottom of the hill the car's total energy is slightly less than it was at the top of the hill, because some of its energy was transformed into thermal energy. And the car won't quite make it to the top of the second hill, since all along its path part of its energy is slowly being transformed to thermal energy. When the car comes to rest on the second hill, its potential energy will therefore be less than its initial potential energy.

And so it is with the pendulum. Friction at the point of suspension, and friction on the bob due to air resistance, give rise to a transformation of kinetic energy into thermal energy. In the case of the colliding billiard balls (Fig. 5–4), if the balls are not very hard and highly elastic, a significant amount of energy will be transformed into heat and so kinetic energy alone will not be conserved.

Another example of the transformation of kinetic energy into heat occurs when the rock in Fig. 5–7 strikes the ground. As the rock falls, its original potential energy changes into kinetic energy. But when the rock hits the ground, its kinetic energy is transformed almost instantaneously into thermal energy. Both the rock and the area of the ground where the rock hits become warmer.

EXPERIMENT

To observe the transformation of kinetic energy into thermal energy, strike a nail or a piece of metal several times in quick succession with a hammer and then touch the metal gently. What do you observe?

Other Forms of Energy: Energy Is Conserved in All Processes

When other forms of energy are involved, such as chemical or electrical energy, we find that the law of conservation of energy is still valid. The kinetic energy of rushing water can be transformed into electrical energy, Fig. 5–10. Electrical energy can be transformed into light energy. The chemical energy of gasoline can be transformed into the kinetic energy of an automobile. In all these processes the total amount of energy is conserved. None is lost. However, in any energy transformation some thermal energy is always produced. For example, most of the kinetic energy of the rushing water is transformed into electrical energy, but some goes into heat. The sum of the electric energy and the thermal energy produced in the process is just equal to the loss of kinetic energy of the water. The total amount of energy remains constant.

The lever problem we discussed at the beginning of this chapter, shown in Fig. 5–2, is another example of the conservation of energy. The man does work to push the lever down. To do this work, he has to expend energy, which comes ultimately from the chemical energy of the food he has eaten. The chemical energy that is used up is just equal to the work he does on the lever plus that lost to heat; and the work done on the lever goes into increasing the potential energy of the rock as it is raised. Chemical energy is transformed into potential energy of the rock, plus a small amount of heat energy. No energy is lost in the process.

Figure 5–10

Kinetic energy of water falling over a dam can be transformed into electrical energy by a generator (see Chapter 14).

Why Is the Energy Concept Important?

We speak of energy as something that cannot be created or destroyed, and as something that can be transferred from one object to another. Does this mean that energy is a kind of material, perhaps a fluid, that can flow from one object to another? No. Energy is an abstract concept. Kinetic energy, for instance, is defined as a product of the factors mass and velocity squared. Gravitational potential energy involves the weight and the height of an object above the ground. These combinations of factors are not really tangible. The idea of energy exists only in our minds, but it is a useful idea.

What makes energy so valuable is the fact that it is *conserved*. What is so valuable about a quantity that is conserved? To a scientist who is looking for order in the universe, the existence of a quantity that does not change, no matter what happens, is gratifying. The universe seems in a state of constant flux, one thing interacting with another; but within this complex of interactions there is a quantity that remains constant—energy. It is a unifying principle and helps to bring order to the natural world.

Furthermore, the fact that energy is conserved makes it easier to solve practical problems. For example, you can figure out the velocity of the car in Fig. 5–8 at any point by using the conservation of energy principle. It can be shown that the conservation of energy principle is essentially equivalent to Newton's laws; therefore it gives us an alternate, and sometimes easier, method of analyzing nature and of solving practical problems.

The conservation of energy is not the only conservation law scientists have discovered. A number of others have come to light, and in this and later chapters we will discuss some of them. Today many physicists feel that the conservation laws are the most basic laws in nature. Therefore, it may be that the law of conservation of energy is more basic than Newton's laws themselves.

4. POWER

Power is defined as the rate at which work is done, or as the rate at which energy is transformed:

$$P = \frac{\text{work}}{\text{time}} = \frac{\text{energy transformed}}{\text{time}}.$$

For example, power refers to how much work a horse or an engine can do per second; or to how much electrical energy an electric heater transforms per second into heat energy; or to the rate at which an electric light bulb transforms electric energy into light energy.

In the metric system, power is measured as joules per second. This unit is given a special name, the *watt*: 1 watt = 1 joule/sec. The

123

watt is used most often in electricity, where the electric power transformed by an appliance or a light bulb is given in watts. For mechanical systems such as engines and automobiles, power is often measured in English units. The usual English unit is foot•pounds per second; for most practical purposes a larger unit, the horsepower, is normally used. The horsepower unit was devised by James Watt (1736–1819), who also developed the steam engine. Watt needed some way to compare the work of his engines with the work a horse could do, since most work at the time was done by horses. He made various measurements of the ability of horses to do work and came to the conclusion that a typical horse could work all day at the rate of about 360 ft•lb/sec. But, being a careful and conscientious man, he did not want to overrate his steam engines. So he multiplied his measured value for the power output of an average horse by about 1½ and thus defined the horsepower (hp) unit as 1 hp = 550 ft•lbs/sec. (The capabilities of Watt's early steam engines ranged from about 4 to 100 horsepower.) The metric horsepower is defined as 750 watts and is almost identical to the British horsepower (which equals 746 watts).

To clarify the difference between power and energy, let us consider a simple example. A person can walk quite a distance, or climb many flights of stairs, before becoming tired. But eventually so much energy is used up that the person will have to stop. A person is limited in physical exertion by the amount of energy stored in the muscles that can be used for doing work. But this is not the only limitation. A person may be able to climb 20 flights of stairs slowly; but if the person runs up the stairs very rapidly, exhaustion may occur after only one flight. Thus a human being is limited not only by the total amount of energy required for a task, but also by the *rate* at which the energy is used, or power.

The maximum rate at which energy can be expended puts a limit on the acceleration of an automobile. That is why automobile engines are rated in horsepower. The brightness of a lamp depends on the rate at which energy is transformed. Thus, light bulbs are rated by power—60 watts, 100 watts, and so on.

EXPERIMENT–PROJECT

Measure your maximum power output by timing how many seconds it takes you to run up a flight of stairs. Your power output will be equal to the increase in your potential energy (your weight times the *vertical* height of the staircase) divided by the time it took you to reach the top. To get your power output in horsepower, divide by 550 if you used feet and pounds, and by 750 if you used meters and newtons.

5. MOMENTUM AND ITS CONSERVATION

The law of conservation of energy was not the first conservation law to be formulated. The law of conservation of momentum had already been discovered by the mid-seventeenth century.

Momentum Is Mass Times Velocity

Momentum is defined as the product of the mass of a moving body times its velocity,

$$momentum = mass \times velocity,$$

or in symbols, momentum $= mv$. As we shall see shortly, the momentum of an object is often a more useful quantity than its mass or its velocity alone. Note that because velocity is a vector, so is momentum; and the direction of the momentum of an object is the same as the direction of its velocity.

According to our definition, a moving object will have a large momentum if either its mass or its velocity, or both, is large; this is in accordance with our everyday use of the term momentum. A fast-moving car has more momentum than a slow-moving car; and a heavy truck has more momentum than an automobile moving at the same speed. The more momentum an object has, the greater will be its effect if it strikes a second object. For example, in a collision, a fast-moving car will cause a more serious accident than a slow-moving car, and a fast-moving heavy truck will cause even more damage. Similarly, a football player is more likely to be stunned when he is tackled by a huge opponent running at top speed than by a lighter or slower-moving tackler. It is the product of mass times velocity—*momentum*—that is important.

Momentum is a useful concept because, like energy, it is a conserved quantity.

Momentum Is Conserved in Collisions

When two bodies collide, the sum of the momenta of the two bodies *remains constant*. Although the momentum of each of the two bodies may change as a result of the collision, the sum of the two momenta is the same after the collision as it was before. Figure 5–11 illustrates the conservation of momentum for the head-on collision of two hard elastic balls of equal mass, one of which is initially at rest. In this particular case, the cue ball comes to rest as a result of the collision, and ball 2 moves off with the same speed that the cue ball had initially. The momentum of the cue ball is completely transferred to ball 2. Momentum is conserved in the collision. If ball

125

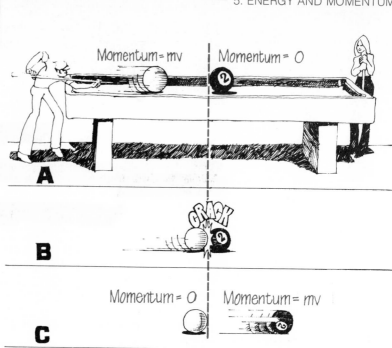

Momentum = mv   **Momentum = 0**

A

B

Momentum = 0   **Momentum = mv**

C

Figure 5–11

Collision of two hard elastic balls of equal mass.

2 had been lighter than the cue ball, it would have moved off at a higher speed to compensate for its smaller mass.

If both balls are found to be moving after the collision, which is generally the case if they have different masses [Fig. 5–12(A)], or if the two balls make a glancing collision and go off at an angle [Fig. 5–12(B)], then we must use vector addition. In either example of Fig. 5–12, the total momentum before the collision is simply the momentum of ball 1. After the collision both balls have momentum, and the momentum of ball 1 is added to that of ball 2, as vectors, to give the total momentum; this total momentum after the collision is equal to the total momentum before the collision. Both the magnitude and the direction of the *total* momentum remain constant.

EXPERIMENT

You can observe collisions of this sort by using billiard balls, or by skidding ice cubes across a wet table. Try to estimate the speed for a number of collisions and see if momentum is conserved. See if you can produce a collision in which momentum is *not* conserved. What happens when one of the balls or ice cubes rebounds at an angle—does the other do so as well? If so, does it rebound to the same side as the first one or to the opposite side? Explain.

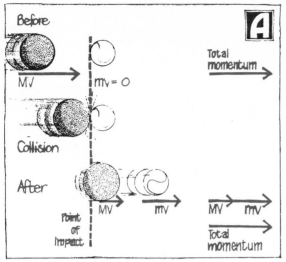

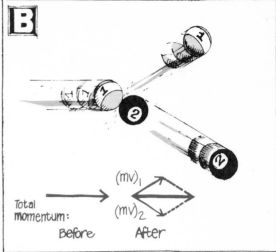

Figure 5–12

Momentum conservation in collisions of two balls:
(A) head-on collision of balls of different masses,
(B) glancing collision.

Momentum is conserved even when two colliding objects stick together, as when two pieces of putty collide or when a football player leaps high in the air to catch a pass, Fig. 5–13. In the latter case, the momentum of the ball before the catch equals the momentum of the ballplayer and the ball together afterward, and so they both move backward, Fig. 5–13(B).

Momentum Conservation Is Related to Newton's Laws

The conservation of momentum principle, like the conservation of energy, is closely related to Newton's laws of motion and, in fact, can be derived from them. Recall from Newton's first law that the velocity and therefore the momentum of an object do not change if no net force acts on the object. However, if a force acts, the velocity changes and therefore the momentum changes as well.

According to Newton's second law, $F = ma$, the greater the force the more rapidly the velocity changes, and therefore the more rapidly the momentum changes. Originally, Newton expressed his second law in just this fashion,[4] that "the rate of change of momentum of an object is equal to the applied force."

[4]Newton's second law says that the applied force equals the mass × acceleration. Since acceleration = $\frac{\text{change in velocity}}{\text{time}}$, we can write

$$\text{Force} = \text{mass} \times \frac{\text{change in velocity}}{\text{time}} = \frac{\text{change in momentum}}{\text{time}},$$

since momentum is mass × velocity. Thus the applied force equals the change of momentum per unit of time.

Now Newton's third law states that when two bodies interact, the forces involved (that each body exerts on the other) have the same magnitude but are in opposite directions. Therefore, the momentum of each of the bodies changes by the same amount but in the opposite direction, and the two changes cancel each other. The change in momentum of one object exactly compensates for the change in momentum of the other body, so that the net change in their combined momentum is zero. The *total* momentum of the two bodies therefore remains constant.

The above statement is true if the only forces involved are those between the two bodies. But if external forces are present—that is, forces produced by other objects—then the sum of the momenta of the two bodies will not be conserved; it will change as a result of those external forces. However, if we include the other objects in our "system," then the total momentum of all the interacting bodies will be conserved. The general statement of the law of conservation of momentum is:

The total momentum of a group of objects is the same after they interact as it was before, so long as no external forces act.

Let's take a simple example. Suppose a railroad car traveling 10 m/sec strikes an identical car at rest. If they hook together as a result of the collision, they will move off with equal speeds. Since momentum is conserved, the momentum before the collision equals the momentum after:

$$(mv)_{before} = (mv)_{after}.$$

Before the collision, only one car is moving with a speed $v = 10$ m/sec. After the collision, twice as much mass is moving; therefore, the velocity must be half as much: 5 m/sec.

The laws of conservation of energy and conservation of momentum give us an alternative to Newton's laws in analyzing motion. But there is a difference between these two approaches. When we use the conservation laws, we can ignore many of the details of the process. We are mainly concerned about the positions and velocities of the objects before and after the interaction. We do not explicitly consider the forces that act during the interaction; in fact, in situations such as collisions it is difficult to measure the strength of the forces between the bodies. This is one reason why the conservation laws are so useful. We can use them to analyze a situation even though we do not know the forces involved in detail.

Rockets and Recoil

The law of conservation of momentum is most valuable in dealing with fairly simple systems, particularly those involving collisions or

Figure 5–13

Momentum is conserved when the player leaps in the air to catch a pass.

128

Figure 5–14

Total momentum is zero both before and after firing a rocket.

A. Total momentum = 0

B. Bullet moves forward & rifle recoils back

Total momentum is still zero

Figure 5–15

(A) Total momentum = 0.
(B) Bullet moves forward and the gun recoils backward. Total momentum is still zero.

certain types of explosion. For example, rocket propulsion, which we explained in Chapter 3 in terms of action and reaction, can also be viewed as arising from the conservation of momentum, Fig. 5–14. The momentum of the burned gases rushing out of the rocket is exactly balanced by the increased forward momentum of the rocket. That is, before the rocket is fired the total momentum is zero. After the firing, the momenta of the rocket and gases add together as vectors to give a *total* momentum that is still zero. Thus, as we saw earlier, a rocket is not propelled because the expelled gases push on the ground or the atmosphere. Rather it is the conservation of momentum, or alternatively Newton's third law, that explains rocket propulsion.

The recoil of a gun is similarly an example of the conservation of momentum, Fig. 5–15. Before firing, the gun and the bullet have zero momentum; they are both at rest. When the trigger is pulled, a minor explosion takes place, propelling the bullet down the barrel. The large forward momentum of the bullet must be compensated by the backward momentum of the gun, its *recoil*. The total momentum, the vector sum of the bullet's forward momentum and the gun's backward momentum, remains zero. The gun's (recoil) velocity is much less than that of the bullet's, since it has a much greater mass.

Which Is More Important—Energy or Momentum?

In the seventeenth century, scientists discovered that the quantities mv and $\frac{1}{2}mv^2$ are conserved in collisions of two hard elastic balls. Soon afterward, a disagreement arose over whether momentum, mv, or kinetic energy, $\frac{1}{2}mv^2$, was the true measure of the "quantity of motion" in a body. This discussion could have no rational resolution, of course, other than that both momentum and energy are important quantities. They are two quite different quantities. Not only does one have the velocity squared and the other simply the velocity; but momentum is a vector, whereas kinetic energy is a scalar. Both kinetic energy and momentum are useful quantities associated with the motion of a body. Both are conserved. One is not "better" than the other. They illuminate different aspects of motion. Though energy is perhaps used more often, momentum can be enlightening as well. For example, in the firing of a gun, the total energy is conserved as well as the momentum. But from energy conservation alone one would have no idea that recoil would occur; only momentum conservation can reveal that.

6. ANGULAR MOMENTUM

The rotational equivalent of ordinary momentum is called **angular momentum**. For a small particle rotating about a fixed point, as in Fig. 5–16(A), the angular momentum is defined as the product of its ordinary momentum, mv, times the distance of the object from the center, r. In symbols, angular momentum = mvr. The angular momentum of an extended body such as a bicycle wheel or a top [Fig. 5–16 (B)] can be calculated by summing up the angular momenta of each little piece that makes up the body. Without going into detailed calculations, we can see that since the angular momentum of each piece of the object equals mvr, the angular momentum is increased (1) if the mass is increased, (2) if the rotational speed is increased, or (3) if more of the mass is concentrated farther from the axis of rotation. Like ordinary momentum, angular momentum obeys a conservation law, the law of conservation of angular momentum:

The angular momentum of an object remains constant so long as no external torques act to change it.

Figure 5–16

(A) Angular momentum of a small rotating object is equal to *mvr*. (B) Angular momentum of a bicycle wheel or a top is proportional to the mass, speed of rotation, and radius.

130

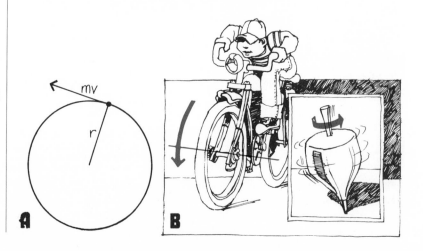

This law is nicely illustrated by a skater doing a spin on ice, Fig. 5–17. When her arms are outstretched, she rotates at a relatively slow rate; when she pulls her arms in close to her body, she begins rotating faster. She has reduced the distance (*r*) of her arms from the axis of rotation; and since angular momentum is conserved, she automatically speeds up to compensate. She doesn't push harder in order to go faster; her speed of rotation increases all by itself when she pulls in her arms. Conservation of angular momentum is evident here. When *r* is large, *v* is small; when *r* is small, *v* is large.

EXPERIMENT

Locate a chair that rotates, such as an office chair. Sit in the chair with your arms and legs extended. Now have someone spin you around. While you are rotating, pull in your arms and legs. What happens to your speed? Try the experiment again, this time holding a couple of books or bricks in your hands. Explain your observations. See Fig. 5–18.

OR

Tie a string around an eraser or some other small object. Hold the other end of the string between your thumb and fingers and start the object rotating. Then put out your forefinger so the string starts wrapping around it. Why does the object's speed increase as the string winds up on your finger? See Fig. 5–19.

Figure 5–17
Conservation of angular momentum. (A) Part of the mass (arms) is far from the axis, and the speed of rotation is slow. (B) All the mass is close to the axis of rotation, and the speed is faster.

What happens when you set a child's top on its point? It falls over. But if the top is spinning, it stays upright. This is another example of angular momentum conservation. If the spinning top were to fall, the axis of rotation would have changed drastically. Now angular momentum is a vector, and although its magnitude might not be affected during the fall, the *direction* would change a great deal.

Figure 5–18
A person rotating on a swivel chair. How does the speed of rotation change?

Only a strong torque would be capable of causing such a large change in the angular momentum.[5] Since such strong external torques are not ordinarily present, the angular momentum stays essentially constant and the top stays upright. However, the forces of friction and gravity do act, so the angular momentum changes slowly; the top gradually slows down and eventually falls over.

The situation is much the same with a bicycle or a motorcycle. If you sit on a bike at rest, it falls over; but if the bike is moving, the angular momentum of the spinning wheels resists any tendency to change and helps to keep the bike upright.[6]

The gyroscope, which is used by mariners to keep on their course, works on the principle of conservation of angular momentum, Fig. 5–20. The rapidly spinning wheel is mounted on a complicated set of bearings so that when the mount is moved, as when a ship pitches in heavy seas, no forces act to change the direction of the angular momentum. Thus the gyroscope wheel remains pointing in the same direction in space no matter where the ship is situated. If initially it is pointing toward the north star, it continues to point in that direction no matter how far the ship travels. Thus a gyroscope helps a captain tell in which direction the ship is moving.

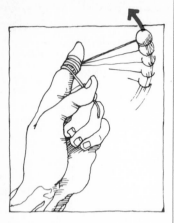

Figure 5–19

The ball's speed increases as the string wraps around the finger.

REVIEW QUESTIONS

1. Work depends on what two factors?
    (a) force and energy
    (b) force and distance
    (c) distance and energy
    (d) energy and momentum

2. What is the difference in the work done between a 100-lb weight carried 100 ft horizontally at constant speed and a 10-lb weight lifted vertically 1 foot?
    (a) more work is done in the first case
    (b) more work is done in the second case
    (c) same amount of work is done in each case
    (d) no work is done in either case

3. A machine can be used to
    (a) increase energy
    (b) transfer energy
    (c) do more work
    (d) save energy

[5]Just as linear momentum (mv) is related to force, so angular momentum is related to torque. That is, the net torque applied to an object is equal to the change of the angular momentum per unit time.

[6]The spinning top and the stability of a moving bike can also be viewed from the point of view of rotational inertia. The rotating top or wheels, because of their rotational inertia, resist any change in their rotational velocity, either in magnitude or direction.

Figure 5–20

A gyroscope.

4. If the mass of a moving object is doubled without changing its speed, its kinetic energy is
   (a) decreased by 1/4
   (b) decreased by 1/2
   (c) the same
   (d) doubled
   (e) increased fourfold

5. When a bird's speed is doubled, its kinetic energy is
   (a) half as large
   (b) the same
   (c) twice as large
   (d) four times as large

6. Potential energy is
   (a) energy of motion
   (b) energy that depends on the position or shape of the body
   (c) always conserved
   (d) the same as heat energy

7. A pile driver is raised to a height of 20 ft. A pile driver of only half the mass must be raised to what height to have the same amount of potential energy?
   (a) 5 ft
   (b) 10 ft
   (c) 20 ft
   (d) 40 ft

8. Energy is
   (a) only associated with motion of molecules
   (b) a form of momentum
   (c) always conserved
   (d) lost if heat is produced
   (e) not convertible from one form to another

9. A "superball" is dropped from a height of 5 ft above the floor. It will rebound to a height of
   (a) less than 5 ft
   (b) 5 ft
   (c) more than 5 ft

10. As a pendulum slows down, it
    (a) gains kinetic energy
    (b) gains potential energy
    (c) loses energy
    (d) transforms kinetic and potential energy into thermal energy

11. The rate at which energy is transformed is called
    (a) work
    (b) momentum
    (c) power
    (d) energy consumption

12. Which of the following requires the most power?
    (a) running down 3 flights of stairs in 14 seconds
    (b) climbing 3 flights of stairs in 12 seconds
    (c) climbing 2 flights of stairs in 6 seconds
    (d) climbing one flight of stairs in 4 seconds

13. Momentum is
    (a) a scalar quantity
    (b) the ability to do work
    (c) the force a body exerts on another during a collision
    (d) equal to *mv*

14. A heavy truck has more momentum than a passenger car moving at the same speed because the truck
    (a) has greater mass   (c) has greater speed
    (b) is not streamlined   (d) has a large wheelbase

15. A gun recoiling when it is fired is an example of
    (a) conservation of momentum   (c) conservation of energy
    (b) conservation of angular   (d) none of these
         momentum

16. An ice skater starts spinning with his arms extended. As he brings his arms in next to his body, he
    (a) spins more slowly   (c) spins at the same rate
    (b) spins faster   (d) stops spinning

17. A tether ball makes more turns per second around the pole as the rope gets shorter. This is an example of
    (a) conservation of momentum   (c) conservation of energy
    (b) conservation of angular   (d) conservation of rotational
         momentum   inertia

EXERCISES

1. How does the meaning of the term "energy" as used in this chapter differ from everyday use of the term?

2. The pyramids of Egypt were built with the help of inclined planes, or ramps. Explain why an inclined plane can be considered a simple machine. Does it have a mechanical advantage?

3. A moving billiard ball strikes a second ball and imparts some of its kinetic energy to the latter. Did it do work on the second ball?

4. In which case is more work done: when a 50-lb bag of groceries is lifted 4 feet, or when a 50-lb crate is pushed 10 feet across the floor with a force of 20 lb?

5. In the old West, a rifle with a long barrel was considered extremely valuable. Explain why, using the fact that the gases from the exploding shell act on the bullet over a longer distance than they do in a rifle with a short barrel.

6. A man exerts a 30-lb force to pull a loaded wagon 200 feet. How much work does he do? How much energy does he use?

7. Why do the stopping distances of a car when the brakes are applied increase with the square of the car's speed?

8. Which has more kinetic energy, a 1000-kg car traveling 80 km/hr or a 2000-kg car traveling 40 km/hr?

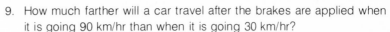

Figure 5–21
A roller coaster.

9. How much farther will a car travel after the brakes are applied when it is going 90 km/hr than when it is going 30 km/hr?

10. A compressed spring has potential energy. How do you know this? Why don't we call it "gravitational" potential energy?

11. Does a stretched rubber band have potential energy? How do you know?

12. An overloaded car sometimes has difficulty climbing a steep hill. Explain.

13. Describe all the energy transformations that take place when you throw a ball into the air, when the ball reaches its maximum height and descends, and finally, when you catch it.

14. Describe all the energy transformations that occur when a sled accelerates as it slides down a steep hill, reaches soft snow halfway down, and slowly comes to rest before it reaches the bottom.

15. Describe the transformations of energy that occur when a child hops around on a pogo stick.

16. What is the purpose of a spring that must be wound up in a pendulum clock?

17. A skier at the top of a hill is ready to descend. Does she have kinetic or potential energy? What happens to this energy as she descends the hill?

18. What does "conservation" mean?

19. A roller coaster is usually pulled by a chain drive to the top of the first rise. It then coasts on its own (see Fig. 5–21). Where does it reach maximum kinetic energy? Where does it reach maximum potential energy? Assuming little or no friction, will the roller coaster make it over all the rises shown? What will happen if a large amount of friction is present?

20. When a car is braked, why do the brakes heat up?

21. To move a car along on a level road at 60 mi/hr requires a force of about 100 to 200 pounds, depending on the car and road conditions. Approximately how much horsepower is required? Do you think high-powered engines capable of 200 or more horsepower are justified?

22. A toy rocket is fired into the air. What kind of energy does it have just before it hits the ground on its return? What happens to this energy when it crashes into the ground?

23. When you are on a long hike and come to a log across your path, it is supposed to be better to step over the log than it is to step on top of it and then jump down on the other side. Why is this so?

24. A stone that is thrown with a particular speed from the top of a cliff will enter the water below with the same speed whether it is thrown horizontally or at an angle. Why?

25. A pendulum bob is swung so that it reaches a maximum height h above its lowest point. Suddenly a peg is placed as shown in Fig. 5–22. How high will the pendulum rise on the right side now? Show this on a diagram.

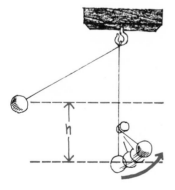

Figure 5–22
Pendulum cord strikes the peg.

26. Can an electric heater be rated in horsepower as well as in watts? Can an automobile be rated in watts as well as in horsepower?

27. One horse does 1200 joules of work in 2 hours. A second horse does 1000 joules in 1½ hours. For which horse is the power output greater?

28. A heavy object and a light object have the same momentum. Which has the greater speed? Which has the greater kinetic energy?

29. Name three everyday experiences that illustrate the conservation of momentum principle.

30. When you jump down from a tree, what happens to your momentum as you hit the ground?

31. Describe how a fish moves forward by swishing its tail back and forth. (Hint: Use conservation of momentum.)

32. According to legend, in ancient times an old miser carrying a bag of gold coins was stranded on the frictionless surface of a frozen lake and froze to death. Although he could not walk or crawl because the ice was frictionless, he still could have got to shore. Explain how.

33. When a ball is dropped to the ground, it is said that the earth comes up to meet the ball. Explain on the basis of the conservation of momentum.

34. A car traveling 20 mi/hr strikes a second identical car from the rear and the two cars lock bumpers. Assuming that the brake of the second car was off and that momentum is conserved in the collision, what is the velocity of the two attached cars after the collision?

35. It is not very useful to consider energy conservation in collisions like the one in the above problem. Why not?

36. Explain on the basis of the conservation of momentum why it is that when a person throws a package out of a boat, the boat moves off in the opposite direction (see Fig. 3–8).

37. Suppose you find yourself teetering on the edge of a cliff or the roof of a tall building (Fig. 5–23) with a heavy physics book in your hand. As you are about to fall, you suddenly remember how physics can save you. What law did you remember and how did you apply it?

38. When you release a balloon that has just been inflated, why does it fly across the room?

39. Why does putting a spiral on a football increase the accuracy of a pass?

40. Rifle barrels generally have a spiral groove in them so that bullets will be spinning when they emerge. Why is this advantageous?

41. Suppose you are standing on the edge of a freely rotating merry-go-round. What would happen to the speed of rotation of the merry-go-round if you started walking along the edge in the opposite direction?

42. Why does a helicopter have more than one propeller?

43. When a quarterback leaps into the air to throw a pass, he rotates the upper part of his body as he throws the ball. If you look carefully, you will notice that the lower part of his body rotates in the opposite direction. Explain.

Figure 5–23

You are falling off the edge. What do you do?

6

RELATIVITY

Figure 6–1
Albert Einstein.

The theory of relativity was largely developed by one amazing man, Albert Einstein (Fig. 6–1). He was an unknown 26-year-old patent clerk in Switzerland when, in 1905, he published three famous papers in different areas of physics. In the first of these papers he presented his "special theory of relativity." The other two dealt with the quantum theory of light and Brownian movement, which we will discuss later. Within a short time Einstein became a well-known person, for these papers were largely responsible for a second revolution in physics—the first, you will remember, was due to the work of Galileo and Newton.

Besides being one of the great scientific minds of history, Einstein was a great humanitarian. Until his death in 1955, he was a constant advocate of world peace and spoke out strongly against all forms of political repression. He wrote extensively on philosophy and the human condition, and his many essays and other works make for stimulating and rewarding reading.

1. REFERENCE FRAMES AND THE ADDITION OF VELOCITIES

The theory of relativity carefully examines how we observe physical *events*. It thus deals with the very foundation of scientific investigation. The first question we must answer is "how do we specify an event?" The answer is that we designate both the time of the event and the place at which it occurred. For example, the statement "The first shot of the American Revolution was fired in Concord on April 19, 1775" specifies both the place and the time of this event. If we don't know where Concord is, we can find its coordinates on a map, namely its latitude and longitude. The theory of relativity is concerned with the times and places at which events take place as observed from different points of view, or different "reference frames."

Measurements Are Made Relative to a Frame of Reference

Whenever we describe an event we have observed, we describe it from a particular frame of reference. Let's take an example. You are traveling on a train and observe a bird flying above you in the sky. You measure its speed and say, "That bird is flying at 35 miles per hour." A friend who is traveling with you replies, "Your statement is not really complete: The bird's speed is 35 miles per hour relative to what? Is it 35 miles per hour with respect to the earth or is it 35

Figure 6–2

A train moving at 60 mi/hr. The waiter walking up the aisle has a different speed according to the two observers. The man seated on the train observes the waiter passing him with a speed of 3 mi/hr. The man on the ground observes the waiter passing him with a speed of 63 mi/hr.

miles per hour with respect to the train? What's your frame of reference?" You decide to ignore your friend and take a seat in the train's club car. You note that the train is traveling 60 mi/hr. A waiter walks past rather swiftly—according to your estimate, about 3 mi/hr with respect to the train. To an observer standing on the ground, however, the waiter's speed is considerably greater. If the waiter is walking toward the front of the train, his speed is 60 mi/hr plus 3 mi/hr, or 63 mi/hr *with respect to the earth*, Fig. 6-2. If he is walking toward the back of the train, his speed is 60 mi/hr minus 3 mi/hr, or 57 mi/hr with respect to the earth. The waiter's speed with respect to the earth is certainly much different from what it is with respect to the train.

Clearly, then, whenever we talk about a speed or a velocity, we must specify the reference point or **reference frame** from which it is measured. In the examples just described, we mentioned two reference frames: the earth and the train. The velocities we spoke of were measured with respect to one or the other of these reference frames. In everyday speech we seldom specify the reference frame, since it is understood from the context. Usually the reference frame we have in mind is the earth.

Addition of Velocities

Let's follow that waiter walking on the train. Suppose we know his velocity with respect to the train, call it v_1, and the velocity of the train with respect to the earth, call it v_2. To find the velocity, v, of the waiter with respect to the earth, we simply *add* the two velocities if they are both in the same direction: $v = v_2 + v_1$. If the two velocities are in opposite directions, we subtract them: $v = v_2 - v_1$. This simple procedure is known as the "addition of velocities."

In a similar way, for a boat traveling up or down a river we can calculate the velocity with respect to the river bank by adding the

Figure 6–3

A boat moving upstream moves more slowly, with respect to the earth, than a boat moving downstream.

velocity of the river current to the velocity of the boat with respect to the water, Fig. 6–3. Suppose the boat can go 10 km/hr in still water. If the speed of the river is 3 km/hr, how fast will the boat be going with respect to the earth?

If the boat is going downstream, its velocity with respect to the earth will be: $v = 10$ km/hr $+ 3$ km/hr $= 13$ km/hr. If the boat is going upstream, it will be fighting the current and its velocity will be less: $v = 10$ km/hr $- 3$ km/hr $= 7$ km/hr. These results make sense, for obviously it takes longer to travel the same distance upstream than downstream.

Here's another example. A ballplayer standing still throws a baseball with a velocity of 20 m/sec. A second player runs toward the ball at a speed of 5 m/sec with respect to the ground and catches it, Fig. 6–4. The ball strikes the second player's glove harder than it would have had he been at rest, because the velocity of the ball with respect to the second ballplayer is $v = 20$ m/sec $+ 5$m/sec $= 25$ m/sec. Now if the second ballplayer were running *away* from the ball at 5 m/sec, the velocity of the ball with respect to the player would be $v = 20$ m/sec $- 5$ m/sec $= 15$ m/sec, and it would be a "softer" catch.

Figure 6–4

The velocity of the ball with respect to the fielder is the sum of the velocity of the ball as thrown plus the velocity of the fielder.

What happens if the player who throws the ball is moving, say at 5 m/sec, and the player who catches the ball is at rest, Fig. 6–

Figure 6–5

The velocity of the ball with respect to the fielder is the sum of the velocity of the ball as thrown by the thrower plus the velocity of the thrower.

5? The thrower still throws the ball at 20 m/sec, but the velocity of the ball with respect to the second player will be either 25 m/sec or 15 m/sec depending on whether the thrower is moving toward or away from the second player. We might say that the ball acquires the additional velocity of the thrower's frame of reference.

Clearly, then, the velocity of an object is different when viewed from different reference frames. In other words, velocity is a **relative concept**. And we say that "motion is relative." This, of course, is not new. It was understood centuries ago by Galileo and Newton.

Position and Time Are Relative

The position of an object is also a relative concept. San Francisco is 600 kilometers from Los Angeles but 4000 kilometers from New York City. It all depends on your point of reference. If someone were to say that Chicago is 1000 kilometers away, you would need to know the speaker's point of reference: 1000 kilometers away from where?

What about time? Is time a relative or an absolute concept? Consider this statement: "A football game was played between Michigan and California on October 23, 1978, beginning at 1:30 P.M., PST; exactly 53 seconds after the game began, California scored a touchdown." Presumably, you will accept this as a complete statement. Whether the observer is in the stands or in an airplane flying overhead, experience leads us to believe that he would observe that the touchdown occurred 53 seconds after the start of the game. In other words, we think of time as an absolute concept, not a relative one. Until the twentieth century physicists were firmly convinced of the absolute nature of time. That conviction is actually built into Newtonian mechanics. In the early part of this century, however, Einstein argued in his special theory of relativity that the absolute concept of time is not correct. For example, the observer in the airplane would not agree that the touchdown occurred 53 seconds after the game began. He would observe that it took somewhat longer, perhaps 54 seconds or maybe 67 seconds—the exact value depends on the velocity of his reference frame, the airplane. Strange, you say? We will soon see how this comes about.

141

2. THE TWO PRINCIPLES ON WHICH RELATIVITY IS BASED

Einstein's special theory of relativity is based on two principles: the so-called "principle of relativity" and the "principle of the constancy of the speed of light."

The Relativity Principle

The first principle, which had in essence been stated earlier by Galileo and others before him, is the **relativity principle:**

The laws of nature are the same in all reference frames that move uniformly (i.e. at constant velocity) with respect to each other.

In other words, the laws of physics have the same form on the earth as they do, say, on a train moving in a straight line at 60 mi/hr. For example, playing table tennis or billiards on a train or a ship is no different than on earth, unless the train or ship lurches, in which case it is not (for that moment) moving uniformly but is accelerating. You can feed yourself normally in a uniformly moving jet plane; and if you drop a fork, it falls just as it does on the earth. Moving objects follow the same laws on a uniformly moving train, ship, or plane as they do on earth. One way to convince yourself is to do the following experiment:

EXPERIMENT

Take a ride in an automobile (not a convertible) with someone else driving. Close the windows to avoid any wind inside the car. With the car going at a constant speed, hold an object (such as a ball or a coin) above you and let it drop. Notice its path. Is it any different from what it would be if you were at rest on the earth when you dropped the object?

When you drop an object it falls in a straight line whether you are at rest on the earth or at rest in a uniformly moving automobile, ship, or train. (However, if you drop an object out the window of a moving vehicle, its path is curved, because air—wind—exerts an extra force on the object.)

We said that if you dropped the object inside the car it would fall straight down. And it does—with respect to the car. But to an observer standing along the road, the object would follow a curved path, Fig. 6–6(B). The path of the object is different for the two ob-

Figure 6–6

The path of ball dropped inside a moving car: (A) as viewed by the person in the car; (B) as viewed by an observer on the ground. The upper views represent the situation when the ball is released, and the lower views, a few seconds later.

servers. But this does not violate the relativity principle, because it is only the *laws* of physics that are the same in these two reference frames. It is the paths that look different. The curved path as seen by an observer on the earth is in accord with the same laws of physics as is the straight path seen by the observer in the car. To the observer on earth, the object had an initial velocity equal to the velocity of the car and therefore follows a curved path like a projectile.

If you throw a ball straight up when you are sitting in a uniformly moving automobile, it comes straight down, just as it would if you were at rest on the earth. When your car is going 80 km/hr, you feel just the same sitting in your seat as you do when you are at rest with respect to the earth. When you accelerate or decelerate, you are pulled forward or backward—you feel unusual forces—but in this case your reference frame is *not* moving uniformly so the laws of physics are different. "Special" relativity deals only with uniformly moving reference frames. Einstein dealt with accelerating reference frames in his "general theory of relativity," but we will not discuss that theory here.

The Relativity Principle Asserts That There Is No Preferred Reference Frame

The relativity principle tells us that one reference frame is as valid as any other that moves uniformly with respect to the first for specifying the laws of physics. Furthermore, an observer on a train moving uniformly with respect to the earth is perfectly justified in saying that he is at rest and the earth is moving beneath him. What we

143

really mean by "relativity" is that one reference frame is as valid as another. *There is no preferred reference frame.* There is no reference frame (or object) that we can claim as be...ıg at absolute rest, and which therefore would be a preferred one to make measurements from.

Scientists before Einstein assumed that there must be a "preferred" reference frame, one that is truly at rest. Early in history it was thought that the earth was this absolute reference frame at rest, and that the planets, the sun, and the stars revolved around the earth. Later, Copernicus and Galileo found that the motion of the heavenly bodies was most easily explained by assuming that the earth moves about the sun. So perhaps we could claim that the sun is absolutely at rest? Alas, that assumption is not any good either, since the sun moves with respect to the center of our galaxy. And the galaxy moves with respect to . . . well, let's say the distant stars (or do the distant stars move with respect to our galaxy?) As Galileo said, "Everything moves." Everything moves with respect to everything else. We cannot tell by experiment whether our galaxy moves with respect to the distant stars or the stars move with respect to our galaxy. That is what relativity tells us: the laws of physics are the same in every uniformly moving reference frame, and so we cannot tell which is the moving one and which is the one at rest. All we can say is that two reference frames are moving *with respect to each other.*

Even in the nineteenth century most physicists believed that there was an absolute reference frame that is at rest. In searching for it, they discovered something startling, whose explanation came only in 1905 with Einstein's theory of relativity. Here is what they found.

Prelude to the Second Principle: The Michelson-Morley Experiment

Measurements of the speed of light played a crucial role in the search for an absolute reference frame. Indeed, light plays a fundamental role in relativity. There's nothing surprising about that. Physics in general, and relativity in particular, requires careful observation, and we make most of our observations through our sense of sight. Light from objects in our environment enters our eyes, and our eyes form an image that we can "see." (We will discuss light in Chapters 11 and 12.) In everyday life, light seems to travel from one place to another instantaneously, although actually it travels with the very high speed of about 300,000 km/sec or 186,000 mi/sec.

In the 1880s two Americans, A. A. Michelson and E. W. Morley, conducted a series of experiments to find the sought-after "absolute

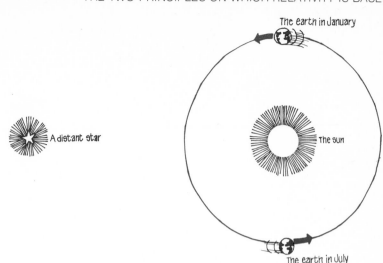

The earth in January

A distant star

The sun

The earth in July

Figure 6–7

The orientation of the sun, earth, and a distant star at two different times; in January the earth is moving toward the star, in July away from it.

reference frame" and to determine the velocity of the earth with respect to it. The results of their experiments had a far greater significance than they could have realized when they began. Their experiments were based on the work of James Clerk Maxwell, who in the 1860s had developed a detailed theory, which is still accepted today, of electric and magnetic phenomena, including light. Maxwell's theory predicted that the velocity of light is 300,000 km/sec. Scientists expected that this must be its velocity with respect to the "absolute reference frame." The velocity of light in any other reference frame would presumably be different, since the velocity of that reference frame would have to be added on.

By measuring the speed of light and determining in what reference frame it has the velocity predicted by Maxwell, Michelson and Morley tried to find out whether this absolute frame was fixed to the earth, to the sun, or to what. Similar experiments were subsequently performed by others. We now discuss one set of experiments in a very simplified way. As shown in Fig. 6–7, the light coming from a distant star can be observed at times of the year six months apart, say in January and July. For the purpose of argument assume that the star is at rest with respect to the sun and the velocity of light as it leaves the star is $c = 300,000$ km/sec. Now the earth moves in its orbit with a velocity, call it u, equal to 30 km/sec with respect to the sun. This is very fast indeed, but it is small compared to the velocity of light. Nonetheless, Michelson and Morley had excellent equipment, namely the "interferometer" invented by Michelson, which could detect even small differences in the velocity of light. If the light leaves the star with velocity c and the earth is moving toward the star with velocity u (January, in Fig. 6–7), then an observer on earth would expect to measure the velocity of the star's light to be $c + u$. Six months later, in July, the earth is going away from the star, and hence the star's light should have a velocity with respect to the earth of $c - u$. The velocity of the light from the star should be less in July when the earth is moving away from the star than it is in January when the earth is approaching the star.

To the astonishment of scientists everywhere, Michelson and Morley found that the velocity of the light from the star, with respect to the earth, was exactly the same in July as it was in January! They tested all sorts of different cases, using different sources of light. In every case the measured velocity of the light was the same. Michelson and Morley had set out to find a fairly simple entity, the "reference frame at absolute rest;" but they discovered instead a far more intriguing and complex phenomenon. (This kind of discovery, though not common, is not unusual in science.)

Many attempts were made in the ensuing years to explain the results of the Michelson-Morley experiments. Most were very complicated explanations and were eventually discredited by internal contradictions or by further experimental results. Einstein finally clarified the situation.

The Second Principle: The Constancy of the Speed of Light

Einstein, convinced that there was no reference frame at absolute rest, proposed a second principle called the *principle of the constancy of the speed of light*:

The speed of light in empty space is the same for all observers regardless of their velocity or the velocity of the source of the light.

This second principle of the special theory of relativity is consistent, in a simple way, with the result of the Michelson-Morley experiment.[1] It says that an observer traveling toward a source of light, Fig. 6–8, would measure exactly the same velocity for that light as would an observer at rest with respect to the light. If a ballplayer runs toward a baseball coming toward him, the velocity of the baseball is the sum of the player's velocity with respect to the ground plus the velocity of the ball with respect to the ground, as we saw earlier. But for light, the velocity with respect to the moving observer is the *same* as its velocity with respect to the ground. We do not add in

---

[1]Einstein later said that he was not fully aware of the results of the Michelson-Morley experiment. He was motivated instead by questions relating to light itself. The point here is that Einstein's theory was fully consistent with the Michelson-Morley experiment.

Figure 6–8

Two observers, one at rest and one moving at high speed toward a source of light, both measure the same velocity for that light.

the velocity of the observer! Now this violates common sense. How can this bizarre fact be reconciled with our everyday experience? Part of the problem is that in our everyday experience we do not measure velocities anywhere near the velocity of light, and so we cannot expect our everyday experience to be helpful in situations at very high speeds.

3. THE PREDICTIONS OF RELATIVITY

The special theory of relativity is based entirely on the two principles we have discussed: (1) the relativity principle and (2) the constancy of the speed of light. From these two principles, Einstein derived a number of extremely interesting results.

Simultaneity Is Not Absolute

Perhaps the most striking aspect of the theory of relativity, at least philosophically, is that we can no longer regard time as an absolute quantity. No one doubts that time flows on and never turns back. Yet the time between two events, and even whether two events are simultaneous, depends on the reference frame in which the events are being observed.

147   When we say that two events occur *simultaneously*, we mean

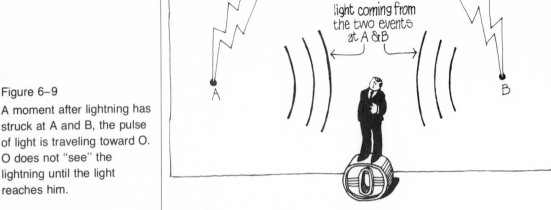

Figure 6–9

A moment after lightning has struck at A and B, the pulse of light is traveling toward O. O does not "see" the lightning until the light reaches him.

that they occur at exactly the same time. But how do you tell precisely whether or not two events are simultaneous?

Before we try to answer that question, we must ask another fundamental question: How do you tell if an event has occurred? Most commonly we *see* it. Light travels from the point of occurrence to our eyes at a finite speed, and only when the light has reached us do we know that the event has occurred.

Now, how do you tell whether or not two events are simultaneous? It's easy to tell that two events are simultaneous if they occur at the same point in space, such as two apples hitting you on the head at the same time. But, if the two events occur at widely separated places, it isn't so easy. An observer would have to take into account the time it takes for the light from each event to reach him and then figure out exactly when the events occurred. If such calculations indicate that the two events occurred at the same time, then they are simultaneous.

Because we are mainly interested in the ideas involved, we will avoid making such calculations and will consider some very simple experiments instead. Einstein himself thought up similar simple experiments that, though not always practical to carry out, do illustrate the important principles. He called these "thought experiments" (in German, "gedanken" experiments).

With the help of a simple thought experiment we will now find an easy way to determine if two events are simultaneous. Then we will try to answer the question: If two events are simultaneous to one observer, will they be simultaneous to a second observer moving with respect to the first? Consider an observer, call him O, situated at point O exactly halfway between two points A and B. For convenience, we use O to refer both to the observer and to the place where he is. At A and B, two events have occurred; let's say that lightning has struck at both A and B, Fig. 6–9. For brief events such as lightning, only short pulses of light will travel outward from A and

B and reach O. O "sees" the lightning when the two pulses of light reach him. Suppose the pulses of light reach O at the same time. The light pulses from the two events travel at the same speed and, since the distance AO equals OB, the time it takes for the light to travel from A to O and from B to O is the same. Therefore, the observer O can definitely state that the two events A and B occurred at the same time; that is, they were simultaneous. In summary, if an observer is situated midway between two events and detects the light coming from each event at the same time, then the two events had to occur simultaneously. If the light from one of the events arrives first, then that event must have occurred first.

Now consider two different observers, call them observer 1 and observer 2 (abbreviated O_1 and O_2), moving with a velocity v with respect to each other. Let's say they are on two different trains, which constitute two reference frames moving with respect to each other. To observer O_2, O_1 is moving to the right with velocity v. On the other hand, to observer O_1, O_2 is moving to the left with velocity v, Fig. 6–10. Both viewpoints are legitimate—remember the relativity principle. There is *no* third point of view that will tell us which reference frame is "actually" moving.

Figure 6–10

Two observers O_1 and O_2 are on two different trains (two different reference frames) moving with a relative velocity v. O_2 claims that O_1 is moving to the right (A), while O_1 claims that O_2 is moving to the left (B). Both viewpoints are legitimate, of course. It all depends on your reference frame.

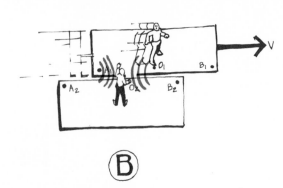

Figure 6–11

Two reference frames as viewed by an observer O_2. To O_2, the other reference frame (O_1) is moving to the right. In (A), one lightning bolt strikes in both reference frames at A_1 and A_2, and a second lightning bolt strikes at B_1 and B_2. According to the observer O_2 the two bolts of lightning strike simultaneously. (B) A moment later the light from the two events reaches O_2 at the same time (simultaneously). But in the other reference frame, the light from B_1 has already reached O_1, whereas the light from A_1 has not yet reached O_1. So in O_1's reference frame the event at B_1 must have preceded the event at A_1. Time is not absolute!

It is a stormy day and lightning strikes, twice in fact, and the lightning makes marks on both trains where it struck—at A_1 and B_1 on O_1's train, and at A_2 and B_2 on O_2's train. O_1 just happens to be midway between A_1 and B_1, and O_2 midway between A_2 and B_2. Before we go further, we must put ourselves in one reference frame or the other. Let us view the situation from reference frame 2, that of O_2. And let us further assume that the two events occurred simultaneously in this (O_2) reference frame, and at exactly the time when O_1 and O_2 are opposite each other, Fig. 6–11(A). As shown in Fig. 6–11(B) [this view was taken a short time after Fig. 6–11(A) was] the light reaches O_2 from A_2 and B_2 at the same time. Since O_2 is midway between A_2 and B_2, he knows that the two events occurred simultaneously in his reference frame.

Now, what does observer O_1 see? The light from the two events takes a finite time to reach O_1, and in this time he has moved to the right. Thus we see in Fig. 6–11(B) that in the reference frame of O_1, the light from B_1 has already reached and passed O_1, while the light from A_1 has not yet reached O_1. Now the reference frame of O_1 is just as good as that of O_2. The speed of light is the same for O_1 as it is for O_2, and the light from A_1 and from B_1 travel at the same speed. Because the light from B_1 arrived before the light from A_1 and since O_1 is midway between A_1 and B_1, observer O_1 claims that event B_1 occurred *before* event A_1. Thus, the two events that are simultaneous to one observer are not simultaneous to the other observer! This is a remarkable conclusion.

"So," you may ask, "which observer, O_1 or O_2, is right?" In other words can we put ourselves in some third reference frame and try to be objective? But this would only make matters more complicated. Anyway, what third reference frame could we choose? Re-

Figure 6–12

Thought experiment that shows how time dilation comes about. The time it takes for the light to travel over and back on the spaceship is longer for the earth observer (B) than for the observer on the spaceship (A).

151

member, there is no "best" reference frame! The only answer is that *both* observers are right. *Time is different in different reference frames. Time and simultaneity are not absolute; they are relative!*

This thought experiment demonstrates that two spatially separated events that are simultaneous to one observer are not simultaneous to a second observer who is moving with respect to the first. The demonstration rests on the two principles of relativity, particularly the second, the constancy of the speed of light. We mentioned it only once in the discussion, but it was crucial.

The two observers disagree as to whether or not the two events are simultaneous. But each has studied relativity and realizes that his own observations are not absolute—that another observer's viewpoint is just as valid even though it is different. For example, O_1 observes that event B occurred before event A but also figures out that observer O_2 will observe that these two events happened at the *same* time.

The relative nature of simultaneity is hard to accept, partly because we are not aware of it in everyday life. Only when the relative velocity of the two reference frames is fairly close to the speed of light, or when the distances involved are huge, will the effect be noticeable.

Time Dilation: Moving Clocks Run Slowly

If two events that are simultaneous in one reference frame are not necessarily simultaneous in a second reference frame, is it possible that clocks would run differently and time would pass differently in two different moving reference frames? Indeed, that is just what Einstein's theory of relativity predicts. Let's try another thought experiment to see how.

In Fig. 6–12 we see a spaceship that is traveling past the earth at a high speed. Fig. 6–12(A) is drawn from the point of view of an observer on the spaceship, and Fig. 6–12(B) from the point of view of an observer on earth. The man on the spaceship flashes a light and measures the time it takes for the light to travel across his spaceship, bounce off a mirror, and return to him, Fig. 6–12(A). He can calculate precisely how long it will take the light to go over and back by using the relation: $velocity = \dfrac{distance}{time}$. Since he knows the distance, which is twice the width of his spaceship, and the velocity, which is the velocity of light, he can readily calculate the time.

An observer on earth, Fig. 6–12(B), observes this same process and she too can calculate how long it takes the light to travel across the spaceship, bounce off the mirror, and return to the sender. But notice in Fig. 6–12(B) that since the spaceship is traveling to the right according to the earth observer, the light itself travels in the diagonal path shown. To the earth observer, then, the light travels a distance *greater* than twice the width of the spaceship. But of course the velocity of the light is the same as it is for the spaceship observer (the second principle of relativity). Therefore the time it takes the light to travel this greater distance must be greater. So the observer on earth calculates that the time needed for the light to travel across the spaceship and back is greater than what the spaceship observer calculates.

Of course both reference frames are valid ones in which the laws of physics hold, and therefore both calculations are in accord with what the watches of the two observers show. Thus the observer on earth sees that the spaceman measures a shorter time between the two events (the sending out and the receiving of the light) than she does. That is, the spaceman's clock seems to run more slowly, and indeed time seems to pass more slowly for him than for the earth observer. This is a general result of the theory of relativity and is known as **time dilation**; it can be stated simply as:

Moving clocks run slowly.

With a little geometry, we could easily calculate how great the time dilation effect would be. However, we will merely give the result here. We let t_0 represent the amount of time between two events that occur at the same place in one reference frame [as in Fig. 6–12(A)] and let t represent the time between the same two events as viewed by an observer in another reference frame; then if v is the relative velocity of the two reference frames, the theory of relativity predicts that

$$t = \frac{t_0}{\sqrt{1 - \frac{v^2}{c^2}}}$$

In this equation c is the speed of light ($c = 300,000$ km/sec). Notice that since the denominator is always less than 1 (except when $v = 0$), t is always greater than t_0.

To illustrate the time dilation effect, let us take a concrete example. Suppose a person is on a train (Fig. 6–13) moving at a very high velocity with respect to the earth, say 0.65 c. This man looked at the clock when he was served dinner; it read 7:00. When he finished eating, it read 7:15. So it took him 15 minutes to eat his dinner. Notice that the two events occur at the same place for this observer. Now observers on the ground would claim that the time between these two events, which to them occurred at different places, was 20 minutes, Fig. 6–13(B), in accordance with the above equation. So the observers on the ground would claim that the clock on the train was running slowly! Of course the train has to be going unrealistically fast, close to the speed of light, in order for this effect to be detected with ordinary watches and clocks.

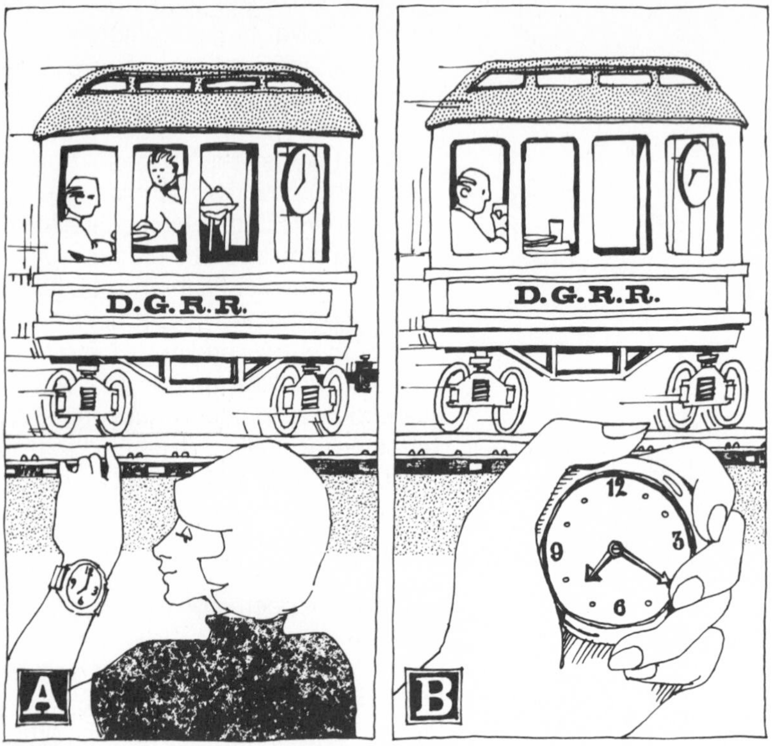

Figure 6–13

A person on a fast-moving train begins dinner at 7:00 (A) and finishes at 7:15 (B) according to an accurate clock on the train. But observers on the ground, who set their watches to correspond with the clock on the train at the beginning of the meal, measure the time needed to eat the meal as 20 minutes.

Incidentally, the football game discussed earlier in this chapter is another example of time dilation. Again, however, the airplane must be going very fast if an observer in it is to measure a noticeably longer time interval between the start of the game and the touchdown than that measured by an observer in the stands.

Space Travel: The Clock Paradox

The time dilation effect has led to some interesting speculations about space travel. Now that humans have walked on the moon, the possibility of going farther seems ever closer to reality. The exploration of the other planets in our solar system is already being planned. Yet, space travel beyond our solar system seems very remote because of the vast distances involved and the long times required for the trips. For example, the nearest star, Proxima Centauri, is four light-years[2] away, so that it would take a spaceship traveling very close to the speed of light about eight years to make a round trip. To reach more distant stars would take even longer.

[2] A light-year is defined as the distance light travels in one year, which, at the rate of 186,000 miles per second, is 5,860,000,000,000 miles.

This is where time dilation comes in. If a spaceship were to travel from the earth to Proxima Centauri and back at a speed of 0.99 c, observers on the earth would measure the trip as having taken about eight years. But because of the time dilation effect, time on the spaceship would appear to pass much more slowly—less than a year would pass for the space traveler. According to an earth observer, not only would the clocks on the spaceship be running more slowly but so would all other processes, including life processes. The clocks do not really slow down (they are perfectly good clocks). They give an accurate measure of how time passes on the spaceship. Indeed, to the people on the spaceship, time passes in a perfectly normal way. To them it seems as if only a year has passed for the whole trip. So, according to observers on earth the trip took some eight years; but because of time dilation, they recognize that the people on the spacecraft age less than a year.

Now consider two twins, each aged 30 when the spacecraft leaves the earth carrying one of them; the other remains on earth. Relativity tells us that the twin on earth ages eight years, while the one on the spacecraft ages only one year. The twin remaining on earth would be 38 when the spacecraft returned, but the traveling twin would just be pushing 31, Fig. 6–14.

What does the twin on the spacecraft say about time? Since everything is relative, any uniformly moving reference frame is as good as any other for applying the laws of physics. So won't the astronaut twin make all the claims that the earth twin does, only in reverse? To the twin in the spaceship, the earth is moving away at a speed of 0.99 c. So the astronaut twin will look back at the earth,

Figure 6–14
The clock paradox.

observe that the clocks on earth are running slowly, and predict that when he eventually returns to the earth, his earthbound twin will have aged less than he has and thus be younger than himself! This is the opposite of what the earth twin expects. They cannot both be right, since after all the spacecraft does come back to the earth and a direct comparison of ages and clocks can be made.

This apparent paradox, known as the "clock paradox," was raised in an attempt to discredit the relativity theory. But the solution was quickly found. There really is no paradox at all. The specific consequences of special relativity, in this case time dilation, can be applied only by observers in uniformly moving reference frames. Remember the first principle of relativity? Whereas the earth is such a uniformly moving reference frame (or very close to it), the spaceship is not. The spaceship starts from rest on the earth, accelerates to high speed, and decelerates on its return. Even more importantly, it turns around at its distant destination and comes back. This turning around, a change in the direction of the velocity, involves a considerable acceleration. Since the reference frame of the spaceship *accelerates* for part of the trip, it is not moving uniformly. So the twin on the spacecraft is not justified in his use of special relativity theory, and his predictions are not valid. The twin on the earth, being in a uniformly moving reference frame, can make the correct predictions. So the original prediction that the astronaut twin will come back younger than the earthbound twin is valid.[3]

Time dilation, it would seem, appears to give us a "fountain of youth." However, this is not quite the same as what Ponce de Leon was looking for, namely an extension of life without the effects of old age. The space traveler we have just considered has aged seven years less than people who remained on the earth. So he is seven years younger than he "ought" to be. During his trip, however, he experienced only one year of life—one year of work, play, reading, what have you. Meanwhile people on earth experienced eight years of life. The space traveler has not really gained youth, or extra life. But he will be younger than his friends when he returns.

On a long enough and fast enough trip, a space traveler could return to earth younger than his children or grandchildren. If he left in the year 1980 and traveled at a speed of $0.99\,c$ for what seemed to him 30 years, he would arrive back on earth in the year 2280. It would be an interesting experience.

Length Contraction: Moving Objects Look Shorter

Since time intervals are different in different reference frames, we

[3]Einstein's "general" theory of relativity, which deals with accelerating reference frames, agrees with this prediction.

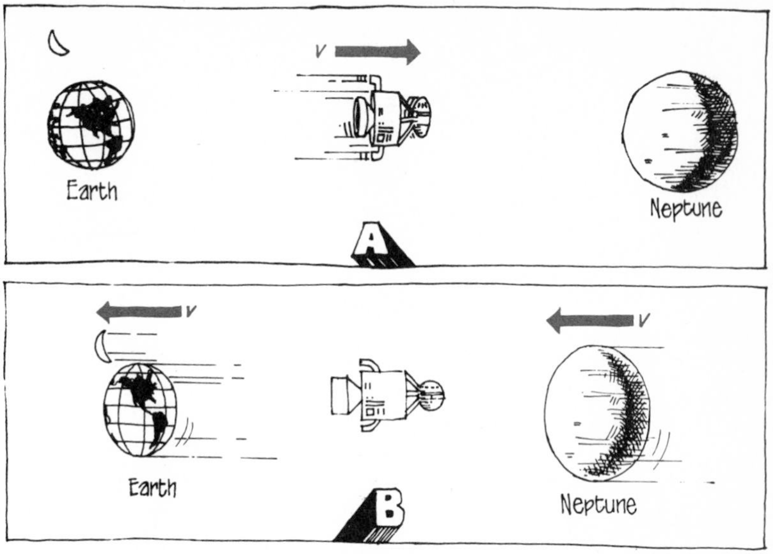

Figure 6–15

(A) Spaceship traveling at very high velocity from earth to Neptune, as seen from earth's frame of reference. (B) The same situation as viewed by an observer on the spaceship; earth and Neptune are moving at the very high velocity *v*.

might expect that space intervals would be different as well. Indeed the special theory of relativity confirms this and specifically predicts that *moving objects are measured to be shorter* than when they are at rest. To see why this is true, let us examine the thought experiment illustrated in Fig. 6–15.

In Fig. 6–15(A) we see an earth observer watching a spaceship heading from earth to Neptune at a very high (but currently unrealistic) speed of 240,000 km/sec or 0.8 *c*. The distance to Neptune is about 5 billion kilometers, and at a speed of 0.8 *c* the trip will take about 6 hours. That is, it will take that long according to an earth observer. However, because of time dilation, to the traveler on the spaceship the trip will take only about 3½ hours. In the space traveler's frame of reference, the earth and Neptune are moving with a speed of 0.8 *c*, as shown in Fig. 6–15(B). According to the space traveler, the earth moves away from him, and Neptune toward him, at a speed of 0.8 *c*. The time difference between departure of earth and arrival of Neptune is only 3½ hours. Because the space traveler measures the *same speed* but *less time*, he must also observe the distance between earth and Neptune as being *less* than what the earth observer sees. (In fact, using the relation distance = velocity × time, we obtain the distance, as viewed by the space traveler, as only about 3 billion kilometers.) To the space traveler, the earth and Neptune are moving and the distance between them is less than what the observer on earth measures.

This is a general result of the theory of relativity. The shortening of distances also applies to the distance between the two ends of an object in motion—that is, it applies to the *length* of a moving object:

The length of an object is measured to be shorter when moving than when at rest.

This **length contraction** occurs only in that dimension of an object that is parallel to the direction of its motion. The other dimensions remain unchanged.

Quantitatively, if the length of an object at rest is L_0, then when it is moving at a velocity v the theory of relativity predicts its length L will be measured to be

$$L = L_0 \times \sqrt{1 - \frac{v^2}{c^2}}$$

However, the amount of contraction will be noticed only if the object is moving very swiftly.[4]

As in the case of time dilation, length contraction will be observed in all moving reference frames. The observers standing on the ground in Fig. 6–13, for example, will observe that the train car is shorter than usual, and that the people on the train look thinner than normal people. In just the same way, a person on the train observes that the people on the ground are thinner than normal people. In fact, all the trees and buildings that pass by the window are thinner than normal. This is, of course, not shown in Fig. 6–13, since this is the view seen by an observer on the ground.

Space–Time: The Fourth Dimension

Let us look again at the situation illustrated in Fig. 6–13. According to the man on the train it took 15 minutes to eat his dinner, which, incidentally, filled a 12-inch-wide plate. According to observers on earth it took the man on the train 20 minutes to eat his dinner from a full plate only 9 inches wide (length contraction). That is, to observers on the ground the food looked smaller but lasted longer.

In a sense these two effects, time dilation and length contraction, balance each other. When viewed from the ground, what the food seems to lose in size it gains in the length of time it lasts. Space, or length, is exchanged for time!

This result led to the formulation of the idea of four-dimensional "space–time": space takes up three dimensions and time is the "fourth dimension." Space and time are intimately connected. Just

---

[4]For example, if v is $0.8\,c$, then $\sqrt{1 - v^2/c^2} = \sqrt{1 - 0.64\,c^2/c^2} = \sqrt{0.36} = 0.6$, so $L = 0.6\,L_0$. If the object is a meter stick, $L_0 = 1.0$ m and $L = 0.6$ m; when traveling at $0.8\,c$, the meter stick will look like it is only 60 cm long.

as when we squeeze a balloon we make one dimension larger and another smaller, so when we view objects and events from different reference frames a certain amount of space is exchanged for time, or vice versa.

This idea of space–time is very satisfying from another point of view. Objects have spatial extent as well as temporal extent. They exist for a certain amount of time at a particular place, such as the food that existed for 15 minutes. Thus, although the term "four dimensions" seems strange, it refers to the fact that any object or event is specified by four quantities—three to describe where in space, and one to describe when in time. The really unusual aspect of four-dimensional space–time is that space and time can intermix: a little of one can be exchanged for a little of the other when the reference frame for viewing the events is changed.

It is difficult for us to understand the idea of four-dimensional time. Somehow we feel, just as physicists did before the advent of relativity, that space and time are completely separate entities. Yet we have found in our "thought experiments" that they are not completely separate. Our difficulty in accepting this is reminiscent of the situation in the seventeenth century at the time of Galileo and Newton. Before Galileo, the vertical direction, that in which objects fall, was considered to be entirely different from the two horizontal dimensions. Galileo showed that the vertical dimension differs only in that it happens to be the direction in which gravity acts. Otherwise, all three dimensions are equivalent, a fact that we all accept today. Now we are asked to accept one more dimension, time, which we had previously thought of as being somehow different. It may, however, take more mental effort to accept time as a fourth dimension on an equal footing with the other three. Perhaps in three centuries this, too, will be a generally accepted idea.

This is not to say that there is no distinction whatsoever between space and time. What relativity has shown is that space and time determinations are *not independent* of one another. Time can no longer be considered totally separate from the three spatial dimensions, any more than the vertical dimension can be considered totally separate from the two horizontal dimensions.

Mass Increase

The three basic quantities we can measure are length, time, and mass. As we have seen, the first two are relative; measurements of length and time intervals are different in different moving reference frames. And the same is true of mass. According to the theory of relativity:

159

Moving objects have increased mass.

Quantitatively (we will not go into details), the theory predicts that if an object has a mass m_0 when it is at rest, then when moving with a velocity v past an observer, the object's mass will be measured to be

$$m = \frac{m_0}{\sqrt{1 - \frac{v^2}{c^2}}}.$$

Since the denominator is always less than 1 (unless $v = 0$), the mass m will always be more than the rest mass m_0.

Relativistic Addition of Velocities

Suppose a rocket ship traveling away from the earth with velocity v sends off a second rocket, as shown in Fig. 6–16. If the velocity of the second rocket as observed by people on the first rocket is u, then we might expect that the velocity of rocket 2 as seen by observers on earth would be $v' = u + v$. (This is like the waiter on the train in Fig. 6–2 who is going 63 mi/hr with respect to the ground.) However, Einstein showed that since both length and time are different in different moving frames of reference, velocity cannot be added up so simply. Instead, Einstein showed that we must use the formula

$$v' = \frac{v + u}{1 + \frac{vu}{c^2}},$$

where c is the velocity of light.

If v and u are each small compared to the velocity of light c, as is usually the case is everyday experience, then $\frac{vu}{c^2}$ is practically zero and $v' \approx \frac{v + u}{1} = v + u$, which is just our common-sense formula.[5] Thus Einstein's formula agrees with everyday experience for ordinary velocities that are much smaller than the velocity of light.

If the velocity of an object is close to the speed of light, rather bizarre effects occur. Suppose for example that rocket 1 in Fig. 6–16 is launched from the earth at a velocity of 180,000 km/sec, which is 6/10 the speed of light, and that rocket 2 is shot from the nose of the first at a velocity of $0.6\,c$ with respect to the first rocket. Common sense tells us that the second rocket should have a velocity of $0.6\,c + 0.6\,c = 1.2\,c$ with respect to the earth; but our "common sense" arises from everyday experience, which does not include objects moving at such high velocities. Instead, we must

[5]The symbol $\approx$ means "approximately equal to."

Figure 6–16

(A) Rocket 1 is launched from the earth with a velocity of 0.6 c. Some time later, (B), rocket 2 is shot from rocket 1 with a velocity relative to rocket 1 of 0.6 c. According to relativity, the velocity of rocket 2 relative to the earth will be 0.88 c.

use Einstein's formula. Substituting $v = 0.6\,c$ and $u = 0.6\,c$ in Einstein's addition of velocities formula, we find[6] that the velocity of rocket 2 with respect to the earth is 0.88 c, or 264,000 km/sec, and not 1.2 c. In general, no matter how close the two velocities are to the speed of light, their sum will not exceed c. Try various velocities (less than c) in the formula and prove it to yourself. Because it is not possible to exceed the speed of light by adding together two velocities that are each less than the speed of light, we have here the first inkling that the speed of light is somehow a speed limit. As we shall see shortly, Einstein's theory of relativity does, indeed, predict that no ordinary object can move faster than the speed of light.

4. THE ULTIMATE SPEED; E = mc²; AND THE EXPERIMENTAL EVIDENCE

The consequences of the theory of special relativity are bizarre indeed. The question is, can they be true? Before we discuss further ramifications of the theory, let us discuss some of the experiments that have been performed to test the theory.

Experimental Evidence

A great deal of experimental evidence confirms the theory of relativity, and none (so far) disputes it. Indeed, nearly all scientists now accept it as an accurate description of nature.

You might think that the relativity effects are much too small to

$$^6 \quad v' = \frac{v + u}{1 + \dfrac{vu}{c^2}} = \frac{0.6\,c + 0.6\,c}{1 + \dfrac{0.36c^2}{c^2}} = \frac{1.2\,c}{1.36} = 0.88\,c$$

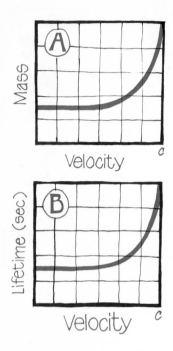

Figure 6–17
(A) Graph of the mass of a particle versus its velocity.
(B) Graph of the average lifetime of muons as a function of their velocity.

be detected at ordinary velocities using ordinary instruments. This is true for the most part. But in 1971, extremely precise atomic clocks were flown around the world in opposite directions by two 600 mi/hr jet planes. These clocks were capable of measuring time to a billionth of a second, and they confirmed experimentally that time dilation does indeed exist.

This was the first time that the predictions of relativity had been tested with ordinary-sized objects and ordinary velocities. But relativity had been experimentally confirmed decades earlier by scientists using very tiny "elementary particles"—electrons and protons—which are the basic constituents of atoms. These elementary particles have extremely small masses and thus require little energy to achieve velocities near that of light. At these high velocities, the relativistic effects of mass increase, for example, should be readily observed. Indeed, experiments with elementary particles have clearly demonstrated that, as velocity increases, the mass of the particle increases exactly as Einstein's formula predicts, Fig. 6–17(A).

An equally impressive experiment is the confirmation of the time dilation formula using high-speed elementary particles. A number of the elementary particles are unstable and decay after a certain time into smaller particles. One of these is a muon (or mu meson), whose average lifetime when at rest is about two-millionths of a second. Muons are commonly formed high in the earth's atmosphere by the collision of high-energy particles coming from the sun and other extraterrestrial sources. In fact, muons make up a large fraction of the "cosmic rays" that reach the surface of the earth and hence are ideal particles for this study.

Experiments show conclusively that the faster the muons move, the longer they live; and the time differences can be readily measured at these high speeds. In other words, the average lifetime of muons increases as the velocity of the muon increases. And it does so exactly as Einstein's formula predicts, Fig. 6–17(B). Of course, if you were to jump onto a muon, or travel along beside it, you would measure its lifetime as being simply its lifetime at rest, since you would be at rest with respect to it.

These are just three of the many experiments that attest to the validity of the special theory of relativity.

The Speed of Light: The Ultimate Speed

That the speed of light is the maximum velocity for objects in the universe can be seen from any of the three formulas given in the last section. Perhaps it is easiest to use the mass increase formula, $m = \dfrac{m_0}{\sqrt{1 - (v^2/c^2)}}$. As a body is accelerated to higher and higher speeds, its mass becomes greater and greater. Therefore more and more energy is required to increase its speed. Indeed, when $v = c$, the denominator in the above mass formula is zero and the mass would become infinite. Thus to accelerate an object to the speed of light would take an infinite amount of energy. Since infinities are not attainable, we see that an ordinary object cannot reach the speed of light.

Furthermore, if the velocity v were greater than c, the denominator, $\sqrt{1 - (v^2/c^2)}$, would be the square root of a *negative* number, and there is no such thing, at least not in real numbers. Mathematicians call such quantities "imaginary numbers." So the mass increase formula tells us that velocities greater than the speed of light are not possible, since in this case the mass would become imaginary. That is, the speed of light seems to be a speed limit in the universe.

This last statement was accepted until the late 1960s, when it was pointed out that Einstein's formula for mass increase does not rule out particles that *always* go *faster* than the speed of light. If such particles existed (scientists proposed the name "tachyons," meaning fast in Greek), their rest mass m_0 would have to be an imaginary number. In this way the mass m would be the ratio of two imaginary numbers when v is greater than c. Now the ratio of two purely imaginary numbers happens to be real. Thus, as long as v is greater than c, the mass m will be real. But if the velocity should drop below c, then the mass would become imaginary. So tachyons would have a *lower* speed limit: they could not go *less* than the speed of light, only faster. All of this is very theoretical. Experimental physicists have conducted experiments to detect tachyons, but they haven't found any. Relativity does not tell us whether they exist or not, only that they are possible. The evidence today indicates that the speed of light is still the ultimate speed in the universe.

E = mc²

The equation $E = mc^2$ is perhaps the most famous equation in all of physics. What does it mean? If a force is applied to an object of rest mass m_0 the object moves faster and faster. Since the force is being applied through a distance, work is being done on the object and its energy is increasing. As the velocity of the object approaches the speed of light, the velocity cannot increase very much because the speed of light is the limiting velocity. But the mass of the object has been increasing as the velocity increased, and as the speed of light is approached the mass increases quite rapidly. Thus, the work being done on the object goes into increasing its *mass*. Normally when work is done on an object, it increases the energy of that object. But relativity predicts that the work can go into increasing the mass as well. This fact leads us to believe that mass is a form of energy. Indeed, Einstein showed that it is a con-

163

sequence of the theory of relativity that mass and energy are equivalent. They are different aspects of the same thing! Einstein deduced from the two principles of relativity that the energy and the mass of an object are related by the equation:

$$E = mc^2.$$

In this equation, E is the energy of a body, m is its relativistic mass, $m = \dfrac{m_0}{\sqrt{1 - (v^2/c^2)}}$, and c is the speed of light. Thus even when an object is at rest it has an energy, called its "rest energy," $E_0 = m_0 c^2$. It has this energy by virtue of its having mass. A moving object has both kinetic energy and rest energy; the sum of these two is known as the "total energy."

The law of conservation of energy remains valid in the theory of relativity but only as long as mass is considered to be a form of energy. Indeed, Einstein predicted that mass can be changed into energy and energy can be changed into mass according to $E = mc^2$. This prediction has been amply verified by experiment. Light energy has been transformed into tiny elementary particles that have mass, such as muons or electrons. The reverse has been observed as well. Perhaps the most graphic illustration occurs in an atomic bomb or a nuclear reactor. In both cases, mass is transformed into energy on a large scale. Indeed, because of the c^2 in the mass-energy equation, we see that even a small mass "contains" an incredible amount of energy. For example, one kilogram of a substance is equivalent to an amount of energy: $E = mc^2 = 1$ kg $\times$ (300,000,000 m/sec)^2 = 90,000,000,000,000,000 joules. This amount of energy would supply the needs of a city of a million people for several years. It is not easy, however, to obtain this energy. The conversion of mass into energy does not occur spontaneously except in certain specific situations. In Chapter 16 we will discuss how it is done in the bomb and in nuclear reactors. Note that the equation $E = mc^2$ does *not* say that an object must travel at c in order to have this much energy. It has this energy even when at rest. The c^2 is merely a sort of "proportionality constant." (See Chapter 1.) By multiplying the mass m by c^2, we determine how much energy is released if the mass m is converted into energy.

Actually, the interchange of mass and energy on a very small scale is a common phenomenon. For example, when you heat a pot of water on the stove, you are putting heat energy into the water; and indeed, the mass of the water increases a small amount. When electric energy flows into a light bulb or an electric frying pan, the mass of the bulb or pan increases. In a chemical reaction in which heat energy is gained or lost, the masses of the reactants and of the products will be different. Ordinarily, however, the change in mass is too small to be detected.

5. MR. TOMPKINS

A little book by George Gamow, *Mr. Tompkins in Wonderland*, delightfully illustrates the results of relativity and what it would be like if ordinary speeds were close to the speed of light. In the excerpt[7] that follows, a Mr. Tompkins has fallen asleep while listening to a lecture on the theory of relativity. He finds himself dreaming of a world in which the natural speed limit, the speed of light, is only 20 mi/hr:

. . . to his surprise there was nothing unusual happening around him; even a policeman standing on the opposite corner looked as policemen usually do. The hands of the big clock on the tower down the street were pointing to five o'clock and the streets were nearly empty. A single cyclist was coming slowly down the street and, as he approached, Mr. Tompkins' eyes opened wide with astonishment. For the bicycle and the young man on it were unbelievably shortened in the direction of the motion, as if seen through a cylindrical lens. The clock on the tower struck five, and the cyclist, evidently in a hurry, stepped harder on the pedals. Mr. Tompkins did not notice that he gained much in speed, but, as the result of his effort, he shortened still more and went down the street looking exactly like a picture cut out of cardboard. Then Mr. Tompkins felt very proud because he could understand what was happening to the cyclist—it was simply the contraction of moving bodies, about which he had just heard. "Evidently nature's speed limit is lower here," he concluded, "that is why the bobby on the corner looks so lazy, he need not watch for speeders." In fact, a taxi moving along the street at the moment and making all the noise in the world could not do much better than the cyclist, and was just crawling along. Mr. Tompkins decided to overtake the cyclist who looked a good sort of fellow, and ask him all about it. Making sure that the policeman was looking the other way, he borrowed somebody's bicycle standing near the kerb and sped down the street. He expected that he would be immediately shortened, and was very happy about it as his increasing figure had lately caused him some anxiety. To his great surprise, however, nothing happened to him or to his cycle. On the other hand, the picture around him completely changed. The streets grew shorter, the windows of the shops began to look like narrow slits, and the policeman on the corner became the thinnest man he had ever seen.

"By Jove!" exclaimed Mr. Tompkins excitedly, "I see the trick now. This is where the word *relativity* comes in. Everything that moves relative to me looks shorter for me, whoever works the pedals!" He

[7]George Gamow, *Mr. Tompkins in Wonderland* (Cambridge, England: Cambridge University Press, 1940). Reprinted by permission of the publisher. This work has been published along with *Mr. Tompkins Meets the Atom* in paperback form as *Mr. Tompkins in Paperback* (Cambridge University Press, 1967).

was a good cyclist and was doing his best to overtake the young man. But he found that it was not at all easy to get up speed on this bicycle. Although he was working on the pedals as hard as he possibly could, the increase in speed was almost negligible. His legs already began to ache, but still he could not manage to pass a lamp-post on the corner much faster than when he had just started. It looked as if all his efforts to move faster were leading to no result. He understood now very well why the cyclist and the cab he had just met could not do any better, and he remembered the words of the professor about the impossibility of surpassing the limiting velocity of light. He noticed, however, that the city blocks became still shorter and the cyclist riding ahead of him did not now look so far away. He overtook the cyclist at the second turning, and when they had been riding side by side for a moment, was surprised to see the cyclist was actually quite a normal, sporting-looking young man. "Oh, that must be because we do not move relative to each other," he concluded; and he addressed the young man.

"Excuse me, sir!" he said, "Don't you find it inconvenient to live in a city with such a slow speed limit?"

"Speed limit?" returned the other in surprise, "we don't have any speed limit here. I can get anywhere as fast as I wish, or at least I could if I had a motor-cycle instead of this nothing-to-be-done-with old bike!"

"But you were moving very slowly when you passed me a moment ago," said Mr. Tompkins. "I noticed you particularly."

"Oh you did, did you?" said the young man, evidently offended. "I suppose you haven't noticed that since you first addressed me we have passed five blocks. Isn't that fast enough for you?"

"But the streets became so short," argued Mr. Tompkins.

"What difference does it make anyway, whether we move faster or whether the street becomes shorter? I have to go ten blocks to get to the post office, and if I step harder on the pedals the blocks become shorter and I get there quicker. In fact, here we are," said the young man getting off his bike.

Mr. Tompkins looked at the post office clock, which showed half-past five. "Well!" he remarked triumphantly, "it took you half an hour to go this ten blocks, anyhow—when I saw you first it was exactly five!"

"And did you *notice* this half hour?" asked his companion. Mr. Tompkins had to agree that it really seemed to him only a few minutes. Moreover, looking at his wrist watch he saw it was showing only five minutes past five. "Oh!" he said, "is the post office clock fast?" "Of course it is, or your watch is too slow, just because you have been going too fast. What's the matter with you, anyway? Did you fall down from the moon?" and the young man went into the post office.

After this conversation, Mr. Tompkins realized how unfortunate it was that the old professor was not at hand to explain all these strange

166

events to him. The young man was evidently a native, and had been accustomed to this state of things even before he had learned to walk. So Mr. Tompkins was forced to explore this strange world by himself. He put his watch right by the post office clock, and to make sure that it went all right waited for ten minutes. His watch did not lose. Continuing his journey down the street he finally saw the railway station and decided to check his watch again. To his surprise it was again quite a bit slow. "Well, this must be some relativity effect, too," concluded Mr. Tompkins; and decided to ask about it from somebody more intelligent than the young cyclist.

The opportunity came very soon. A gentleman obviously in his forties got out of the train and began to move towards the exit. He was met by a very old lady, who, to Mr. Tompkins's great surprise, addressed him as "dear Grandfather." This was too much for Mr. Tompkins. Under the excuse of helping with the luggage, he started a conversation.

"Excuse me, if I am intruding into your family affairs," said he, "but are you really the grandfather of this nice old lady? You see, I am a stranger here, and I never . . . " "Oh, I see," said the gentleman, smiling with his moustache. "I suppose you are taking me for the Wandering Jew or something. But the thing is really quite simple. My business requires me to travel quite a lot, and, as I spend most of my life in the train, I naturally grow old much more slowly than my relatives living in the city. I am so glad that I came back in time to see my dear little granddaughter still alive! But excuse me, please, I have to attend to her in the taxi," and he hurried away leaving Mr. Tompkins alone again with his problems.

REVIEW QUESTIONS

1. Reference frames are
   (a) necessary for certain measurements
   (b) used for all measurements
   (c) used mainly for adding velocities
   (d) used only in the theory of relativity

2. A gun is capable of shooting a bullet at a speed of 600 mi/hr. The gun is taken onto a train traveling 60 mi/hr and fired toward the front of the train. The velocity of the bullet, as seen by an observer on the ground, is
   (a) 600 mi/hr
   (b) 60 mi/hr
   (c) 660 mi/hr
   (d) 540 mi/hr

3. A flashgun that puts out light, which travels 186,000 mi/sec, is fired forward from a train traveling 100 mi/hr. The speed of the light as seen by an observer on the ground is
   (a) 186,000 mi/sec
   (b) 100 mi/hr
   (c) 186,000 mi/sec + 100 mi/hr
   (d) 186,000 mi/sec − 100 mi/hr

4. The *principle of relativity* states that
   (a) the laws of physics are the same in all reference frames that move at constant velocity with respect to each other
   (b) the laws of physics are relative
   (c) all quantities are relative
   (d) velocity is relative

5. The theory of relativity is based on two principles. One is the relativity principle, the second is
   (a) the law of conservation of momentum
   (b) Newton's second law
   (c) the constancy of the speed of light
   (d) the addition of velocities

6. "Simultaneity is not absolute" means
   (a) two events can never be simultaneous
   (b) two events that are simultaneous to one observer may not be simultaneous to a second observer standing next to him
   (c) two events that are simultaneous to one observer may not be simultaneous to a second observer moving with respect to the first
   (d) one cannot make simultaneous measurements

7. Which of the following statements is true according to the theory of relativity?
   (a) Time passes the same for all people and is measured accurately by clocks
   (b) Time is absolute
   (c) Time passes differently in different moving reference frames
   (d) Time only seems to pass more slowly in some reference frames

8. The theory of relativity says that moving objects are measured
   (a) to be shorter when moving than at rest
   (b) to be longer when moving than at rest
   (c) to always have the same length
   (d) to have the same length as when at rest but appear to be shorter

9. According to a traveler in a high-speed spaceship, the distance from earth to Pluto as compared to measurements made from earth is observed to be
   (a) longer   (c) the same
   (b) shorter   (d) unmeasurable

10. The idea of four-dimensional space–time means that
    (a) space and time are two distinct quantities
    (b) space and time can be interchanged
    (c) space takes up time
    (d) it is impossible to distinguish space from time

168

11. As an object increases in speed, its mass
    (a) increases   (c) is always the same
    (b) decreases

12. A space traveler is moving away from the earth with a speed of ½ c. He observes a meteor overtake and pass him and measures its speed as ½ c. What is the velocity of the meteor with respect to the earth?
    (a) ½ c
    (b) greater than ½ c but less than c
    (c) less than ½ c
    (d) c

13. According to the theory of relativity, an ordinary material body
    (a) cannot have a speed equal to, or in excess of, the speed of light
    (b) can travel at the speed of light but no faster
    (c) can travel faster than the speed of light if enough energy is given to it

14. Einstein's formula $E = mc^2$ deals with
    (a) the energy produced by burning a certain mass of fuel
    (b) the energy required to move a mass m at speed c
    (c) the energy produced when converting mass to energy
    (d) the energy in light waves

EXERCISES

1. A man is standing on the top of a moving railroad car. He throws a heavy ball straight up (it seems to him) in the air. Ignoring air resistance, will the ball land *on* the car or *behind* it?

2. A boat whose speed in still water is 7 mi/hr is traveling along a river whose current is 2 mi/hr. What is the net speed of the boat when going upstream? When going downstream?

3. How long will it take the boat in the above problem to make a round trip of 60 miles (30 miles upstream and 30 miles back again)?

4. Would Newton's first law be invalid for an observer moving at uniform velocity? If so, how would it differ? How would it differ if the observer is accelerating uniformly?

5. An airplane having a maximum speed of 600 mi/hr flies into a head-wind of 100 mi/hr (with respect to the earth). What is the speed of the airplane with respect to the earth?

6. The theory of relativity rests on two basic principles. What are they?

7. The predictions of the theory of relativity seem to contradict certain of our everyday intuitive notions and therefore don't seem to "make sense." What are those notions? Examine them in detail and determine if any physically measurable contradiction exists.

8. According to the principle of relativity, it is just as legitimate for a person riding in a uniformly moving automobile to say that the automobile

169

is at rest and the earth is moving as it is for a person on the ground to say that the car is moving and the earth is at rest. Do you agree, or are you reluctant to accept this? Discuss the reasons for your response.

9. Does the earth *really* go around the sun? Or is it also valid to say that the sun goes around the earth? Discuss in view of the first principle of relativity (that there is no preferred reference frame).

10. If you were on a space vehicle traveling at half the speed of light away from a bright star, with what speed would the star's light go past you?

11. Suppose you were placed in a laboratory that could "float" on a layer of air just above the ground and there was one small window you could look out. If you saw a row of trees going past you outside the window, could you do an experiment to prove (a) that the trees were at rest and you were moving uniformly, or (b) that you were at rest and the trees were moving uniformly? Explain.

12. Does the theory of relativity show that Newtonian mechanics is wrong?

13. As Mr. Tompkins pedals along the street in Wonderland, does he notice his bicycle getting shorter as he pedals faster? Do people standing on the street notice his bicycle getting shorter?

14. You are sitting in a rowboat on a lake. A rocket ship passes by at a speed of 0.8 c. A person in the rocket ship claims the rocket ship is 100 m long and that your rowboat is 5 m long. What would you say about these two lengths?

15. Does time dilation mean that time actually passes more slowly in moving reference frames, or that it only *seems* to pass more slowly?

16. Suppose you were to journey to a distant star 100 light-years away. If you traveled at a very high speed, would the distance you had to travel still be 100 light-years? Explain.

17. Today's subways are said to age people prematurely. Suppose that in the future subway trains could be designed that traveled very close to the speed of light. How do you think this would affect the aging process?

18. A mu meson lives for about two-millionths of a second when at rest. How long will it appear to live when it travels at a speed of 0.8 c?

19. It takes light four years to reach us from the nearest star. Is it possible for an astronaut to travel to that star in one year?

20. If you were traveling in a very high-speed spaceship away from the earth, would you notice a change in your heartbeat? Would your mass or width change? Would someone on earth (using a telescope) observe any such changes in you?

21. A young-looking astronaut has just arrived home from a long trip. She rushes up to an old gray-haired man and in the ensuing conversation refers to him as her son. Is this possible?

22. Is it possible for a person to live long enough to travel to a star 10,000 light-years away? (A light-year is the distance that light travels in one year.)

23. Consider an object of mass m to which is applied a constant force for an indefinite period of time. Discuss how its velocity and mass change with time.

24. Suppose the speed of light were infinite. What would happen to the relativistic predictions of length contraction, time dilation, and mass increase?

25. Why don't we notice the special effects of the theory of relativity in everyday life? Be specific.

26. A neutrino is a massless elementary particle that is believed to travel at the speed of light. Could you ever catch up with a neutrino that passed you? Explain. If you were traveling at half the speed of light away from a source of neutrinos, how fast would they pass you?

27. Do length contraction, time dilation, and mass increase occur at ordinary speeds, say 50 mi/hr?

28. What is the fourth dimension?

29. A person on a rocket traveling at half the speed of light (with respect to the earth) observes a meteor come from behind and pass him at a speed he measures as half the speed of light. How fast is the meteor moving with respect to the earth?

30. Two spaceships leave the earth in opposite directions, each with a speed of one-half the speed of light with respect to the earth. (a) What is the velocity of spaceship 1 relative to spaceship 2? (b) What is the velocity of spaceship 2 relative to spaceship 1?

31. A vehicle at rest is 100 feet long; if it moves past an observer at 0.8 c, what will the observer measure its length to be?

32. Show that the relativistic addition of velocities formula is consistent with the second principle of relativity; that is, if the object is light so that $v = c$, then show that $v' = c$ also.

33. Does the equation $E = mc^2$ conflict with the law of conservation of energy? Explain.

34. Does $E = mc^2$ apply only to objects traveling at the speed of light?

35. Does an object's density (mass/volume) change when the object is traveling near the speed of light? If so, how?

36. A red-hot iron bar is cooled to room temperature. Does its mass change?

37. Mr. Tompkins finds that the clock in the post office advanced 30 minutes while his own watch advanced only 5 minutes. Since the post office was moving with respect to him, shouldn't it have been the reverse? Explain.

FLUIDS

By experience, everyone knows the difference between the three states of matter—solid, liquid, and gas—at least in most cases. No one would argue that at room temperature iron is a solid, water is a liquid, or oxygen a gas. But what precisely distinguishes the three states of matter?

We can characterize and distinguish them in the following way, Fig. 7–1. **Solids** maintain a fixed shape and a fixed volume. Even if a force is applied to a solid, it does not readily change its shape or volume. A **liquid** does not maintain a fixed shape but takes on the shape of its container. However, a liquid does maintain a fixed volume and thus, like solids, is practically incompressible.[1] When water or milk is poured from one pitcher into another, the liquid maintains its volume, but it takes on the shape of the new pitcher. A **gas** maintains neither a fixed shape nor a fixed volume. A gas expands to fill whatever container it is in and is easily compressible. When air (which is a gas) is pumped into an automobile tire, it does not all run to the bottom of the tire as a liquid would; it fills the whole space inside the tire.

Because liquids and gases do not maintain a fixed shape, they both have the ability to flow. Because of this property, liquids and gases are sometimes referred to collectively as **fluids**.

[1] Both external forces and temperature affect the volume of liquids and solids, but the effect is much smaller than for gases.

Figure 7–1
(A) A solid, (B) a liquid, (C) a gas.

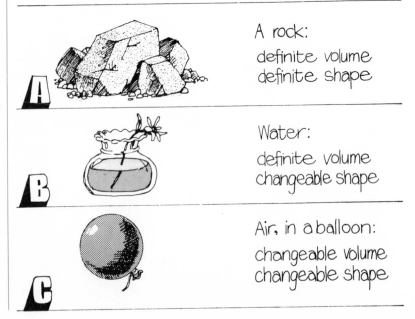

A rock:
definite volume
definite shape

Water:
definite volume
changeable shape

Air, in a balloon:
changeable volume
changeable shape

174

In earlier chapters we studied the motion of solid bodies and some of their properties at rest. In this chapter, we will study fluids.

1. DENSITY AND SPECIFIC GRAVITY

It is often said that lead is "heavier" than wood. Yet we can have a large piece of wood whose mass and weight are greater than that of a small piece of lead. What we should say is that lead is more *dense* than wood.

Density is another important property of materials. The **mass density** of a substance is defined as its mass per unit volume:

$$\text{mass density} = \frac{\text{mass}}{\text{volume}}.$$

We know that if we have twice the size or volume of a particular material, the mass will be twice as great as well. For example, several pieces of aluminum may each have a different mass and a different volume, but each will have the same ratio of mass to volume; that is, the density will be the same for each. Objects made up of the same material have the same density. The densities of some common materials are shown in Table 7–1.

Weight density is also a useful quantity. It is defined as weight per unit volume:

$$\text{weight density} = \frac{\text{weight}}{\text{volume}}.$$

The weight densities of various substances are also shown in Table 7–1. The densities of liquids and gases are defined in the same way as for solids.

If the term "density" is used by itself, it is generally understood to mean mass density and not weight density.

The **specific gravity** of a substance is defined as the ratio of the density of that substance to the density of water. Specific gravity is easily determined in the metric system since the density of water is exactly 1000 kg/m³. Thus we need only divide the density of a substance (in kilograms per cubic meter) by 1000. For example, from the first column of Table 7–1 we see that the density of aluminum is 2700 kg/m³. Therefore the specific gravity of aluminum is (2700)/(1000) = 2.7. Similarly, the specific gravity of alcohol is 0.8, and so on (see Table 7–1).

Wine and beer makers keep a sharp eye on the specific gravity of their fermenting fluids, since it steadily decreases as alcohol, with its low specific gravity, is formed from the sugar and starches of the original materials. An instrument known as a **hydrometer** is used

175

Table 7–1
Densities and Specific Gravities of Various Substances
(at 0° C and atmospheric pressure)

SUBSTANCE	MASS DENSITY kg/m³	WEIGHT DENSITY lb/ft³	SPECIFIC GRAVITY
Solids			
Aluminum	2700	170	2.7
Concrete	2300	140	2.3
Gold	19,000	1200	19
Ice	920	58	0.92
Iron	7800	480	7.8
Lead	11,000	700	11
Platinum	21,500	1300	21.5
Silver	10,500	650	10.5
Wood, balsa (approx.)	200	12	0.2
Wood, pine (approx.)	400	25	0.4
Wood, oak (approx.)	800	50	0.8
Liquids			
Alcohol (ethyl)	800	48	0.8
Gasoline	680	42	0.68
Mercury	13,600	830	13.6
Oil	900	55	0.9
Water, pure	1000	62	1.0
Water, sea	1030	64	1.03
Gases			
Air	1.3	0.08	0.0013
Carbon dioxide	2	0.12	0.002
Helium	0.2	0.012	0.0002
Hydrogen	0.09	0.0054	0.00009
Nitrogen	1.3	0.077	0.0013
Oxygen	1.4	0.087	0.0014

to measure specific gravity. It consists of a weighted glass tube (Fig. 7–2) and floats at the 1.000 mark when placed in water.

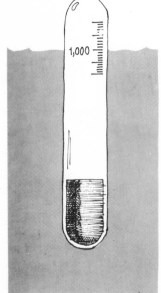

Figure 7–2
A hydrometer.

2. PRESSURE

Pressure Is Force per Unit Area

When we discussed solids in motion or at rest, we found the concept of force very useful. Because fluids do not have a definite shape, when dealing with them it is more convenient to use the concept of pressure,

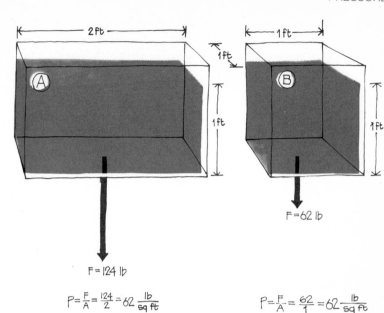

Figure 7–3
Pressure is equal at equal depths.

$$P = \frac{F}{A} = \frac{124}{2} = 62 \frac{lb}{sq\,ft}$$

$$P = \frac{F}{A} = \frac{62}{1} = 62 \frac{lb}{sq\,ft}$$

rather than force. **Pressure** is defined as the force exerted per unit area:

$$pressure = \frac{force}{area}.$$

If a force is exerted over a small area, the pressure will be greater than if the same force is exerted over a large area. To see more clearly the distinction between pressure and force, consider a simple example. Push a pin head against your finger; now turn the pin around and push the sharp point of the pin against your finger with the same force. Because the cross sectional area of the pin tip is much smaller than that of the pin head, the force per unit area, or pressure, is much greater. It is the *pressure* that determines whether your skin is punctured or not.

A fluid exerts a pressure against the walls and bottom of its container. Let us consider two containers, one with a bottom that has twice the area of the other; each is filled to the *same height* with the same liquid, Fig. 7–3. The larger container contains twice the weight of liquid as the smaller one, and therefore the liquid exerts twice the force on the bottom. But the pressure (force per unit area) due to the weight of the liquid is the same on both, since the container with twice the bottom area has twice the force acting on it. This is a general result: *the pressure at equal depths in the same liquid is everywhere the same.*

Of course at different depths the pressure is different, as the following experiment shows.

177

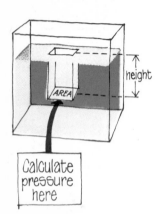

Figure 7–4

Calculating the pressure at a given depth in a liquid.

Divers are well aware that pressure increases as one goes deeper in water. Since water pressure at any depth is due to the weight of water above that point, the pressure is directly proportional to the depth. In fact, it is easy to show[2] that the pressure of a liquid at any depth is equal to the product of the weight density of the liquid times the depth in the liquid:

$$\text{pressure} = \text{weight density} \times \text{depth}.$$

This is illustrated in Fig. 7–5. Notice that the pressure is exerted not only against the bottom of the vessel but against its sides as well. In fact, the pressure at any depth is the same in all directions; and it always acts at right angles to the surface of any submerged object, including the walls of the vessel.

That the liquid exerts a pressure in all directions is obvious to swimmers and divers who feel the water pressure on all parts of their bodies. Fig. 7–6 illustrates why the pressure must be the same in all directions at a given depth. Consider a tiny cube of the liquid that is so small that we can ignore the gravitational force on it. The pressure on one side of it must equal the pressure on the opposite side of it, as shown. If the pressure weren't equal, that tiny piece of liquid would move (i.e., the liquid would flow) until the pressure did become equal. Therefore, when a liquid is still, the pressure must have the same magnitude in all directions at the same depth.

Buoyancy in Fluids and Archimedes' Principle

Objects seem to weigh less when submerged in water, as the following experiment shows.

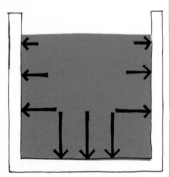

Figure 7–5

Pressure increases with depth and is exerted on the sides of the container as well as on the bottom.

[2]The pressure at any depth is just the weight of the liquid above that point divided by the area. Thus (see Fig. 7–4):

$$\text{pressure} = \frac{\text{force}}{\text{area}} = \frac{\text{weight}}{\text{area}} = \frac{\text{wt density} \times \text{volume}}{\text{area}} = \frac{\text{wt density} \times \text{area} \times \text{depth}}{\text{area}}$$

$$= \text{wt density} \times \text{depth},$$

where we have used the fact that $\text{wt density} = \frac{\text{weight}}{\text{volume}}$, or weight = wt density × volume, and volume = cross-sectional area × depth.

Use a pool or nearby lake or pond; or fill your bathtub with water. Take a heavy rock, one that is rather hard for you to lift, and submerge it in the water. Be sure it is completely covered with water. Now slowly lift it out of the water. Compare its weight when out of the water with its apparent weight when submerged.

Figure 7–6

Pressure is the same in all directions in a fluid; otherwise the fluid would be in motion.

Figure 7–7

(A) Water pressure increases with depth, so the water pressure pushing up on the bottom of a rock is greater than the water pressure pushing down on the top. (B) The net effect of this difference of pressure is a net *buoyant force,* up.

A rock that you would have difficulty lifting off the ground can be readily lifted from the bottom of a lake or stream. But when the rock reaches the surface, it suddenly seems much heavier. Some objects, like wood, actually float in water. These are two examples of the effect called **buoyancy**.

Why does a rock seem to weigh less in water? Fig. 7–7(A) shows the forces on a rock; these are the gravitational force, which is its normal weight, and the pressure of the water pushing on the sides, top, and bottom of it. The water pressure on the left and right sides of the rock cancel each other and do nothing to make it seem heavier or lighter. But the water pressure on the top and bottom of the rock do make a difference. Since the water pressure increases with depth, the water pressure on the bottom of the rock is greater than the water pressure on the top. Thus, the water exerts a net upward

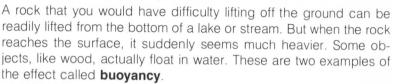

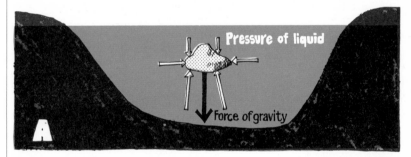

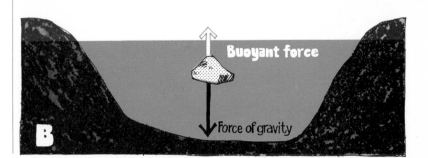

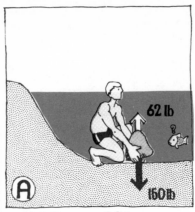
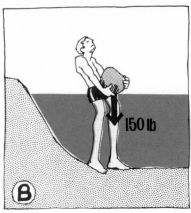

Net force = 88 lb   Net force = 150 lb

Figure 7–8

Because of the buoyant force, it is easier to lift a rock when it is under water than when it is out of the water.

force on the rock. This upward force is the net effect of the water pressure on all parts of the rock and is called the **buoyant force**; it is shown in Fig. 7–7(B). The total net force on the rock is the gravitational force downward minus the buoyant force upward. Thus the net downward force on the submerged rock is less than when it is not submerged, which is why it takes less force for you to lift it.

How large is this buoyant force? The answer to this question was discovered by the Greek philosopher and mathematician Archimedes (287–212 B.C.), and is known as Archimedes' Principle:

The buoyant force on a body immersed in a fluid is equal to the weight of the fluid displaced by that object.

To see what Archimedes' Principle means, we must first examine what is meant by the phrase "the weight of the fluid displaced by that object." If you drop a stone into a glass of water, the water level rises. The water above the original level represents the water displaced by the stone. Of course, the volume of water displaced is equal to the volume of the rock,[3] and the "weight of the fluid displaced" is the weight of this amount of water.

Let's take a specific example and use Archimedes' Principle to find the buoyant force B on the rock of Fig. 7–7. If the rock weighs 150 lb and has a volume of one cubic foot, the submerged rock will displace one cubic foot of water. The weight of one cubic foot of water is about 62 lb (see Table 7–1). Thus, in this case, the "weight of the water displaced" by the rock is 62 lb. According to Archimedes' Principle, then, the buoyant force on the rock is 62 lb.

180

[3]Incidentally, this is a good way to measure the volume of an irregular solid object, like a rock. If the glass or beaker has a scale on the side, the change in volume can be read right off the scale and equals the volume of the rock.

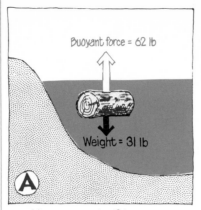
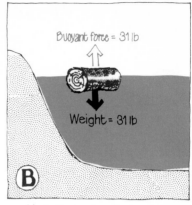

Net force is 31 lb up.　　Net force is O.

Figure 7–9

(A) A log rises to the surface because the buoyant force is greater than the force of gravity. (B) The log reaches equilibrium when partially submerged; the weight of water displaced equals the log's total weight.

The net force acting on the rock is 150 lb − 62 lb or 88 lb downward, Fig. 7–8. To lift the rock up through the water, you would have to exert a force of 88 lb; but once it is out of the water, you would have to exert a force of 150 lb.

What happens to an object whose density is *less than* that of the liquid in which it is placed, for example a piece of wood in water? It floats, of course. But let's see why. Suppose the piece of wood has a density that is half that of water. This means that if the volume of the wood is one cubic foot, it would weigh only 31 lb. If you were to submerge this piece of wood in water and then let go of it, it would immediately rise to the surface and float. This happens because when the wood is submerged [Fig. 7–9(A)], the buoyant force upward is 62 lb (remember Archimedes' Principle: the wood displaces one cubic foot of water, which weighs 62 lb) whereas the downward force of gravity is only 31 lb. Thus, there is a net upward force of 31 lb on the wood, and it is forced to move upward to the surface. It comes to rest when the net force on it is zero. This will happen when the buoyant force upward equals the gravity force downward. At this point, according to Archimedes' Principle, the wood displaces 31 lb of water, which is ½ cubic foot of water. Thus the log is half submerged, Fig. 7–9(B).

In general, a body floats when its density is less than that of the fluid in which it is immersed, and it sinks if its density is greater.

The density of ice is 0.9 that of water (900 kg/m³), and hence ice floats on water. A piece of ice, or an iceberg in the ocean, must displace a volume of water equal to 0.9 of its own volume if it is to float, Fig. 7–10. Thus only one-tenth of an iceberg shows above the surface of the water. How do steel ships float? After all, steel is denser than water. Steel ships are not solid steel—they have lots of empty space filled with air, so the density of a ship as a whole is less than that of water.

181

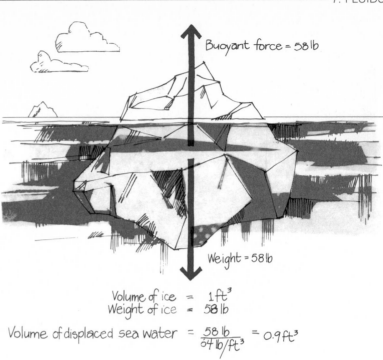

Buoyant force = 58 lb

Weight = 58 lb

Volume of ice = 1 ft³
Weight of ice = 58 lb

$$\text{Volume of displaced sea water} = \frac{58\ \text{lb}}{64\ \text{lb/ft}^3} = 0.9\ \text{ft}^3$$

Figure 7–10

Small iceberg of volume 1 cubic foot displaces 0.9 cubic foot of sea water. Therefore it is 9/10 submerged.

Denser fluids exert a greater buoyant force than less dense fluids. This is true because the buoyant force of an object is equal to the weight of fluid it displaces. A ship will therefore float higher in sea water (wt density 64 lb/ft³) than it will in fresh water. An ice cube that floats in water will sink in alcohol (density 790 kg/m³, or 48 lb/ft³) since the ice is denser than the pure alcohol. The density of an alcohol-water mixture lies in between, according to the proportion of each that is present. If someone gives you a drink in which the ice cubes have sunk, beware.

EXPERIMENT

Put an egg in a glass of water. The egg will probably rest on the bottom. Now slowly add salt to the water while stirring. Can you get the egg to float? Why does this happen?

Pascal's Principle:
Transmission of Force Through a Fluid

The French philosopher and scientist Blaise Pascal (1623–1662) discovered another important principle at work in fluids. It is known as Pascal's Principle:

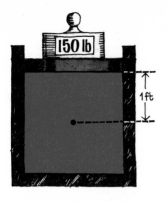

Figure 7–11

Pressure at a depth of 1 foot is the sum of the ordinary water pressure plus the pressure due to the external weight applied on top.

Pressure applied to an enclosed fluid is transmitted throughout the fluid and acts in all directions.

In Fig. 7–11 a piston is pushing down on a cylinder of water with a force of 150 lb. If the cross-sectional area of the cylinder is 3 square feet, then the piston exerts a pressure of 150 lb/3 ft² = 50 lb/ft². At a depth of one foot the water pressure is normally 62 lb/ft², but because of the external pressure, Pascal's Principle tells us that the total pressure one foot below the surface is 62 lb/ft² + 50 lb/ft² = 112 lb/ft². Pascal's Principle makes sense since we expect the pressure at any point in a fluid to be due to the weight of everything above it, fluid plus piston.

Pascal's Principle applies to many practical situations, including the hydraulic brakes in an automobile and the hydraulic press or lift, Fig. 7–12. In the latter case, a small force can be used to exert a large force by making the area of one piston (the output) larger than that of the other (the input). Although the force can be effectively multiplied in this way, energy is still conserved, as we saw in Chapter 5, since the input force must act through a greater distance than the output.

Figure 7–12

Pascal's Principle finds use (A) in the hydraulic brakes of an automobile, (B) in a hydraulic lift.

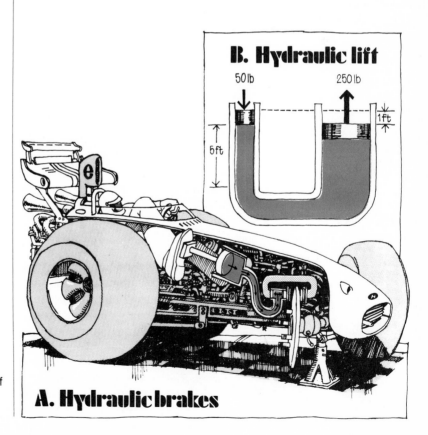

B. Hydraulic lift

50 lb

250 lb

5 ft

1 ft

A. Hydraulic brakes

3. GASES AND THE ATMOSPHERE

We are less aware of gases in everyday life than we are of liquids and solids. Perhaps it is because many common gases are colorless or nearly so. Yet gases are all around us. The air we breath, the earth's atmosphere, is a mixture of gases. The odors of food, flowers, and perfumes are due to gaseous substances. Many gases—for example, chlorine, helium, hydrogen, argon, and carbon dioxide—have important commercial uses.

Properties of Gases

Gases, as we mentioned earlier, expand to fill their containers and are readily compressible. Like liquids, they flow easily and hence are fluids. Many of the macroscopic concepts and principles that apply to liquids apply equally well to gases. The concept of pressure, for example, is valid and is very important when dealing with gases. Archimedes' and Pascal's Principles are valid as well. The rising of helium-filled balloons and the "floating" of dirigibles or blimps are examples of Archimedes' Principle applied to gases. Some hydraulic lifts [Fig. 7–12(B)], which operate in accordance with Pascal's Principle, use air instead of a liquid.

Scientists understand gases better than the other states of matter, and in the next chapter we will examine the behavior of gases in more detail. Now we will examine some of the more practical aspects of gases, particularly as they relate to the most familiar gas of all—the earth's atmosphere.

The Atmosphere

The atmosphere of the earth is a mixture of gases. Near the ground level, air is made up of about 78 percent nitrogen, 21 percent oxygen, slightly less than 1 percent argon, and very tiny amounts of carbon dioxide, neon, and other gases. Water vapor is also present, but the amount varies widely from almost none to as high as 4 percent.

The amount of each gas present, particularly gases present in small amounts, varies slightly from place to place. This is quite noticeable in areas of heavy smog. The pollutants that make up smog include not only gases such as carbon monoxide, nitrogen oxides, hydrocarbons, and sulfur dioxide, but large globs of molecules—they are almost droplets—known as aerosols, along with tiny smoke particles. The smoke particles are mainly responsible for poor visibility under smoggy conditions.

The molecules that make up our atmosphere are moving at speeds that average about 1000 mi/hr (see Chapter 8). Now a gas normally expands to fill its container. But the earth's atmosphere has no outside walls; why then doesn't all the air escape into outer space? It stays on earth because gravity holds it here. In Chapter 3 we saw that an object must attain a speed of about 25,000 mi/hr to escape from the earth's gravity. Very few molecules in our atmosphere reach that speed and so they don't escape.

We are fortunate that the force of the earth's gravity is as strong as it is. The force of gravity on the moon, for example, is too weak to hold an atmosphere. The moon may once have had a gaseous atmosphere, but during the past few billion years the gas has all escaped into space.

Because of the pull of gravity, the gases of the earth's atmosphere tend to accumulate near the surface of the earth. Thus the air is densest at sea level and becomes thinner and thinner at higher elevations. At an elevation of 18,000 ft, the air is only half as dense as it is at sea level. At 29,000 ft it is only one-third as dense. This is why airplanes are pressurized and why Himalayan mountain climbers often use tanks of oxygen to help them breathe at high elevations. At 29,000 ft, for example, a lungful of natural air contains only a third as much oxygen as at sea level. This variation in density of the earth's atmosphere can be compared to liquids, such as water, which maintain an essentially constant density regardless of height or depth (although, of course, pressure increases with depth). There is no distinct upper limit to our atmosphere. It just fades out gradually into space.

Atmospheric Pressure

We tend to think of air and other gases as being weightless, but this is not so. A cubic meter of air at sea level has a mass greater than 1 kg. The air in an average living room weighs more than a hundred pounds.

Just as water pressure is caused by the weight of water, the weight of all the air above the earth causes an atmospheric pressure of about 14.7 lb per square inch, or 1×10^5 newtons per square meter, at sea level.

At higher elevations, air pressure is less since there is less air pushing down from above. For example, at 18,000 ft the air pressure, like the density, is only about half what it is at sea level. At any given altitude, atmospheric pressure varies slightly according to climatic conditions. Climate, in turn, is affected by differences in air pressure at different places on the earth's surface. Wind, for example, is a result of air rushing from a region of high pressure to a region of low pressure.

We can find the total force exerted on an object due to atmospheric pressure by rearranging the definition of pressure as force/area. Thus, force = pressure × area. Ordinarily the atmosphere exerts a pressure of 14.7 lb on every square inch of surface in contact with it. There must, therefore, be an enormous total force on our bodies. Why don't we notice it? Why aren't we crushed by this huge force? The answer is that the cells of our body and those of other terrestrial organisms maintain an equal pressure within them; thus the pressure inside the cells equals the atmospheric pressure outside. When placed in a vacuum, many cells actually explode because of this internal pressure. Biological organisms have evolved in this high-pressure atmosphere of ours; in order to survive they had to be able to internally balance the outside pressure of the atmosphere.

We are not always aware of the pressure exerted by the air, but the following experiments attest to its existence.

EXPERIMENT

Fill a drinking straw with water and place your fingers over both ends. Now hold the straw vertically and remove your finger from the bottom end, but keep your finger tightly over the top end. Why doesn't the water run out of the straw? What happens when you release your finger from the top?

AND

Put about two inches of water in the bottom of an empty gallon can that has a tight-fitting lid, Fig. 7–13. With the top of the can removed, heat the water to boiling. When steam emerges from the opening, put the cap back on and immediately remove the can from the burner. The steam has forced most of the air out of the can. As the can cools, the water condenses to the liquid state and leaves a partial vacuum. What happens to the can?

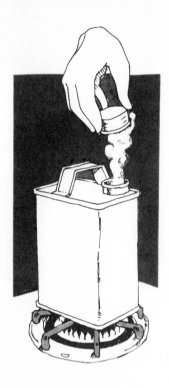

Figure 7–13

Put the lid on the can when the small amount of water inside is boiling vigorously, and immediately remove the can from the burner.

Suction Is Not a Force

In the above experiment, is the can crushed by the air pressure or by the suction "created" by the vacuum? If we say that suction is pulling the walls of the can inward, we're implying that the vacuum is exerting an inward force. But a vacuum is the *absence* of matter; a vacuum is *nothing*. How can nothing exert a force? Thus the can is not crushed by sucking; the air pressure on the outside forces it to cave in. Normally the can is open to the air and the pressure on the inside is the same as on the outside. But when some of the air is removed from the inside, the pressure is lower on the inside than on the outside and therefore a net force pushes in on the can. Air pressure can be very powerful.

A suction cup works on the same principle. When it is pushed

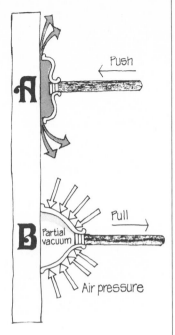

Figure 7–14

(A) Pushing air out from under a suction cup.
(B) Removal of suction cup is difficult because the air pressure outside is greater than pressure underneath.

onto a flat surface such as a wall, most of the air is pushed out from under it, Fig. 7–14(A). When you try to pull the suction cup away from the wall, the space between the wall and the cup increases slightly; because there is very little air present, the pressure is reduced, Fig. 7–14(B). Since the air pressure is greater outside than inside, there is a net force pushing the suction cup back to the wall. To remove it you must exert a force that can overcome the force of the air pressure, or bend the suction cup so that leaking develops.

The action of a vacuum pump or of a vacuum cleaner and the drinking of a soft drink through a straw are also examples of suction; but again suction does not act as a force. A vacuum pump works by making available empty spaces into which air can flow—remember, a gas tends to fill its container. A vacuum pump repeatedly increases the size of the space into which gas molecules can flow because of their high velocity, and then the pump pushes these molecules through another opening, Fig. 7–15.[4] When you suck on a straw you enlarge your lungs slightly and allow more volume for the air to fill. This of course means that the air in your lungs and mouth is less dense and exerts less pressure than the air outside. The ordinary atmospheric pressure on the surface of the drink is thus greater than the air pressure at the top of the straw, which is why the drink is forced up the straw. The drink is not sucked, it is *pushed* up the straw.

[4]The "pump" in a vacuum cleaner is usually nothing more than a high-speed fan. Can you explain, using the principles described in this chapter, how a vacuum cleaner works?

Figure 7–15

Diagram of one type of pump. When the piston moves up, the intake valve opens, and air (or fluid to be pumped) rushes in to fill the empty space. When the piston moves down, the outlet valve opens and the fluid is pushed out.

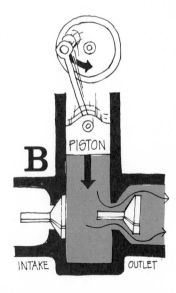
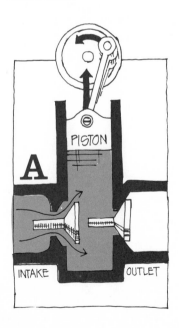

Our lungs work because we expand them with muscles inside our body. When we breathe in, we expand our rib cage and lower our diaphragm. This action is illustrated by the following experiment.

Figure 7–16
Model of a lung.

Figure 7–17
(A) A mercury barometer,
(B) an aneroid barometer.

EXPERIMENT

Find a bottle whose bottom has been cut off or use the glass "chimney" from a kerosene lamp. Cover the broad opening with rubber sheeting (from a balloon, for example) and secure it with a rubber band. Attach a balloon to the end of a glass tube (a drinking straw may work) and insert it through a cork into the smaller opening (or use modeling clay), as shown in Fig. 7–16. Now pull down on the rubber sheet at the bottom (which represents your diaphragm). This is easier to do if, before stretching it over the opening, you wrap a rubber band around a small section so there is something to pull on. What does this do to the pressure in the bottle? What happens to the balloon (which represents one of your lungs)? Why?

Barometers Measure Air Pressure

Instruments for measuring air pressure are called barometers. The oldest form of barometer is the mercury barometer shown in Fig. 7–17(A). A tube is filled with mercury and is then inverted into a bowl of mercury; some of the mercury runs out into the dish. At standard atmospheric pressure, 14.7 lb/in², a column of mercury 76 cm or 30 inches high remains in the tube. Since there is no gas pressure on the top of the tube (there is a vacuum), the pressure

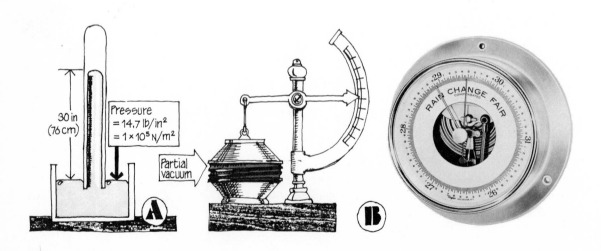

at the base of the tube of mercury due to its own weight must be equal to atmospheric pressure. It is atmospheric pressure that keeps the mercury up in the tube. A column of air that reaches all the way to the top of the atmosphere, and that has the same diameter as the tube of mercury, weighs the same as the 30-inch column of mercury. When the air pressure rises, the mercury will rise in the tube; when the air pressure drops, the mercury drops.

Because the mercury barometer was devised many years ago, air pressures are still often given as "inches of mercury" or "centimeters of mercury," rather than as pounds per square inch. Thus "30 inches of mercury" corresponds to 14.7 lb/in², and so on.

Mercury is a very dense liquid; it is 13.6 times denser than water. If a barometer using water were constructed, standard atmospheric pressure would maintain a column of water 10 meters, or 34 feet, high; that is, 13.6 times higher than a mercury column. The column of water would be 10 meters high only if there were a perfect vacuum at the top of the tube, as there is for the column of mercury in Fig. 7–17(A). If you were to try to drink water through a straw longer than 10 meters, no matter how strong your lungs or how perfect a vacuum pump you used, the water would not rise higher than 10 meters.

A modern barometer that is handier than the mercury barometer is the aneroid barometer shown in Fig. 7–17(B). A partially evacuated metal box whose volume is very sensitive to changes in external pressure is linked to a dial. The dials of many aneroid barometers are calibrated in inches or centimeters of mercury.

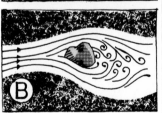

Figure 7–18
Fluid flow: (A) streamline, (B) turbulent.

4. FLUID FLOW

So far we have dealt mainly with fluids at rest. Now we examine fluids that are in motion—that is, fluids that are flowing. We usually distinguish two types of flow: streamline and turbulent. If the flow is smooth so that neighboring layers of fluid slide by each other smoothly, the flow is **streamline**, Fig. 7–18(A); each particle of the fluid follows a smooth path, called a "streamline." If a fluid flows very fast, or if there are obstacles in its path, the flow becomes **turbulent**; the streamlines are distorted into small, irregular, whirling **eddies**, Fig. 7–18(B). We will be interested mainly in streamline flow.

In general, fluids flow whenever there is a difference in pressure from one area to another. We saw examples of this phenomenon in the last section (water being pushed up a straw, air rushing into your lungs). Now you might think that once water was flowing in a horizontal tube or pipe it would continue to flow so long as no outside force was applied, as Newton's first law suggests. However, there is always some friction present between the fluid and the walls

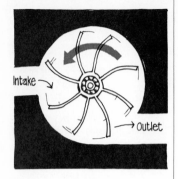

of the tube and also between layers of the fluid. This internal friction is called **viscosity**. Different materials have different amounts of viscosity. Syrup is more viscous than water, for example. And gases are in general much less viscous than liquids. Because of viscosity, there must be a difference in pressure between the two ends of a horizontal tube if a fluid is to flow at a constant speed. Thus pumps are needed to force water or oil through level pipes as well as when the pipes go uphill.

Pumps

The purpose of a **force pump** or a **circulating pump** is to exert pressure on a fluid to make it flow. We can use pumps to lift a fluid (such as water from a well) or to force a viscous fluid (such as oil) through a pipe. The diagram of a pump in Fig. 7–15 can in principle apply to a force pump (one that increases pressure to force a fluid to flow) as well as a vacuum pump (one that creates a vacuum, as discussed with reference to Fig. 7–15); if used as a force pump, fluid comes in through the intake valve and then is forced out through the outlet valve. Another type of pump is the **centrifugal pump** shown in Fig. 7–19. It, too, can be designed as either a vacuum pump or a force pump; it is commonly used in vacuum cleaners and as a water pump in automobiles. In the latter case, it is called a **circulation pump**, since it keeps water circulating in a closed path through the engine to cool it.

The human **heart** (and the hearts of other animals) is basically a circulation pump. The pumping action is caused by heart muscles that move the walls of the heart so that the several cavities of the heart change in volume. When a cavity contracts, the increased

Figure 7–19

A centrifugal pump. The rotating blades force fluid through the outlet pipe. This type of pump is used in vacuum cleaners and to circulate the water in automobile engines.

Figure 7–20

Action of human heart.

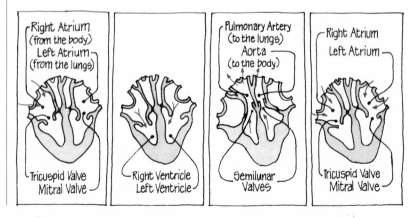

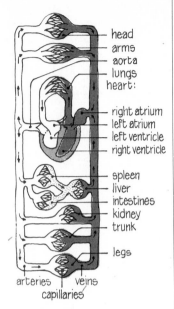

Figure 7–21

Human circulation system.

head
arms
aorta
lungs
heart:

right atrium
left atrium
left ventricle
right ventricle

spleen
liver
intestines
kidney
trunk

legs

arteries | veins
capillaries

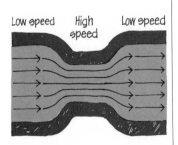

Figure 7–22

Fluid flow rate (mass of fluid passing by a given point per second) is the same at all points in the tube. Hence the fluid has greater speed where the tube is narrower.

pressure forces the blood out. When a cavity expands, blood flows into it because of the reduced pressure. Figure 7–20 shows the operation of the heart in detail. The heart pumps blood around two different paths as shown in Fig. 7–21. First the blood is pumped by the contracting right ventricle to the lungs, where it picks up oxygen. The oxygen-laden blood returns to the heart and is then pumped to the rest of the body through the arteries to the capillaries, where the oxygen is transferred to tissues (and where the blood picks up waste products). Finally, the blood returns to the heart through the veins to begin another pumping cycle.

Continuous Flow

The **flow rate** of a fluid is defined as how much mass (or volume) of the fluid passes a given point per second. For a fluid flowing smoothly in a continuous tube whose diameter varies, as shown in Fig. 7–22, the flow rate must be the same all along the tube. To get through the narrower portion of the tube, the fluid must therefore speed up. This is what happens along the course of a river. Where a river is wide, the water meanders slowly; but through a narrow gorge it rushes at high speed. Thus, *where the cross-sectional area is large, the fluid speed is small; and where the area is small, the fluid speed is large*.

There is an interesting parallel here to the flow of blood through our bodies. As shown in Fig. 7–21, blood flows from the heart through the aorta and into the arteries, which branch into tiny capillaries that bathe our tissues with blood and oxygen. Although the capillaries are tiny, there are billions of them, and their total cross-sectional area is far greater than that of the aorta. Thus, although the speed of blood in the aorta is about 30 cm/sec, in the capillaries it is only about 0.05 cm/sec. At this slow rate, there is plenty of time for oxygen (and other materials) to be exchanged between the blood and the tissues.

Bernoulli's Principle

Have you ever wondered why smoke goes up a chimney or a thrown baseball curves? These are just two of many interesting, and sometimes surprising, examples of a principle worked out by Daniel Bernoulli (1700–1782). Bernoulli's Principle, in a simplified form, states that *where the velocity of a fluid is high, the pressure in the fluid is low; and where the velocity is low, the pressure is high*. This may at first seem surprising to you, but the following experiments will show you that it is true.

Figure 7–23

Experiments showing
Bernoulli's Principle.

EXPERIMENT

Hold a sheet of paper horizontally with one edge just below
your mouth, as shown in Fig. 7–23(A). If you blow strongly
across the top of the paper, would you be surprised if the paper
moved *up*? Try it and see.

AND

Hold two pieces of paper vertically as shown in Fig. 7–23(B).
If you blow strongly between them, do they move apart or come
closer together? Why?

In each case, you confirmed Bernoulli's Principle that the pressure
is low where the velocity is high; the higher pressure of the still air
forced the paper into the region where the air was moving swiftly
and thus had lower pressure. You might have expected that where
the speed is highest, the pressure would be highest. But remember
that you are not blowing *at* the paper but rather *across* it.

To see why Bernoulli's Principle is consistent with Newton's laws,
consider the flow of fluid in a tube, as shown in Fig. 7–24. Where
the tube narrows, the speed is higher, as we saw earlier. Since the
fluid accelerates, we conclude that the pressure in the wide part of
the tube, P_1, must be greater than the pressure in the narrow part
of the tube, P_2.

Bernoulli's Principle can be used to explain many everyday phenomena, some of which are illustrated in Fig. 7–25. Airplane wings
and other "airfoils," such as that shown in Fig. 7–25(A), are designed to deflect more air upward than downward so that the
streamlines are closer together above the wing than below it. We

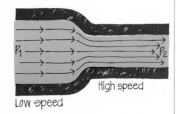

Figure 7–24

Flow of fluid in a tube. Where
the speed is higher, the
pressure must be lower.

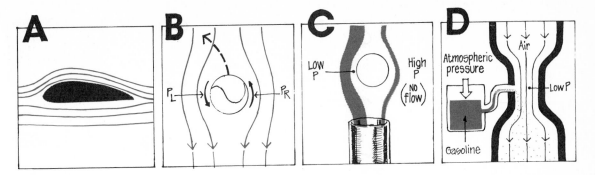

Figure 7–25
Examples of Bernoulli's
Principle.

saw in Fig. 7–22 that where the streamlines are forced closer together, the speed is faster; so it is with the wing—the air flows faster over the top of the wing than underneath. Thus the pressure is greater below the wing, and this greater pressure helps push the wing upward. (This, however, produces only part of the *lift*; the wing is also tilted, which means the onrushing air is deflected downward off the bottom of the wing and thus exerts an upward force on the wing as it rebounds.)

When a baseball or tennis ball flies through the air, the air moves past it with a certain speed. If the ball is spinning, as in Fig. 7–25(B), some air is dragged along with the ball. So the speed of the air on the left side of the ball is greater than on the right side. The resultant higher pressure on the right side pushes the ball to the left. This is the secret behind throwing curve balls in baseball. And it's the spin that makes tennis balls and golf balls curve.

A Ping-Pong ball can be made to float above a blowing jet of air, as shown in Fig. 7–25(C). If the ball starts to leave the jet of air, the higher air pressure outside the jet pushes the ball back in.

A Venturi tube is a pipe with a narrow section, called the "throat." One use of a Venturi tube is in the carburetor of an automobile, Fig. 7–25(D). As air flows into the carburetor, it must speed up in the narrow throat and therefore the pressure there is lower. Gasoline is then pushed by the higher atmospheric pressure in the fuel reservoir of the carburetor into the Venturi, where it mixes with the air before entering the cylinders.

Finally, smoke goes up a chimney because wind blows across the top of the chimney, making the pressure there less than it is inside the house. The higher pressure inside pushes the smoke up. Even on a still night there is usually enough air flow, particularly four or five meters above the ground, to allow the upward flow of smoke. It should be added that ordinary convection ("hot air rises"—see Chapter 9) also plays a role.

REVIEW QUESTIONS

1. Density is
   (a) the same as mass
   (b) the same as volume
   (c) mass/volume
   (d) volume/mass

2. A pound of cotton compared to a pound of iron
   (a) weighs more
   (b) weighs less
   (c) has greater density
   (d) has less density
   (e) has the same density

3. Pressure is defined as
   (a) force per unit area
   (b) force times area
   (c) area/force
   (d) the force due to a liquid or gas

4. The pressure at the bottom of a liquid is *not* proportional to
   (a) the depth of the liquid
   (b) the area of the liquid's surface
   (c) the density of the liquid
   (d) any of these

5. The pressure of water on a submerged object is greatest
   (a) on its top surface
   (b) on its bottom surface
   (c) on its sides
   (d) is equal on all surfaces of the object

6. A 70-lb rock has a volume of ½ ft³. What is its apparent weight when submerged? (Assume 1 ft³ of water weighs 60 lb.)
   (a) 100 lb
   (b) 70 lb
   (c) 40 lb
   (d) 30 lb
   (e) 10 lb

7. A block of ice is 10 cm on a side. When placed in a glass of water, the ice cube sinks down (assume the density of ice is 0.9 that of water)
   (a) all the way beneath the surface
   (b) 1 cm
   (c) 5 cm
   (d) 9 cm

8. An empty 60-lb canoe displaces
   (a) no water
   (b) 60 lb of water
   (c) less than 60 lb of water
   (d) an amount of water that depends on its volume.

9. The rising of a helium balloon is an example of
   (a) Archimedes' Principle applied to gases
   (b) random motion
   (c) antigravity forces
   (d) suction

10. Hydraulic brakes are an example of
    (a) Pressure increasing with depth
    (b) Archimedes' Principle
    (c) Pascal's Principle of transmission of pressure through a fluid
    (d) Brownian movement

11. You can't drink on the moon by sucking through a straw because
    (a) the straw will collapse
    (b) you don't have as much breath
    (c) there is no atmospheric pressure
    (d) the force of gravity is too small

12. A fluid flowing through a pipe of variable diameter
    (a) has its greatest speed where the pipe is widest
    (b) has its lowest speed where the pipe is narrowest
    (c) flows with the same speed at all points in the pipe
    (d) has its highest speed where the pipe is narrowest

13. The pressure in a moving fluid is
    (a) greatest where the speed is greatest
    (b) least where the speed is greatest
    (c) the same at all points
    (d) the same at equal depths independent of the speed

14. Smoke goes up a chimney partly because
    (a) smoke is lighter than air
    (b) the air pressure is less at higher elevation
    (c) the moving air outside sucks it up the chimney
    (d) the higher pressure of still air inside the house pushes it up

EXERCISES

1. Which has the greater volume, 10 pounds of lead or 1 pound of water?
2. What is the specific gravity of a material whose density is 600 kg/m³?
3. If one material has a greater density than another, does this mean that molecules of the first are necessarily heavier? Explain.
4. The hydrometer shown in Fig. 7–2 floats in a solution of grape juice, whose specific gravity is greater than that of water. Is the fluid level above or below the 1.000 mark? As the grape juice ferments, producing alcohol (see Table 7–1), does the hydrometer float higher or lower in the liquid?
5. When you blow up a balloon, what makes it stretch?
6. What is the pressure beneath a 300-newton orange crate 0.4 m wide by 0.7 m long?
7. The tires of a 2400-lb car carry 30 lb/in² pressure. What area of each tire is in contact with the ground?

195

Figure 7–26

Water stands at the same level in each of these interconnected containers.

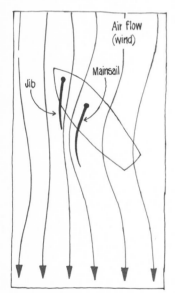

Figure 7–27

Use Bernoulli's Principle to explain how a sailboat can move forward against the wind.

8. What is the pressure due to water 10 ft beneath the surface of a lake?

9. Where is the pressure greater, at the bottom of a 1-cm diameter glass tube 2 m high filled with water, or at the bottom of a lake 1 m deep?

10. In which is the pressure greatest, 20 cm deep in a bottle of water, a bottle of alcohol, or a bottle of mercury? In which is it lowest?

11. It is sometimes said that heavy objects sink. Is this an accurate statement? Explain.

12. Why is it easier to float in salt water than in fresh water?

13. Legend has it that a Dutch boy held back the whole North Sea by plugging a hole in a dike with his finger. Is this possible and reasonable?

14. How do you suppose a submarine is able to sink and later to float on the surface of the sea?

15. Does the buoyant force on an object depend on its weight? Does it depend on its volume?

16. An overloaded ship barely stays afloat in the Atlantic Ocean and sinks when it moves up the Hudson River. Why?

17. Discuss what is meant by the quantities 10 lb, 10 lb/ft^2, 10 lb/ft^3.

18. It is often said that "water seeks its own level" (see Fig. 7–26). Explain.

19. It is harder to pull the plug out of the drain of a full bathtub than an empty one. Why? Is this a contradiction of Archimedes' Principle?

20. Will iron float on mercury? If so, what fraction of the iron will lie above the surface?

21. When you drink through a straw, is the liquid being pushed up or sucked up? Explain.

22. When you are wading on a rocky beach, why do the rocks hurt your feet less when you are in deeper water?

23. Draw a diagram showing all the forces (and their relative strengths) acting on an ice cube as it floats in water, and also when it is held submerged beneath the water's surface.

24. Why does a helium-filled balloon rise only to a particular height in the atmosphere and go no higher?

25. Estimate the air pressure on the Tibetan plateau, 16,000 ft above sea level.

26. Tire pressure gauges are calibrated to read the excess pressure in the tire over and above atmospheric pressure. What is the total pressure inside a tire when the gauge reads 25 lb/in^2.

27. What keeps the mercury from running out of a barometer (Fig. 7–17)?

28. Airplanes often travel at high altitudes where the air pressure is low; however, they are pressurized on the inside to keep travelers comfortable. Explain, on this basis, why airplane windows are so small.

29. If a single piece of newspaper is spread across a meter stick lying on a flat table, it will take quite an effort to quickly lift the stick. (Try it!) Explain.

30. What is atmospheric pressure due to?

31. Would a vacuum cleaner pick up dust from a carpet on the moon where there is no atmosphere? Explain.

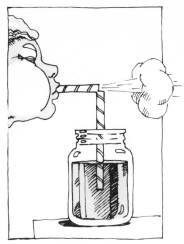

Figure 7–28
Homemade atomizer.

32. Compare the flow of air in a heating duct to the flow of air in a room.

33. Children are told to avoid standing too close to a rapidly moving train because they might get sucked under it. Is this possible? Explain.

34. If two ships traveling in parallel paths approach too closely, they may find themselves crashing into one another. Explain.

35. Why does the canvas top of a convertible bulge out when a car is traveling at high speeds?

36. A sailboat can move forward against a headwind if the mainsail can curve outward and if a small "jib" sail in front of it directs the wind between them, as shown in Fig. 7–27. Explain.

37. Sometimes when you are taking a shower and you turn the water on hard, the shower curtain seems to be pulled in toward you. Explain.

38. Red corpuscles tend to flow in the *center* of blood vessels. Why?

39. Explain how a perfume atomizer works; use Bernoulli's Principle. You can make a simple atomizer yourself with a straw cut halfway through and bent as shown in Fig. 7–28.

40. A tornado or a hurricane does not rip off the roofs of houses by blowing against them. Explain how they are blown off; use Bernoulli's Principle.

ATOMS
AND
TEMPERATURE:
KINETIC
THEORY

What would happen if you cut a piece of iron in half, then cut the halves in half, and continued to subdivide the substance into tinier and tinier pieces? Would you ever reach a "tiniest" piece of iron? Can you subdivide a substance into smaller and smaller pieces indefinitely? Is matter made up of tiny particles, or building blocks, or is matter continuous? People have been asking these questions about the ultimate nature of matter since the time of the ancient Greeks.

The earliest recorded hypothesis that matter is made up of tiny particles is credited to the Greek philosopher Democritus, who lived in the fifth century B.C. According to Democritus, there is a limit to how far matter can be subdivided. The tiniest pieces of matter beyond which no further subdivision is possible were called **atoms**, from the Greek word for "indivisible." However, most ancient philosophers, including Aristotle, believed that matter was continuous, and that it consisted of four basic elements: earth, air, water, and fire. This view predominated until the Renaissance. Although Democritus and his atomic theory had a number of followers, his role in history is not unlike that of Aristarchus and the heliocentric theory (see Chapter 2).

For many centuries, acceptance of either the continuous theory or the atomic theory of matter was based more on intuition than on experimental evidence. However, toward the end of the Renaissance, philosophers began to notice that the atomic theory made certain aspects of the behavior of matter more intelligible. Galileo and Newton were avowed atomists. But the atomic theory did not really have much support until experiments were made in chemistry in the eighteenth and early nineteenth centuries.

We will begin this chapter with a discussion of the atomic theory and follow that with a discussion of temperature—the hotness or coldness of an object. Although this may seem at first like a strange sequence, we will see that these two concepts are actually closely related. And they led to one of the great scientific theories—the "kinetic theory."

1. THE ATOMIC THEORY

Elements, Compounds, and Mixtures

Chemists have long been concerned with how different materials could be combined to form other materials, and with whether or not a given substance could be broken down into simpler substances. They found that certain materials, such as iron, gold, and silver, could not be broken down by any chemical means. They called these materials **elements**. They found that other substances were made up of combinations of elements, and these they called **compounds**. For example, water is a compound; it can be broken down into the elements hydrogen and oxygen. Ordinary table salt, or sodium chloride, is a compound; it is made up of sodium and chlorine.

Compounds are not the same as mixtures. A compound is a wholly new substance, completely different from the substances of which it is made. In a mixture, on the other hand, the separate materials are easily recognizable. For example, the two gases hydrogen and oxygen can be mixed together to form a mixture that is still gaseous. If, however, a spark is applied to this mixture, an explosion occurs and a new substance, water, is formed. Water is a compound of hydrogen and oxygen but it is a liquid that has very different properties from a mixture of hydrogen and oxygen. Many of the solid materials we deal with in everyday life are mixtures of elements and compounds: concrete, glass, pottery, wood, food, and all living matter.

The Law of Definite Proportions Supports the Atomic Theory

Just before 1800, chemists discovered an experimental law that relates the weights of the different substances required to make a given compound. This is the *law of definite proportions*. Briefly stated, this law says that

When two or more elements combine to form a compound, they always do so in the same proportions by weight.

For example, salt is always formed from 23 parts by weight of sodium and 35 parts by weight of chlorine; water is formed from one part hydrogen and 8 parts oxygen; hydrochloric acid is formed from one part hydrogen and 35 parts chlorine; and so on.

The English scientist John Dalton (1766–1844) quickly recognized that the law of definite proportions lent significant support to the atomic theory. A continuous theory of matter could not readily account for the definite proportions of each element required to make a compound. Dalton showed that the atomic theory explains it very simply. He reasoned that the weight proportions of each element required to make a compound correspond to the relative weights of the combining atoms. For example, to explain why 23 parts by weight of sodium always combine with 35 parts by weight of chlorine to make the compound sodium chloride, Dalton argued that each atom of sodium combines with a single atom of chlorine,

Gold

Iron

Water (H_2O)

Salt (NaCl)

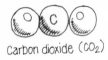

Carbon dioxide (CO_2)

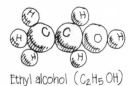

Ethyl alcohol (C_2H_5OH)

Figure 8–1

A simple model of some atoms and molecules. Chemical symbols (Table 8–1) are used for each atom.

and that one atom of sodium weighs 23/35 times as much as one atom of chlorine. He called the combination of one chlorine atom with one sodium atom a **molecule** (from Latin, meaning "little mass"). A molecule is thus a fixed combination of atoms.

By measuring the relative amounts of each element required for the formation of a great variety of compounds, a table of the relative weights of atoms was soon established. Hydrogen, which was found to be the lightest of all the atoms, was arbitrarily given the relative weight of 1; on this scale sodium was about 23, chlorine 35, iron 56, and so on. From the various compounds that oxygen formed, the relative weight of oxygen was found to be 16. This was hard to reconcile with the fact that in water the weight ratio of oxygen to hydrogen is 8:1. This difficulty was explained, however, by assuming that two hydrogen atoms combined with one oxygen atom to form the water molecule. A good many other molecules were also found to contain two or more atoms of the same kind, Fig. 8–1.

The atomic theory neatly describes the difference between elements, compounds, and mixtures. Substances that contain only one kind of atom, such as iron or aluminum, are **elements**. Substances that are made up of one kind of molecule, such as water, carbon dioxide, salt, or ethyl alcohol, are **compounds**. Substances that are a mixture of different kinds of atoms and molecules, such as air, milk, concrete, wood, and other organic substances, are **mixtures**. Today, there are only 105 known elements. Most of the substances we know in our everyday lives are compounds or mixtures.

A list of the known elements, their relative weights, and the chemical abbreviation or symbol for each are given in Table 8–1.

Brownian Movement

So much evidence was collected in support of the atomic theory of matter following Dalton's findings that by the early twentieth century most scientists accepted it as true.[1] It is one of the most important and influential theories in all of science.

Perhaps the most direct evidence in support of the atomic theory of matter was a phenomenon discovered accidentally by the biologist Robert Brown in 1827. Brown noted that tiny pollen grains suspended in water under his microscope followed a tortuous path even though the water appeared to be perfectly still, see Fig. 8–2(A). This phenomenon is now called Brownian movement. Tiny

[1]The last major holdout was Ernst Mach (1838–1916), an Austrian physicist and philosopher. His positivist philosophy led him to the conclusion that since atoms cannot be directly sensed, it was meaningless to believe in them.

Table 8–1
The Elements

ELEMENT	SYM-BOL	ATOMIC WEIGHT	ELEMENT	SYM-BOL	ATOMIC WEIGHT	ELEMENT	SYM-BOL	ATOMIC WEIGHT
Actinium	Ac	(227)	Hafnium	Hf	178.5	Praseodymium	Pr	140.9
Aluminum	Al	27.0	Hahnium	Ha	(262)	Promethium	Pm	(145)
Americium	Am	(243)	Helium	He	4.0	Protactinium	Pa	(231)
Antimony	Sb	121.8	Holmium	Ho	164.9	Radium	Ra	226.1
Argon	Ar	40.0	Hydrogen	H	1.0	Radon	Rn	(226)
Arsenic	As	74.9	Indium	In	114.8	Rhenium	Re	186.2
Astatine	At	(210)	Iodine	I	126.9	Rhodium	Rh	102.9
Barium	Ba	137.3	Iridium	Ir	192.2	Rubidium	Rb	85.5
Berkelium	Bk	(249)	Iron	Fe	55.8	Ruthenium	Ru	101.1
Beryllium	Be	9.0	Krypton	Kr	83.8	Rutherfordium	Rf	(261)
Bismuth	Bi	209.0	Lanthanum	La	138.9	Samarium	Sm	150.4
Boron	B	10.8	Lawrencium	Lw	(257)	Scandium	Sc	45.0
Bromine	Br	79.9	Lead	Pb	207.2	Selenium	Se	79.0
Cadmium	Cd	112.4	Lithium	Li	6.9	Silicon	Si	28.1
Calcium	Ca	40.1	Lutetium	Lu	175.0	Silver	Ag	107.9
Californium	Cf	(249)	Magnesium	Mg	24.3	Sodium	Na	23.0
Carbon	C	12.0	Manganese	Mn	54.9	Strontium	Sr	87.6
Cerium	Ce	140.1	Mendelevium	Md	(256)	Sulfur	S	32.1
Cesium	Cs	132.9	Mercury	Hg	200.6	Tantalum	Ta	180.9
Chlorine	Cl	35.4	Molybdenum	Mo	95.9	Technetium	Tc	(99)
Chromium	Cr	52.0	Neodymium	Nd	144.2	Tellurium	Te	127.6
Cobalt	Co	59.0	Neon	Ne	20.2	Terbium	Tb	158.9
Copper	Cu	63.5	Neptunium	Np	(237)	Thallium	Tl	204.4
Curium	Cm	(245)	Nickel	Ni	58.7	Thorium	Th	232.0
Dysprosium	Dy	162.5	Niobium	Nb	92.9	Thulium	Tm	168.9
Einsteinium	Es	(253)	Nitrogen	N	14.0	Tin	Sn	118.7
Erbium	Er	167.3	Nobelium	No	(253?)	Titanium	Ti	47.9
Europium	Eu	152.0	Osmium	Os	190.2	Tungsten	W	183.8
Fermium	Fm	(255)	Oxygen	O	16.0	Uranium	U	238.0
Fluorine	F	19.0	Palladium	Pd	106.4	Vanadium	V	50.9
Francium	Fr	(223)	Phosphorus	P	31.0	Xenon	Xe	131.3
Gadolinium	Gd	157.2	Platinum	Pt	195.1	Ytterbium	Yb	173.0
Gallium	Ga	69.7	Plutonium	Pu	(242)	Yttrium	Y	88.9
Germanium	Ge	72.6	Polonium	Po	(210)	Zinc	Zn	65.4
Gold	Au	197.0	Potassium	K	39.1	Zirconium	Zr	91.2

Figure 8–2

(A) Brownian movement: the straight lines connect successive positions of a single pollen grain suspended in water observed under a microscope at 10-second intervals.
(B) Brownian movement is explained by continual random motion of molecules.

pieces of dust or oil droplets suspended in water, and smoke particles suspended in still air, also undergo random movement.

The atomic theory explains Brownian movement if the further reasonable hypothesis is made that the atoms or molecules in any substance are continually in motion, Fig. 8–2(B). Thus the tiny pollen grains of Fig. 8–2(A) are buffeted about by a barrage of rapidly moving molecules of water. A piece of wood floating in still water is too large and has too much inertia to be affected by the minuscule molecular bombardment; but a tiny pollen grain is small enough so that if several molecules chance to strike it simultaneously on the same side, it will rebound. The random bombardment of water molecules from all directions thus gives rise to the observed erratic motion of Fig. 8–2(A). The pollen grain is jostled about on a vast molecular sea.

In 1905 Einstein explained Brownian movement from this atomic point of view, and his quantitative studies made it possible to estimate atomic sizes and weights. It was found that atoms were extremely small, on the order of 0.00000001 (1×10^{-8} cm) in diameter. In the wake of Einstein's analysis, few scientists persevered in opposition to the atomic theory.

A Brief Look Inside the Atom

The historical development of our understanding of the atom is a fascinating subject, and we will investigate it in detail in Chapters 15 and 16. For now we will summarize the contemporary view of the atom very briefly.

Atoms are not indivisible, as was once thought. Physicists early in the twentieth century found that an atom consists of a very tiny but heavy **nucleus**; tiny **electrons** revolve around the nucleus, much like the planets orbit the sun, Fig. 8–3. However, it is not the gravitational force that keeps the electrons in their orbits; it is instead the "electrical force," a force we will discuss in Chapter 13. The electrons are "negatively charged" and the nucleus is "positively charged." A positive charge always attracts a negative charge, and it is this electrical attraction that holds the electrons in orbit around the nucleus.

Why is iron hard and lead soft? Why is mercury a liquid at room temperature and helium a gas? Why does oxygen and not helium react readily with most metals? These and other properties of an element are determined by the number of electrons in an atom of the element. The number of electrons in an atom (called the **atomic number**) thus distinguishes one kind of atom from another. Thus hydrogen has an atomic number of one because it has only one

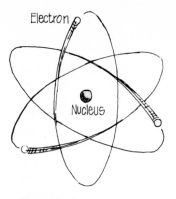

Figure 8-3
Model of an atom, showing electrons revolving around the nucleus.

electron orbiting its nucleus; helium has two, carbon six, oxygen eight, and uranium ninety-two electrons.

The Periodic Table of the Elements

The 105 elements known today correspond to atomic numbers from one through 105. But as early as 1871 enough elements had been studied to enable the Russian chemist Dmitri Mendeleev (1834–1907) to find some order among them. By arranging the elements in a table according to increasing atomic weight, Mendeleev found a regular repetition (or periodicity) of chemical and physical properties. The modern version of Mendeleev's *periodic table of the elements* is shown in Table 8–2. Each vertical column contains elements with similar properties. For example, the first column contains the very reactive metals lithium, sodium, potassium, rubidium, cesium, and francium, whereas the last column on the right contains the "inert" gases, helium, neon, argon, krypton, xenon, and radon, that chemically react only very slightly if at all.

Why this periodicity of the elements should exist was not understood until the quantum theory was conceived in the early twentieth century (see Chapter 15).

In Table 8–2, the periodic table, each square contains the atomic number and then the symbol of the element; below the symbol is the relative atomic weight of that element.

Solids, Liquids, and Gases

The distinctions we made at the beginning of Chapter 7 between solids, liquids, and gases were based on **macroscopic**, or large-scale, properties. Distinctions at the atomic or molecular level are also useful. We refer to the latter as **microscopic** properties, although submicroscopic would perhaps be a better word since atoms cannot actually be seen with a microscope.

Microscopically, the differences between solids, liquids, and gases can be attributed to the strength of the forces existing between the atoms or molecules. That molecules exert forces on one another is apparent, since otherwise an object—such as this piece of paper—would fall apart into tiny pieces. The forces between molecules are electrical in nature as we shall see later.

In all three states, solid, liquid, and gas, the molecules (or atoms, in the case of an element) are in rapid motion. In a solid, the forces between the molecules are so strong that each cannot move very

Table 8–2
The Periodic Table

1	2	3	4	5	6	7	8	9	10	11	12	13	14	15	16	17	18
1 H 1.0																	2 He 4.0
3 Li 6.9	4 Be 9.0											5 B 10.8	6 C 12.0	7 N 14.0	8 O 16.0	9 F 19.0	10 Ne 20.2
11 Na 23.0	12 Mg 24.3											13 Al 27.0	14 Si 28.1	15 P 31.0	16 S 32.1	17 Cl 35.4	18 Ar 39.9
19 K 39.1	20 Ca 40.1	21 Sc 45.0	22 Ti 47.9	23 V 50.9	24 Cr 52.0	25 Mn 54.9	26 Fe 55.8	27 Co 58.9	28 Ni 58.7	29 Cu 63.5	30 Zn 65.4	31 Ga 69.7	32 Ge 72.6	33 As 74.9	34 Se 79.0	35 Br 79.9	36 Kr 83.8
37 Rb 85.5	38 Sr 87.6	39 Y 88.9	40 Zr 91.2	41 Nb 92.9	42 Mo 95.9	43 Tc (99)	44 Ru 101.1	45 Rh 102.9	46 Pd 106.4	47 Ag 107.9	48 Cd 112.4	49 In 114.8	50 Sn 118.7	51 Sb 121.8	52 Te 127.6	53 I 126.9	54 Xe 131.3
55 Cs 132.9	56 Ba 137.3	57–71 see below	72 Hf 178.5	73 Ta 180.9	74 W 183.8	75 Re 186.2	76 Os 190.2	77 Ir 192.2	78 Pt 195.1	79 Au 197.0	80 Hg 200.6	81 Tl 204.4	82 Pb 207.2	83 Bi 209.0	84 Po (210)	85 At (210)	86 Rn (222)
87 Fr (223)	88 Ra (226)	89–103 see below	104 Rf (261)	105 Ha (262)													

57 La 138.9	58 Ce 140.1	59 Pr 140.9	60 Nd 144.2	61 Pm (147)	62 Sm 150.4	63 Eu 152.0	64 Gd 157.2	65 Tb 158.9	66 Dy 162.5	67 Ho 164.9	68 Er 167.3	69 Tm 168.9	70 Yb 173.0	71 Lu 175.0
89 Ac (227)	90 Th 232.0	91 Pa (231)	92 U 238.0	93 Np (237)	94 Pu (242)	95 Am (243)	96 Cm (247)	97 Bk (247)	98 Cf (251)	99 Es (254)	100 Fm (253)	101 Md (256)	102 No (254)	103 Lw (257)

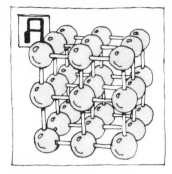

Figure 8–4
Arrangement of atoms in
(A) a crystalline solid,
(B) a liquid, (C) a gas.

far, and therefore they vibrate about nearly fixed positions. Most solids occur as crystals, although the crystals are often very tiny and hard to see, as in metals; such crystalline solids have an orderly arrangement of molecules, as shown in Fig. 8–4(A).[2]

In a liquid, the forces between the molecules are not strong enough to hold them in fixed positions, and so they are free to roll over one another. However, the forces are sufficiently strong so that the molecules cannot get very far apart. In a gas, the molecules are moving so rapidly, or the forces are so weak, or both, that the molecules don't even stay close together. They move every which way, filling any container they are in, and occasionally they collide with one another. On the average, the velocities of gas molecules are sufficiently high so that when two of them collide, the forces of attraction are too weak to keep them close together and the two molecules fly off in new directions, Fig. 8–4(C). A gas thus expands to fill its container simply because the molecules are moving rapidly in random directions. The speed of air molecules averages about 1000 mi/hr at room temperature. Because the molecules of a gas are usually far apart on the average, gases are easy to compress compared to liquids and solids.

2. TEMPERATURE

Temperature has meaning for us mainly through our sense of touch. When we touch something that feels hot—whether it's a hot stove or the hot air of a summer day—we say it has a high temperature. A cold object on the other hand has a low temperature.

Properties of Matter Change with Temperature

Nearly all materials expand when they are heated. An iron rod is longer when it is hot than when it is cold. Concrete sidewalks expand and contract with changes in temperature, which is why compressible spacers are placed at regular intervals. We can remove the metal lid from a tightly closed jar by warming it under hot water, because the lid expands more than the glass. Glass itself expands, and if one part of a glass container is heated or cooled more rapidly than adjacent parts, the glass may break. The same thing can happen to an automobile engine block; water is added to an overheated

[2]The molecules in a few solids—such as glass, rubber, and sulfur—have no crystal structure. These *amorphous* solids have some properties characteristic of liquids. Rubber is flexible and glass flows (slowly) as can be seen in old windows, which are thicker at the bottom than at the top. Hence the division of matter into solids, liquids, and gases is not always precise.

automobile engine slowly and with the engine running to avoid "cracking" the block.

Another property of matter that changes with temperature is color. At very high temperatures solids become red. An example is iron; you have probably seen the burner of an electric stove turn red when it is hot. At still higher temperatures, iron and other solids turn orange, and then white. The white light from an incandescent light bulb is emitted by an intensely hot tungsten wire.

Temperature also affects the "state" of a material—whether it is solid, liquid, or gas. For example, water at low temperatures is a solid (ice). At higher temperatures it is a liquid; and at still higher temperatures it is a gas (steam). Other materials pass through these same three states as temperature increases. However, the temperature at which the transition from one state to another occurs is different for each material.

We are all familiar with how biological organisms, particularly humans, respond to temperature. When we touch something that has either a very high or a very low temperature, it hurts. This response of our nervous system was a very important development in biological evolution, because biological tissue is readily damaged by temperature extremes. Biological organisms can tolerate a relatively limited range of temperatures. Even bacteria are sensitive to temperatures and grow well only over a small range. In fact, the pasteurization of milk products is a heating process that kills certain harmful bacteria. Viruses are much less sensitive to temperature.

Thermometers

In order to measure temperature quantitatively, we must use some property of matter that changes with temperature. A device that does this is called a **thermometer.** Most thermometers use the expansion property of a material. The first thermometer, invented by Galileo, employed the expansion of a gas. Common thermometers today are glass tubes that contain liquid mercury or alcohol colored with a red dye. Since either liquid expands more rapidly than the glass, the liquid level rises with temperature, Fig. 8–5. An interesting early fever thermometer is shown in Fig. 8–6.

The length of a metal rod might conceivably be used to measure temperature, but the change in length would be too small to be measured accurately at normal temperatures. However, a useful thermometer can be made from two dissimilar metals bonded together, as in Fig. 8–7(A). Because the two metals expand or contract by different amounts when heated or cooled, the bimetallic strip bends. Bimetallic strips in the shape of a coil are used in many

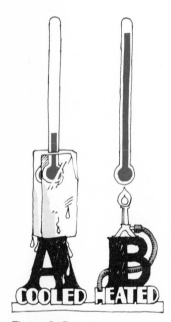

Figure 8–5
Liquid in glass
thermometer.

Figure 8–6

Replica of a seventeenth-century clinical thermometer (the original is in the Museum of the History of Science, Florence, Italy). The glass frog, containing spheres of slightly different density suspended in alcohol, was strapped to a patient's wrist. When the patient had a fever, one or more balls would sink as a result of the decreased density of alcohol with increased temperature.

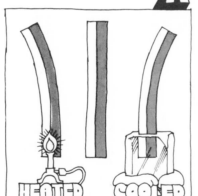

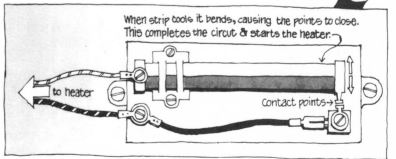

When strip cools it bends, causing the points to close. This completes the circuit & starts the heater.

to heater

contact points→

Figure 8–7

(A) Bimetallic strip: one metal expands or contracts more than the other when heated or cooled. (B) Common thermometer made from coiled strip. (C) Bimetallic strip used in a thermostat: at a certain temperature, the strip bends sufficiently to touch the contact that closes the electric circuit and starts the heater.

common thermometers [Fig. 8–7(B)], including oven thermometers. One end of the coil is fixed and the other is attached to a pointer. Bimetallic strips are also used in thermostats [Fig. 8–7(C)] and in the automatic choke control in automobiles.

EXPERIMENT

Construct a simple thermometer by pushing a glass tube (or a sturdy drinking straw) through a cork with a hole in it. Put the tube and cork into a small bottle filled with water, Fig. 8–8. If you can't find a cork, use plasticene or clay to seal the opening. The water level will be more visible if you put a few drops of food coloring in the water. Note the water level in the tube and then place the thermometer in a pan of boiling water. Note the water level now. Finally, put the thermometer on ice or in a refrigerator and see what happens.

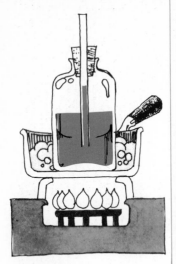

Figure 8–8

A homemade thermometer.

Temperature Scales

If a thermometer is to be of any use, it must be calibrated—that is, it must be marked off with some sort of numerical scale. The two most common temperature scales are: the Fahrenheit scale and the Celsius, or centigrade, scale.

To define a temperature scale, we assign arbitrary values to two easily reproducible temperatures. The two points most commonly chosen are the freezing point of water and the boiling point of water (at atmospheric pressure). When we calibrate a thermometer, we first place it in a mixture of ice and water; then we make a mark where the mercury level or the pointer comes to rest. Next we put the thermometer in boiling water and make a second mark, Fig. 8–9. On the Celsius scale, the freezing point of water is chosen arbitrarily to be 0°C (read "zero degrees Celsius" or "zero degrees centigrade"), and the boiling point is chosen to be 100°C. The distance between these two points is divided into 100 equal intervals, or degrees; hence the name "centigrade scale." For temperatures below 0°C and above 100°C, the scale is extended with equally spaced intervals. On the Fahrenheit scale, the freezing point of water is labeled 32°F and the boiling point of water 212°F. The scale in between is divided into 180 equal intervals. The Fahrenheit scale is in everyday use in the United States, although the Celsius scale is becoming more common. For most scientific work the Celsius scale is used. Most other countries in the world use the Celsius scale, which is part of the metric system.

Figure 8–9
Calibrating a thermometer.

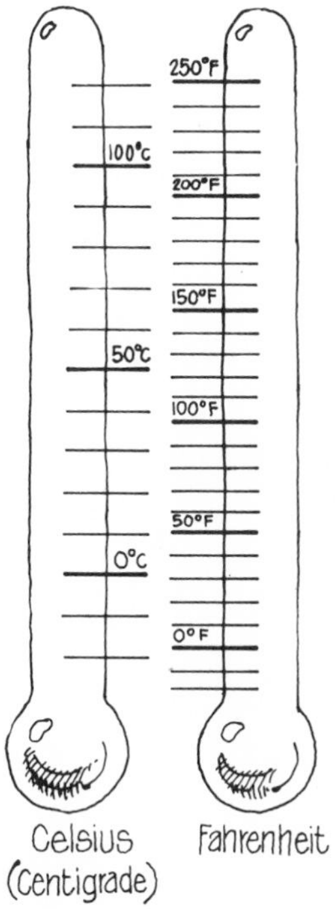

Figure 8–10

Comparison of Celsius (centigrade) and Fahrenheit scales.

Every temperature on the Celsius scale corresponds to a particular temperature on the Fahrenheit scale, and it is easy to convert from one scale to the other with a little arithmetic, or by using a table like the one in Fig. 8–10.

3. THE GAS LAWS

As we have just seen, temperature affects the properties of liquids and solids. But the effect of temperature on *gases* is even more striking. We will now investigate the behavior of gases in detail, and we will see that early investigations eventually led to a deeper understanding of temperature and heat.

Volume of a Gas Is Inversely Proportional to Pressure—Robert Boyle

Robert Boyle (1627–1691), who conducted some of the early experiments on gases, discovered a law that is consistent with everyday experience. Boyle's law states that

When the temperature of a gas is not changed, the volume of the gas is inversely proportional to the pressure applied to it.

For example, if the pressure on a confined gas is doubled (Fig. 8–11), the volume of the gas decreases to half of what it was; if the pressure is tripled, the volume decreases to a third of what it was; and so on. By using abbreviations or symbols, Boyle's law can be written

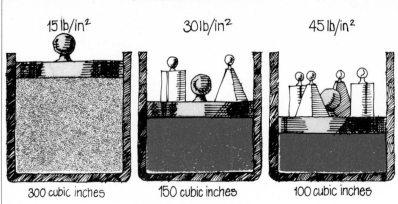

$$V \propto \frac{1}{p} \text{ [at constant temperature]},$$

where V refers to volume and p to pressure. Boyle's law can also be stated as follows: at constant temperature, if either the volume of a gas or the pressure applied to the gas is varied, the other varies as well, such that the product of the two remains constant. That is,

$$pV = \text{constant [at constant temperature]}.$$

**Volume is Directly
Proportional to the Absolute
Temperature—Jacques Charles**

Temperature also affects the volume of a gas, as the following experiment shows.

EXPERIMENT

Take an "empty" bottle (which of course is filled with air) and attach a balloon or a small plastic bag to its neck, Fig. 8–12. Heat the air in the bottle by placing the bottle in boiling water or over a flame. What happens to the balloon or the plastic bag?

OR

Blow up a balloon, tie its end firmly, and measure its diameter. Put it in a freezer for 15 to 30 minutes and then measure its diameter again. What do you find?

We all know that a gas expands when it is heated. Yet the quantitative relationship between the volume of a gas and its temperature was not discovered until more than a century after Boyle perceived

Figure 8–11
Volume is inversely proportional to pressure.

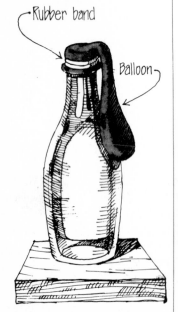

Figure 8–12
"Empty" bottle with a balloon over its neck, ready to be heated.

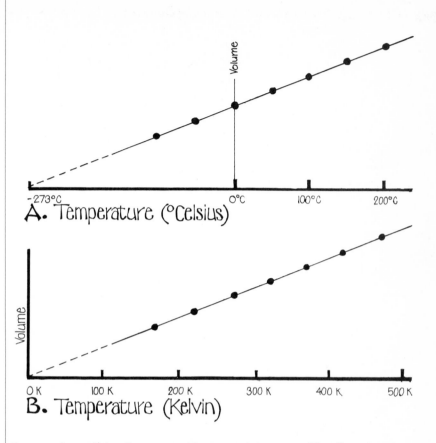

A. Temperature (°Celsius)

B. Temperature (Kelvin)

Figure 8–13
(A) Volume of an enclosed gas as a function of values of the volume at temperature in degrees Celsius (at constant pressure). The dark dots represent the measured values of the volume at various temperatures.
(B) Volume of an enclosed gas, at constant pressure, is directly proportional to the absolute temperature.

the precise relation between volume and pressure. The French scientist Jacques Charles (1747–1823) found that, when the pressure on a gas is kept constant, the volume increases with temperature at a constant rate, Fig. 8–13(A). However, since gases liquefy at low temperatures (air at −196°C), the graph of Fig. 8–13 cannot be carried to temperatures below the liquefaction point.

Nonetheless, when the straight line of the graph is projected to lower and lower temperatures, as shown by the dotted line of Fig. 8–13(A), it crosses the axis at a temperature of −273°C. This kind of graph can be drawn for all gases, and the graph always projects to zero volume at −273°C. This seems to imply that if a gas could be cooled to −273°C it would have no volume! It also implies that at temperatures lower than −273°C the volume would be negative, which of course makes no sense. This suggests that perhaps −273°C is the lowest temperature attainable. A good many other experiments indicate that this is indeed the case. Therefore, −273°C is called the **absolute zero** of temperature. (This is −460°F on the Fahrenheit scale.) Although scientists have been able to reach within 0.00001°C of absolute zero, absolute zero itself apparently cannot be reached precisely.

Absolute zero is the basis of a third temperature scale, which is called the absolute, or Kelvin, scale. The size of a degree on this scale is the same as on the Celsius scale, but absolute zero, −273°C, is chosen as zero, or 0 K, on the absolute scale (note that no degree sign is used with the K, which stands for "kelvins"); the freezing point of water is 273 K and the boiling point of water is 373 K. When a temperature is given in degrees Celsius, we can translate it to the absolute scale by adding 273: K = °C + 273.

If we graph the volume of a gas against the absolute temperature [Fig. 8–13(B)], we see that the straight line passes through the origin of the graph. In other words,

If the pressure on a gas is not allowed to change, the volume of the gas is directly proportional to its absolute temperature.

This is Charles's law. If we let T represent the absolute temperature, Charles's law can be written

$$V \propto T \text{ [at constant pressure]}.$$

In other words, if the absolute temperature of a gas were doubled, its volume would double as well, so long as the pressure remained constant.

At Constant Volume, Pressure Increases with Temperature

What happens when we keep the volume of a gas constant and change the temperature? This is the situation when a gas is heated in a closed (inelastic) container. We find that

The pressure is directly proportional to the absolute temperature.

If the absolute temperature is doubled, the pressure of the gas inside the container is doubled. This was discovered by Joseph Gay-Lussac (1778–1850), a contemporary of Charles's, and is known as Gay-Lussac's law. In symbols, Gay-Lussac's law is written

$$p \propto T \text{ [at constant volume]}.$$

A familiar example of the increase in pressure with an increase in temperature when volume remains constant is what happens when we throw a closed jar into a fire, or place it on a hot stove burner: the air pressure that builds up inside the jar causes it to break.

The Ideal Gas Law

Boyle's law, Charles's law, and Gay-Lussac's law have to do with the relations among the three important variables of a gas: the pressure, the volume, and the temperature. Collectively, these laws are known as the **Gas Laws**. Each law relates two of these variables while the third variable is kept constant. This technique of keeping one or more variables constant in order to observe the effect of changing one of the other variables is a very useful tool for scientists. However, since it is not always practical to hold one of the variables constant, we can combine the three separate laws into a single law that relates all three variables. This single law can be stated in symbols as

$$pV \propto T.$$

When any two of the quantities—pressure, volume, and temperature—are changed, this relation tells us how the third will change. Thus it is more general than Boyle's, Charles's, and Gay-Lussac's laws separately. Note, however, that at constant temperature the above relation reduces to Boyle's law, pV = constant; at constant pressure it reduces to Charles's law, $V \propto T$; and at constant volume it reduces to Gay-Lussac's law, $p \propto T$.

In addition to volume, pressure, and temperature, a gas has a fourth, equally important variable, which up to now we have assumed to be constant. This is the total mass of the gas, or the total number of molecules in the gas. If we keep the temperature and pressure of a gas constant but change the volume of the gas by adding more molecules, we find that the volume increases directly with the number of molecules, N. That is $V \propto N$ when temperature and pressure do not change. If you have ever blown up a balloon, you have discovered this for yourself, Fig. 8–14. If you double the amount of air in the balloon, the volume doubles.

Similarly, if we keep the volume rather than the pressure constant, we find that the pressure is directly proportional to the amount of gas present: $p \propto N$.

This fourth variable, the number of molecules, can be combined with the relation linking pressure, volume, and temperature and can be stated in symbols as

$$pV \propto NT.$$

Finally, putting in a constant of proportionality, k (see Chapter 1), we obtain

$$pV = NkT.$$

Figure 8–14

Volume is proportional to the amount of gas at constant pressure and temperature.

This is known as the **ideal gas law**. It is called "ideal" because real gases do not follow it precisely, particularly at very high pressures, or when the gas is near the liquefaction point. (This is also true of Boyle's, Charles's, and Gay-Lussac's laws.) However, under most ordinary circumstances real gases follow the ideal gas law quite closely.

The constant k in the ideal gas law is found experimentally to be the same for *all* gases, whether they be hydrogen, oxygen, argon, or whatever. This amazing fact was discovered by Amadeo Avogadro de Quarenga (1776–1856) and is known as Avogadro's law. Avogadro stated it in the following simple way:

Equal volumes of gases at the same pressure and temperature contain equal numbers of molecules.

That Avogadro's discovery should be true is not at all obvious. That it *is* true is a remarkable expression of simplicity in nature. Avogadro's law was to play an important role in the acceptance of the atomic theory.[3]

Kinetic Theory of Gases

4. KINETIC THEORY

Kinetic theory refers to the idea that matter is made up of molecules and that these molecules are in constant motion. The theory is based on the assumption that, on the average, the molecules of a gas are rather far apart (the average distance between molecules is perhaps ten or more times greater than the diameter of a molecule) and that they frequently collide with one another and with the walls of the container. Because of these collisions, a particular molecule is assumed to be moving randomly, sometimes in one direction and sometimes in another, sometimes at high speed and sometimes at low speed, or somewhere in between.

As we mentioned earlier, this view of a gas explains why gases are so easily compressible. It is also easy to see why gases expand to fill any container and why they leak through tiny openings—it is because the molecules are constantly moving, and at rather high speeds.

By the mid-nineteenth century many scientists had come to ac-

[3]Avogadro had no way of measuring how many molecules there are in a given volume of gas. In fact, accurate measurements were not possible until the twentieth century. These measurements indicate that there are about 2.7×10^{22} molecules in one liter of an ideal gas, or 2.7×10^{25} in a cubic meter, at STP (Standard Temperature and Pressure, meaning 0°C and 1 atmosphere).

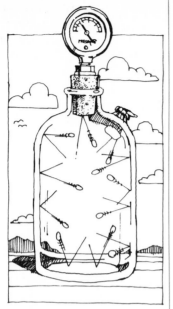

Figure 8–15

Pressure on the walls of a container is due to the bombardment of the rapidly moving gas molecules.

cept the kinetic (or atomic) theory of matter and were beginning to use it to explain the behavior of gases, particularly the ideal gas law.

Temperature Is a Measure of the Kinetic Energy of Molecules

A crucial development in the kinetic theory of gases was an idea put forward by Rudolf Clausius (1822–1888) in 1847. Clausius showed that the entire behavior of a gas, including the behavior specified by the gas laws, could be explained by the hypothesis that

The temperature of a gas is proportional to the average kinetic energy of its molecules.

This can be written in symbols as $T \propto \overline{KE} = \overline{\frac{1}{2} mv^2}$, where the bar (——) means average.[4]

How does this relation explain the ideal gas law? Let's take a simple illustration. The pressure on the walls of a vessel containing a gas results from the constant bombardment of the gas molecules against the walls, Fig. 8–15. According to the above relationship, when the temperature of a gas is increased, the molecules have more kinetic energy. They move faster and strike the container walls with more force; this means that the pressure is higher. Thus, an increase in temperature corresponds to an increase in pressure, which is just what we observe experimentally (Gay-Lussac's law).

By using kinetic theory and the relation between temperature and average molecular kinetic energy we can also explain the observed behavior of gases on which Charles's and Boyle's laws are based. If we keep the pressure on a gas constant—by allowing the volume of its container to change (as with a balloon)—an increase in temperature means that the molecules will strike the walls of the container harder, forcing the container to enlarge. This is just what Charles found experimentally. Finally, at constant temperature, if a gas is compressed so that its volume is smaller, the pressure increases (Boyle's law). According to kinetic theory, the molecules have been pushed closer together, and more of them will therefore strike the walls of the vessel in any time interval; thus the force on the walls, and therefore the pressure, will be increased.

[4]This is not to imply that *all* the molecules of a gas have this same amount of kinetic energy. Rather, it is the *average* kinetic energy of all the molecules that is proportional to the absolute temperature. At any given instant, roughly half the molecules have a kinetic energy greater than the average and the other half have a kinetic energy less than the average.

Kinetic Theory of
Solids and Liquids

The kinetic theory, and the idea that temperature is a measure of the average molecular kinetic energy, can accurately explain the observed behavior of gases. Applying it to solids and liquids is a more difficult task, because the forces between molecules are much more important. For example, temperature is not in precise direct proportion to the average kinetic energy of molecules in the solid and liquid phases. Nonetheless, it is still a measure of the motion of the molecules. Thus, in a liquid, an increase in temperature means that the molecules roll over one another more rapidly; and in a solid it means the atoms or molecules are vibrating faster about their fixed position in the crystal.

Kinetic theory can account for the expansion of liquids and solids as temperature increases: at higher temperatures the molecules are moving faster and, on the average, they will be a little farther apart. Thus the object being heated will take up more space. Kinetic theory is also useful in explaining the processes of melting and boiling—the change of state. We will return to this in Chapter 9.

REVIEW QUESTIONS

1. Elements are different from compounds in that
   (a) elements can be broken down but compounds cannot
   (b) compounds can be broken down chemically but elements cannot
   (c) elements can contain oxygen but compounds cannot
   (d) none of the above

2. Which of the following does not specifically support the atomic theory?
   (a) law of definite proportions   (c) Archimedes' Principle
   (b) Brownian movement   (d) all of the above

3. Brownian movement refers to
   (a) the motion of small particles of pollen or smoke due to collisions with molecules of a gas or liquid
   (b) the motion of molecules in a gas
   (c) the vibration of molecules in a gas
   (d) the motion of electrons in an atom

4. What distinguishes one element from another?
   (a) size of the atom
   (b) number of electrons in the neutral atom
   (c) size of the electrons and their orbits
   (d) number of nuclei in the atom

5. The periodic table refers to
   (a) periodic waves

(b) a specific way to list times

(c) a multiplication table

(d) a specific way to arrange the elements

6. *Macroscopic* means

   (a) at the molecular level   (c) large-scale everyday world

   (b) visible under a microscope   (d) a particular kind of solid

7. At the molecular level, the distinction between solids, liquids, and gases at room temperature is attributed to

   (a) the speeds of molecules

   (b) the number of electrons the molecules have

   (c) the mass of the constituent molecules

   (d) the strength of the forces between molecules

8. Which of the following properties does *not* change with temperature?

   (a) volume   (d) the state of a material

   (b) length   (e) all of the above change

   (c) color

9. Thermometers mainly make use of which property of materials?

   (a) color   (c) kinetic energy

   (b) expansion   (d) fluidity

10. As we heat a gas at constant pressure, its volume

   (a) increases   (c) stays the same

   (b) decreases

11. The volume of a gas is directly proportional to its

   (a) pressure   (c) Celsius temperature

   (b) Fahrenheit temperature   (d) Kelvin temperature

12. A gas is maintained at constant temperature. If the pressure on the gas is doubled, the volume is

   (a) increased fourfold   (d) reduced by half

   (b) doubled   (e) decreased by a quarter

   (c) the same

13. Blowing up a balloon is an example of

   (a) volume increasing with pressure

   (b) pressure increasing with volume

   (c) volume increasing with increased number of molecules

   (d) none of the above

14. The kinetic theory of matter

   (a) can be derived from the law of conservation of energy

   (b) is based entirely on experimental data with no arbitrary assumptions

   (c) contains assumptions accepted because the resulting theory agrees with experimental data

   (d) cannot be considered valid because molecules cannot be readily seen or measured

219

15. Temperature is a measure of the
    (a) potential energy of molecules
    (b) amount of heat in a body
    (c) average kinetic energy of molecules
    (d) total energy of all molecules in a body

EXERCISES

1. Explain how the law of definite proportions supports the atomic theory.
2. How does Brownian movement support the atomic theory?
3. There are about 3.2×10^{25} molecules in a quart of water, which is a volume of about 1000 cm³ (more precisely, 946 cm³). Estimate the size of one water molecule.
4. An atom of copper (Cu) is how many times heavier than an atom of sulphur (S)?
5. What is the principal factor that determines whether the atoms or molecules of a particular substance will form a solid, liquid, or gas at ordinary temperatures?
6. When a mercury thermometer is warmed, the mercury expands. What happens when the mercury expands? Do more molecules of mercury appear, or does the distance between molecules increase?
7. Room temperature is often defined as 68°F. What temperature is this on the Celsius scale? On the absolute scale?
8. What is 30°C on the Fahrenheit scale? On the Kelvin scale?
9. Why is it sometimes easier to remove the lid from a tightly closed jar after warming it under hot running water?
10. A closed bottle placed in a fire will suddenly explode. Why?
11. On a hot day do you think air molecules are on the average closer together, farther apart, or the same distance apart as on a cold day? Assume that atmospheric pressure is the same on both days.
12. If a gas is compressed to one-fifth its original volume but the temperature is kept constant, what happens to the pressure?
13. How can Avogadro's law be true if molecules of different gases have different sizes?
14. If the pressure on a gas is tripled, what happens to the volume if the temperature is kept constant?
15. The temperature of a gas in a glass container is increased from 300 K to 600 K. How does the pressure change?
16. If the pressure on a gas is kept constant, by what factor will its volume change if the temperature is increased from 27°C to 177°C?
17. Explain in words how Charles's law follows from the relation between average kinetic energy of molecules and absolute temperature.
18. If the average kinetic energy of molecules in a gas were to double, how would this affect the temperature?
19. If the temperature of a gas is raised from 20°C to 100°C, by what factor does the kinetic energy of the molecules increase?

220

20. If a copper penny has a higher temperature than a steel paper clip, does this necessarily mean that the average speed of molecules in the penny is greater than that of molecules in the paper clip? Explain.

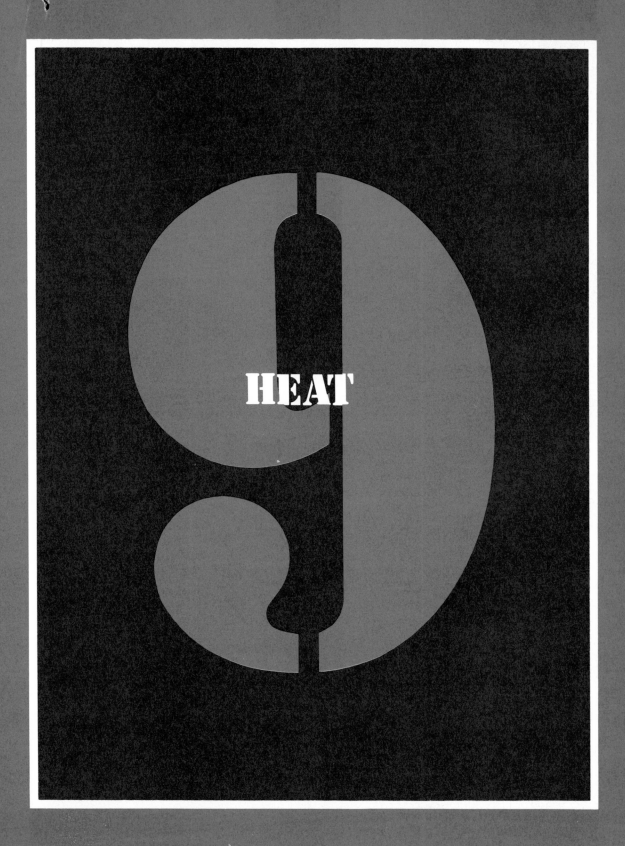

HEAT

We warm our houses with heat from burning coal, gas, or oil, or with heat generated by electricity, and we use the heat from electricity or gas to cook our food. We are all familiar with the phenomenon known as heat, and most of us are aware that heat is related to energy.

1. HEAT AND THE FIRST LAW OF THERMO-DYNAMICS

We all have an intuitive notion of what heat means. Yet it is not easy to define. Heat is often confused with temperature, but the two are different concepts. Just what do we mean by heat?

Heat: A Transfer of Molecular Energy

First, an example. A pot of cold water is heated on a hot stove; the temperature of the water increases. Since temperature is a measure of the kinetic energy of molecules, the kinetic energy of the water molecules must have increased. Thus, energy was transferred to the water from the hot burner. We often say that heat flows from the hot burner to the cold water. This is but one example of the fact that **heat** is a *transfer of energy from one body to another because of a temperature difference between them.* And heat flows only from a hot body (the stove burner) to a cooler body (the water).

The kinetic theory helps us explain how this energy transfer known as heat occurs. Let's go back to the cold water on the stove. The burner of an electric stove or the flame of a gas stove is at a high temperature, which means that its molecules are moving very rapidly. When these high-speed molecules strike the cold pot, they transfer some of their energy to the molecules of the pot (Fig. 9–1); thus the molecules of the pot gain in kinetic energy and the pot's temperature increases. In a similar way, the faster-moving pot molecules collide with the "cooler" water molecules, giving them increased energy, and so the water's temperature increases as well.

Whenever a hot substance is in contact with a cold substance, heat flows from the hotter one to the colder. Unless the process is interrupted, the two objects eventually reach the same temperature. When an ice cube is placed in a warm drink, heat flows from the warm drink to the ice cube; the drink becomes cooler and the ice cube becomes warmer and melts. When a pot is heated on the stove, however, the burner doesn't cool down very much because energy is continually being supplied to it by electricity or gas. The

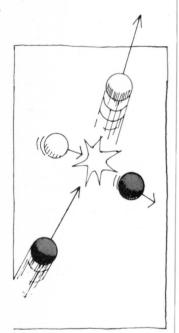

Figure 9–1

Kinetic energy is transferred from a fast-moving molecule (dark) to a slow-moving molecule by means of a collision.

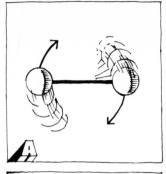

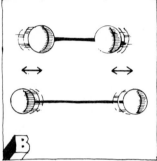

Figure 9–2

Molecules can have
(A) rotational energy and
(B) vibrational energy (the
atoms in a molecule vibrate
within the molecule).

pot gets hotter and hotter. But the stove burner cannot heat the pot to a temperature higher than its own temperature; heat will only flow "downhill" so to speak—only from a high temperature to a low temperature.

When an object is heated, not all the energy it absorbs goes into increasing the velocity of its molecules. Some goes into rotational motion of the molecules and some into internal vibrational energy; this is illustrated in Fig. 9–2. Thus the total energy of the molecules in a substance is the sum of the ordinary kinetic energy, the rotational kinetic energy, and the vibrational energy.

When we want to refer to the *total* energy of all the molecules in a given amount of material we use the term **thermal energy**.

Heat Was Once Thought to Be a Fluid

The modern scientific view of heat is based on kinetic theory. But kinetic theory is a relatively recent development. Although a few scientific thinkers as far back as the seventeenth century had suggested that heat might be associated with the motion of atoms, until the nineteenth century heat was thought to be a fluid called *caloric*.

According to the caloric theory, every object has a certain amount of heat, or caloric, in it. When caloric was added to an object, its temperature increased; and when caloric escaped, the temperature of the object decreased. However, since this caloric substance could not be detected by any means, it was assumed to be odorless, tasteless, massless, and invisible—a somewhat mysterious substance indeed. Nonetheless, the caloric theory satisfactorily explained many observations such as the "flow of heat" from a hot to a cold body.

The unit of heat was named the **calorie** (after "caloric"), which was defined as the amount of heat required to raise the temperature of one gram of water by one degree Celsius. Although we have discarded the caloric theory itself, we still use the calorie as the unit of heat. Its definition, though, has been slightly refined:

One calorie is the amount of heat required to raise one gram of water from 15°C to 16°C.

The reason that this definition specifies a particular temperature range is that the amount of heat necessary to change the temperature of water by one degree is slightly different at different temperatures. A **kilocalorie** of heat, or Calorie spelled with a capital C, is 1000 ordinary calories. The energy value of food—that is, the amount of energy that is released when the food is digested—is measured in Calories.

The caloric theory gave us not only our unit of heat but the expression "flow of heat." We still use this expression, even though we no longer consider heat a fluid.

Heat as Energy

Although the caloric theory successfully explained many aspects of heat, it could not explain them all. This deficiency eventually led scientists to abandon it and to develop a macroscopic view of heat as energy.

One of the main shortcomings of the caloric theory was that it could not account for all the heat generated by friction. By rubbing one's hands together or by rubbing two pieces of metal together one can generate heat indefinitely. That is, the amount of heat that can be generated by the friction of two objects being rubbed together seems limitless. And yet the caloric theory maintained that heat is a substance, and therefore any object can contain only a fixed amount of it. It was an American expatriate, Benjamin Thompson (1753–1814), later Count Rumford of Bavaria, who dealt with this discrepancy. He rejected the idea that heat is a substance and argued instead that heat could only be a kind of *motion*. Rumford claimed that the performance of mechanical work (as when two objects are rubbed together) was responsible for the generation of heat. He was thus at the threshold of interpreting heat as a form of energy. But it was an English brewer, James Prescott Joule (1818–1889), who performed the crucial experiments that settled the question.

Joule was convinced that heat is a form of energy. He performed an exhaustive series of experiments to prove his point.[1] One of these was the paddle wheel experiment illustrated in Fig. 9–3. The slowly falling weight does work on the paddle wheel, causing it to turn against the resistance of the water. The potential energy lost by the falling weight is first converted to kinetic energy of the paddle wheel, which in turn produces heat by friction with the water. By measuring the rise in the temperature of the water, Joule could calculate how much heat was produced by a particular amount of mechanical energy. In this, and in a great many other experiments, Joule found that 4.2 joules of energy always generated one calorie of heat.

The First Law of Thermodynamics

What Joule had shown by his experiments was that if heat is interpreted as energy, then mechanical energy can be converted into heat, or thermal energy, with no energy being lost or created in the process. This result revealed a new and very general conservation law, the conservation of energy. We discussed this law in Chapter

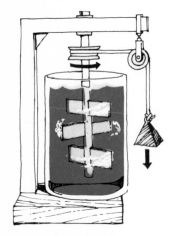

Figure 9–3

Joule's paddle wheel experiment. The falling weight turns the paddle wheel, which heats the water because of "friction" between the paddles and the water.

[1]In recognition of his work the metric unit of energy was named the joule.

5, particularly as it concerns mechanical energy in the absence of friction. But real situations do involve friction and heat. Thus it was only after heat was seen to be related to energy that the principle of the conservation of energy was stated as a general law of nature.

Because it involves heat, the law of conservation of energy is also called the first **law of thermodynamics**. Simply stated, it asserts that

> The total amount of energy—kinetic energy + potential energy + thermal energy + electrical energy + other forms of energy —remains constant.

When the kinetic theory emerged later in the nineteenth century, it reinforced Joule's interpretation of heat as energy and the conservation of energy principle. Joule's view of heat was a macroscopic view. The kinetic theory added a microscopic view of thermal energy by explaining how the transfer of energy as heat takes place at the molecular level.

Heat, Thermal Energy, Temperature

Let us now briefly review what we mean by temperature, heat, and thermal energy.

Temperature is a directly measurable property of an object. At the molecular level, temperature is a measure of the *average* kinetic energy of the individual molecules.

Thermal energy refers to the *total* energy (kinetic energy, vibrational energy, and rotational energy) of *all* the individual molecules in a given object. Thermal energy is what is usually meant by the expression "heat content" of a body. Thus we see that temperature and thermal energy refer to very different properties.

Although heat is often used to mean the same thing as thermal energy, these terms are not synonymous. Strictly speaking, heat refers to energy transferred from one body to another because of a difference in temperature. One object "heats" another object because the former transfers energy to the latter. Heat does not refer to energy itself, but rather to the *transfer* of energy.

Heat is thus akin to work. Heat is the transfer of energy from one object to another because of a temperature difference. Work is the transfer of energy from one body to another by means other than by difference in temperature.

To further clarify the distinctions between temperature, thermal energy, and heat, consider a hot brick. If you cut the brick in half and throw one half away, the remaining half will still have the same temperature. But it will have only half the thermal energy, and it will have only half the ability to heat another object. The following experiment will help make this clear.

227

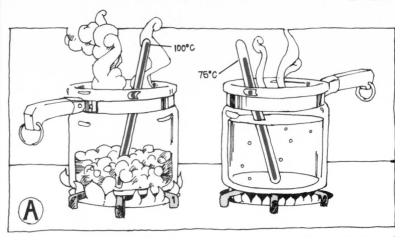

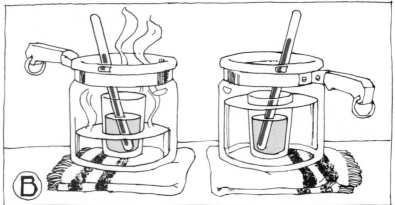

Figure 9-4

(A) The two pots are on the stove; the pot on the left contains 2 cups of water and is heated to boiling; the pot on the right contains 8 cups and is heated only to 75 °C. (B) Both pots are removed from the stove, and identical glasses or tin cans are placed in each (see experiment in text).

EXPERIMENT-PROJECT

Find two kitchen pots of about the same size and made of the same material. Put 2 cups of water in one and 8 cups of water in the other, Fig. 9-4(A). Put them on identical burners and, using a thermometer, heat the first to boiling, 100°C, (212°F), and the second only to 75°C (167°F). Notice which pot takes longer to heat and therefore requires more heat. Remove both pots from the stove and immediately place an identical tin can (or glass) containing one cup of cold water in each pot, Fig. 9-4(B). Wait a few minutes and then check the temperature of the water in each tin can. Which can of water is heated to the higher temperature? Which pot of water, the 2 cups at 100°C or the 8 cups at 75°C, possessed the greater thermal energy? Alternatively, after removing the two pots from the stove, see how many ice cubes each can melt.

Figure 9–5
Heat conduction.

2. HEAT TRANSFER: CONDUCTION, CONVECTION, AND RADIATION

Heat can be transferred from one place to another in three different ways: by **conduction**, by **convection**, and by **radiation**.

Conduction

When you put one end of a metal rod in a fire, or a silver teaspoon in a hot cup of soup, the other end becomes warm even though it is not directly in contact with the source of heat. We say that heat has been *conducted* from the hot end to the cold end, Fig. 9–5.

Conduction is the transfer of energy through the collision of molecules. As one end of the object is heated, the molecules at that end move faster and faster. When they collide with their slower-moving neighbors, some of their energy is transferred to them. These molecules now speed up. When they in turn collide with other molecules farther up the rod (or spoon), they transfer energy to them. And so on.

Heat conduction takes place only when there is a difference in temperature, and the heat always flows from the higher temperature to the lower temperature. The rate at which the heat flows depends on the temperature difference, on the size and shape of the material, and on the kind of material it is. Some materials are good **conductors** of heat, which means that heat flows through them quickly. Other materials are poor conductors, or heat **insulators**; these materials conduct heat only very slowly.

EXPERIMENT

To demonstrate the difference between conductors and insulators, you can use a lighted candle or the burner of a stove. Be careful. Hold one end of a paper clip in your fingers and stick the other end in the flame; drop it quickly if it gets too hot. Now try a glass rod and other materials such as a nail, a silver spoon, a wooden rod (be careful it doesn't burn, though charring won't hurt), and a stone. Which materials are good conductors? Which are good insulators?

Most metals are very good conductors of heat, although there is wide variability. Test this statement by putting a silver spoon and a stainless steel spoon in the same cup of hot water and then touch the handles.

Even the best insulators conduct some heat. Glass, for example, is a much poorer conductor of heat than metals are. Even so, a

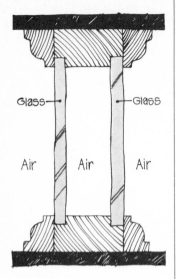

Figure 9–6
A double-glazed window.

good deal of heat is lost through the windows of a house on a cold day; the glass, though a poor conductor, is usually quite thin, so the heat doesn't have to travel far to be lost. The use of "double glazing," that is, two panes of glass separated by a layer of air, Fig. 9–6, helps considerably to reduce heat loss by conduction.

Air and other gases are the best insulators of all, because the molecules of a gas are so far apart on the average that they make fewer collisions with their neighbors than do the molecules in solids; hence they transfer energy much more slowly. A perfect vacuum would be the ideal insulator, because it would contain no molecules at all to transfer energy. The double-glazed window of Fig. 9–6 insulates well because of the air layer between the sheets of glass.

The best insulating materials are those that trap air in pockets or between their fibers: materials like fiberglass and the various types of foam that are used to insulate houses and buildings. Wool and down are very good for insulating people; down is especially effective, because even a small amount of it fluffs up and traps a great deal of air. It is not the material our clothes are made of that keeps us warm but the air trapped in the material.

Convection

Although liquids and gases are generally not very efficient conductors of heat, they can transfer heat quite rapidly by convection. **Convection** is the transfer of heat through the mass movement of molecules from one place to another. In conduction, molecules move only through very small distances; but in convection, molecules move through large distances.

A forced-air furnace, in which air is heated and is then blown by a fan into a room, is an example of **forced convection**. **Natural convection** occurs as well, and one familiar example is the fact that hot air rises. For example, the air above a radiator expands as it is heated; hence its density decreases and it rises. Warm and cold ocean currents, such as the Gulf Stream, display natural convection on a large scale. Wind is another example of convection, and weather in general is the result of convective air currents.

When a pot of water is heated, Fig. 9–7, convection currents are set up as the heated water at the bottom of the pot becomes less dense, rises, and is replaced by cooler water from above. This same principle is used in home heating systems like the hot-water radiator system shown in Fig. 9–8. As water is heated in the furnace its temperature goes up, and the water expands and rises into the radiators. As the water circulates through the system, heat is transferred by conduction to the radiators and thence to the air; the cooled water then returns to the furnace to be heated again. The water circulates because of convection, though pumps are sometimes

Figure 9–7
Convection occurs when water is heated on a stove.

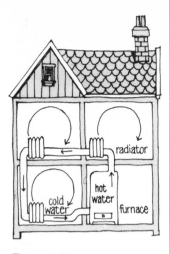

Figure 9–8

A hot water radiator heating system uses convection.

used to speed up its circulation. Convection also brings about the uniform heating of the rooms in which the radiators sit. As the radiators heat the air, it rises and is replaced by cooler air, resulting in convective air currents, Fig. 9–8.

Other types of furnaces also depend on convection. Hot-air furnaces with openings near the floor often depend on natural convection to bring about the distribution of heated air; the cold air returns to the furnace through other openings.

Another example of convection and its effects is given in the following excerpt from Francois Matthes's "The Winds of Yosemite Valley":[2]

It happens to be so ordained in nature that the sun shall heat the ground more rapidly than the air. And so it comes that every slope or hillside basking in the morning sun soon becomes itself a source of heat. It gradually warms the air immediately over it, and the latter, becoming lighter, begins to rise. But not vertically upward, for above it is still the cool air pressing down. Up along the warm slope it ascends, much as shown by the arrows in the accompanying diagram [Fig. 9–9(A)]. Few visitors to the valley but will remember toiling up some never-ending zigzags on a hot and breathless day, with the sun on their backs and their own dust floating upward with them in an exasperating, choking cloud. Perhaps they thought it was simply their misfortune that the dust should happen to rise on that particular day. It always does on a sun-warmed slope.

But again, memories may arise of another occasion when, on coming down a certain trail the dust ever descended with the travelers, wafting down upon them from zigzag to zigzag as if with malicious pleasure. That, however, undoubtedly happened on the shady side of the valley. For there the conditions are exactly reversed. When the sun leaves a slope the latter begins at once to lose its heat by

[2]Reprinted from the *Science Club Bulletin,* June 1911, pp. 91–92.

Figure 9–9

"Winds of Yosemite."

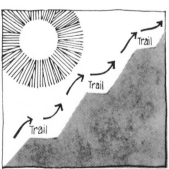

A. Warming air currents

B. Cooling air currents

radiation, and in a short time is colder than the air. The layer next to the ground then gradually chills by contact, and, becoming heavier as it condenses, begins to creep down along the slope [Fig. 9–9(B)]. There is, thus, normally a warm updraft on a sunlit slope and a cold downdraft on a shaded slope—and that rule one may depend on almost any day in a windless region like the Yosemite. Indeed, one might readily take advantage of it and plan his trips so as to have a dust-free journey.

EXPERIMENT

On a cold day, open an outside door in a well-heated room in your house. Light a candle and hold it first near the very top of the door opening, and then near the floor, Fig. 9–10. Does the flame change direction? The hot air, which has expanded and risen, escapes near the top; the cold air, on the other hand, is drawn into the room near the floor and replaces the air that has escaped.

Figure 9–10
Detecting convection currents.

The human body generates a great deal of heat. Only about 20 percent of the food energy transformed by the body is used to do work; some 80 percent appears as heat, and this heat must somehow be transferred to the outside if the body is not to become overheated. But the body is not a very good conductor of heat. How does it get rid of this superfluous heat? The blood acts as a convective fluid and carries the heat to the surfaces of the body where it is transferred to the air by conduction. Our blood serves many purposes!

Radiation

Conduction and convection require the presence of matter. The third form of energy transfer, **radiation**, does not. All life on earth depends on energy from the sun, and this energy is carried to the earth at the speed of light by radiation. The warmth we receive from a fire reaches us mainly through radiation.

As we shall see later, radiation consists of electromagnetic waves; for now, we merely point out that the radiation from the sun includes not only visible light but many other wavelengths as well.

All objects give off radiation. But the higher the temperature of the object, the more energy it radiates. In fact, the rate of radiation of energy is proportional to the fourth power of the Kelvin temperature (T^4). Thus an object at 600 K as compared to one at 300 K has only twice the temperature but radiates $(2)^4 = 16$ times as much energy!

Bodies that glow, such as the sun, a fire, or a light bulb, are radiating energy. Indeed, any object will glow (that is, will give off *visible* radiation) if it is raised to a high enough temperature. But even at lower temperatures objects give off radiation, though it may not be visible.

Consider, for example, two bodies close to one another that are otherwise the same except that one is cold and the other hot. Both objects emit energy, but the hot one emits more energy than the cold one. Thus there is a *net* transfer of radiant energy from the hot one to the cold one. Indeed, there will be a net transfer of energy by radiation between two bodies as long as there is a difference in their temperatures.

Every object both emits energy and absorbs energy. But an object may not absorb all the energy that falls on it; it may reflect some of it. Good absorbers reflect very little of the radiant energy that falls on them and thus appear black; that is why dark-colored objects are relatively good absorbers. Light-colored objects, on the other hand, reflect most of the energy that falls on them and so are poor absorbers. That is why light-colored clothing is more comfortable than dark clothing on a bright hot day.

It is found, too, that good absorbers are good emitters, and poor absorbers are poor emitters. That is, at the same temperature, a black body will radiate energy at a faster rate than a similar white body at the same temperature. For example, hot tea in a dark-colored metal container cools more quickly as a result of radiation than does tea in a brightly polished metal container.

The fact that all objects, including human bodies, emit radiation, has many practical consequences. For example, a person may be sitting in a room where the air temperature is 25°C (77°F), which you might think would be comfortable enough. But the room may feel chilly if the walls are at a somewhat lower temperature (perhaps because they are in contact with cold outside air). Human skin temperature is normally about 33° or 34°C; if the walls are at, say, 15°C, there may be a significant flow of radiant energy away from the person, and the person becomes chilled. Rooms are most comfortable when the walls and floor are warm and the air a bit less warm. Hot-water pipes or electric heating elements could be used to heat walls and floors, but such systems are rare today. Interestingly, though, 2000 years ago the Romans heated the floors of their houses with hot-water and steam pipes, even in the remote province of Britain!

Glass windows cause substantial heat loss through both conduction and radiation. They not only conduct heat out of a room, but at night, when the outside temperature is low, considerable radiation from a body to the outside occurs. That's why it helps to close the curtains on a cold night.

3. SPACE HEATING AND SOLAR ENERGY

A large part of the energy consumed in the United States is used for space heating—that is, for keeping the inside of houses and buildings warm. Most heating systems burn gas, oil, or coal, which are in limited supply. Moreover, the extraction and burning of these fossil fuels create environmental problems: destruction of the landscape, rendering farmland less useful, and air and water pollution. A typical home furnace is only about 60 percent efficient, which means that only 60 percent of the energy released by the burning of the fuel actually heats the house; the other 40 percent goes out the vent or chimney or is wasted as unburned or partially burned fuel.

Electric heating is generally even less efficient and more wasteful, since most electric energy today is generated by the burning of fuel in systems that are only about 30 to 35 percent efficient.

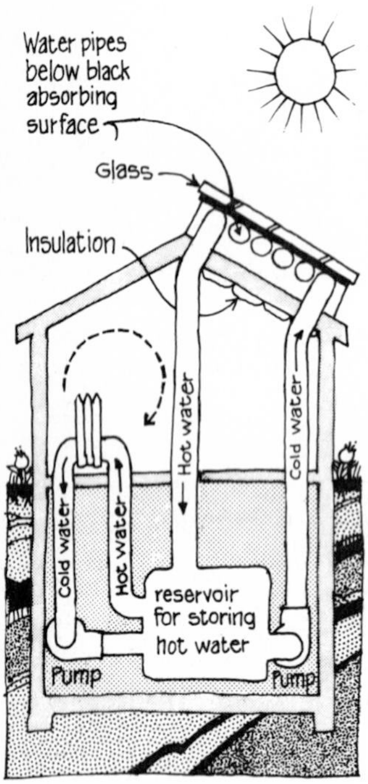

An alternative to space heating is to use solar energy—that is, radiation from the sun. It is now feasible to install solar-energy heating systems in individual houses, although some back-up system is often needed in climates with prolonged periods of cloudiness. In a solar-heating system for a house, Fig. 9–11, water-carrying tubes are located on the roof in contact with a large black surface that absorbs the sun's radiant energy and heats the water. The surface is covered with glass to prevent heat loss by convection, and the tubes are well insulated to reduce heat loss by conduction. The heated water is then circulated to a large, well-insulated reservoir, where it is stored and recirculated to heaters in the house. The reservoir can also supply hot water for other needs.

Greater efficiency and higher temperatures can be achieved by using reflectors to concentrate the radiation from a large area onto a smaller absorbing surface. The possible use of solar energy for generating electricity will be discussed in Chapter 14.

Figure 9–11

A solar energy heating system for an individual house.

4. CHANGE OF STATE

When water is heated to a high enough temperature, it boils and becomes a gas. When water is cooled to a low enough temperature, it freezes and becomes a solid. The temperature at which a liquid boils and changes to a gas is called its **boiling point**; that gas cools and condenses into a liquid at the *same* temperature (the **condensation point**). The temperature at which a solid melts into a liquid is called its **melting point**; that liquid freezes and changes back into a solid

Table 9-1
Melting and Boiling Points of Various Substances

SUBSTANCE	MELTING POINT (°C)	BOILING POINT (°C)
Helium	−272*	−269
Oxygen	−219	−183
Nitrogen	−210	−196
Alcohol (ethyl)	−117	79
Mercury	−39	357
Water	0	100
Lead	327	1744
Gold	1063	2966
Iron	1535	3000

*Under pressure of 26 atmospheres. (Does not condense under one atmosphere.)

at the *same* temperature (the **freezing point**). The melting and boiling points for a variety of common substances are given in Table 9–1.

Heat is Involved in a Change of State

Let's see what happens when we continuously heat a solid substance until it becomes a liquid and then a gas. We'll use water. Suppose we take 1 gram of ice at −40°C and heat it at a continuous rate, Fig. 9–12. As we heat it, its temperature rises. It takes about

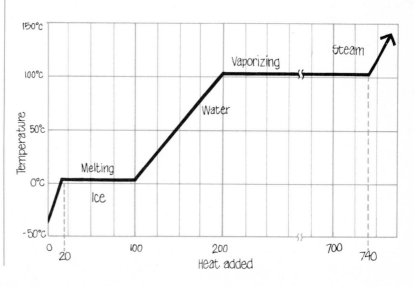

Figure 9–12

Graph showing the heat required to heat 1 gram of ice at −40°C to steam at 100°C.

half a calorie to raise the temperature of one gram of ice by one degree Celsius. When the temperature of the ice reaches 0°C, the ice begins to melt. But it doesn't all melt at once. As more heat is added, the ice melts into water little by little and a water-ice mixture results. Throughout this process, the temperature of the surviving ice does not change; it remains at 0°C until all the ice has melted. Heat is required just to melt ice without any increase in temperature. To melt 1 gram of ice at 0°C to water at 0°C requires 80 calories; this is known as the **heat of melting** for water. Once all the ice has melted, the temperature begins to rise again; since it takes about 1 calorie to raise the temperature of 1 gram of water by 1 degree Celsius, a total of 100 calories is necessary to bring the 1 gram of water to the boiling point at 100°C. At this point a considerable amount of energy is required to change the water from the liquid to the gaseous state (which we call **steam**, or water vapor). This is the **heat of vaporization**, which for water is 540 calories per gram. The temperature of the water remains constant until it has all been converted to steam. With further heating, the temperature of the steam increases. This whole process is graphed in Fig. 9–12.

Evaporation and Condensation

To change water from the liquid to the gaseous state does not necessarily require that the water be heated to boiling. The change occurs even at room temperature, although much more slowly. This is the process of **evaporation**. If you leave a glass of water out overnight, you will find that the water level has gone down by morning. Eventually all the water in the glass will evaporate; the liquid water slowly becomes a gas. Evaporation also occurs in other liquids, of course.

Let us examine the process of evaporation from the microscopic viewpoint, making use of kinetic theory. The molecules of water in a glass at room temperature are rolling over one another in random directions and at various speeds: slow, fast, and in between. At any instant a few of the molecules will be moving so fast that if they are near the surface they can overcome the forces holding them in the liquid and escape into the air above, Fig. 9–13. This is much like a rocket, which, if its speed is great enough, can overcome the gravitational attraction of the earth and escape into space (Fig. 3–34), except that, for molecules, the force is electrical in nature.

Since it is the fastest molecules that escape from the surface, the average kinetic energy of the molecules remaining in the liquid is lowered and hence the temperature drops. Thus evaporation is a cooling process. If you have ever exercised hard enough on a hot day to work up a sweat, you know how the slightest breeze makes you feel cool; you feel cool because the perspiration is evaporating

Figure 9–13
Evaporation. Very fast-moving molecules near the surface of the liquid can escape.

Figure 9–14

Water condensing on a cold glass.

from your skin. Similarly, when a person is rubbed down with alcohol, which evaporates more rapidly than water, the evaporating alcohol has a pleasant cooling effect. (This cooling process can also be viewed from the point of view of energy; the cooling is a result of energy taken from the body and given to the liquid—the heat of vaporization—that causes it to vaporize.)

The reverse of evaporation is **condensation**, which occurs when gas molecules strike the surface of the liquid and are absorbed. What happens when water in the air condenses on the outside of a glass of an iced drink, Fig. 9–14? Water molecules in the vicinity of the cold glass lose kinetic energy when they collide with the slow-moving molecules of the glass and the cool layer of air next to the glass. As more and more of the water molecules slow down, the attractive forces between them pull them together and form a drop of water on the glass.

Fog, clouds, and rain are also the result of the condensation of water into droplets when air is cooled. The cooling of air at night often results in the condensation we know as morning dew.

These explanations in terms of kinetic theory remind us that a theory is only as good as its power to explain phenomena; and kinetic theory works very well here, as elsewhere.

Melting and Boiling According to the Kinetic Theory

What happens at the molecular level when a solid is heated to melting and the resulting liquid is then heated to boiling? That is, what happens during the process graphed in Fig. 9–12? As the solid is heated to higher temperatures, its molecules vibrate faster about their normal positions. At a sufficiently high temperature the vibrations become so vigorous that the attractive forces can no longer hold the molecules in place, and they start rolling over one another. As more heat is added, more and more molecules break loose from their fixed positions in the solid. At this point, melting is occurring. Notice, however, that since the temperature is not changing during the melting process, the heat energy required, which we called the *heat of melting*, is not going into increasing the kinetic energy of the molecules. Where is it going? It is going into potential energy of the molecules. In the orderly array of a crystalline solid, the molecules have less potential energy than when they are in the disorderly array of a liquid. Work must be done against the attractive forces between the molecules in the solid to pull them apart and set them into the more random motion of a liquid. This raises their potential energy, just as work done on lifting a rock from the ground raises the rock's potential energy. Thus, the heat of melting goes into the potential energy of the molecules.

After melting occurs, continued heating increases the kinetic energy of the liquid molecules even further. As the temperature rises, a very small amount of evaporation will occur; but at the boiling point, very rapid evaporation starts to occur. The molecules are moving so fast now that the attractive forces can no longer hold them in a limited region of space, and the molecules move out in all directions, as a gas. Work must be done against the attractive forces, of course, and this raises the potential energy of the molecules. The energy necessary to do this is the heat of vaporization, which also goes into the potential energy of the molecules.

The heat of vaporization is required whether the evaporation takes place at the boiling point or below it. We have already seen that evaporation even at room temperature cools a liquid, and we can interpret this cooling as an absorption of the heat that is needed to evaporate the liquid.

The energy needed to pull liquid molecules far apart into the gaseous state intuitively seems to be more than the energy needed for the smaller reorientation required when a solid becomes a liquid. Indeed, for water, the heat of vaporization is much greater than the heat of melting, 540 calories per gram versus 80 calories per gram.

The Boiling Process Examined

Let's look at the boiling process in more detail.

EXPERIMENT
Put some cold water in a pan and heat it on a stove to boiling. Notice where and how bubbles form in the water as the boiling point is approached. Examine the water carefully just as it is about to boil vigorously.

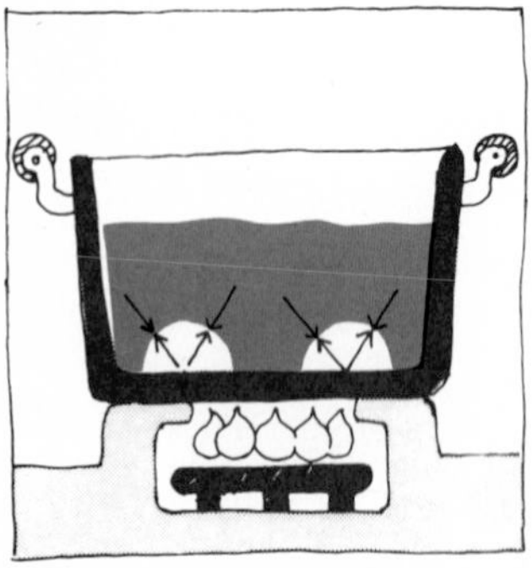

Figure 9–15
Bubbles of steam forming at the bottom of a heated pot just before boiling takes place.

As the water approaches the boiling point, evaporation begins to take place on the bottom and sides of the pan where the temperature is highest. The molecules are moving so fast there that a group of them may enter the gaseous state and start to form a bubble. But for a while any bubble that tends to form will immediately collapse unless the pressure caused by the fast movement of molecules within the liquid is sufficient to balance the downward pressure of the liquid around it, Fig. 9–15. As heating continues, the vapor pressure within the bubbles builds up until they release themselves from the pan and float to the top. Bubbles soon begin appearing throughout the liquid and rise to the top; vigorous boiling has begun.

Boiling occurs when the pressure within the bubble overcomes the outward pressure caused by the liquid and the atmosphere above it. Thus, if the pressure at the surface of the liquid is increased, a higher temperature is necessary before the molecules are moving fast enough to form a bubble. This is how a pressure cooker works. The evaporating water builds up a high pressure within the sealed pot and boiling occurs at a higher than normal temperature. At this higher temperature foods cook much more quickly.

At high elevations, where the air pressure is less, it takes longer to cook foods than it does at sea level. This is because less internal vapor pressure is needed and bubbles can form at a lower temperature. Thus the temperature of boiling water is lower at high elevations. If you apply more heat to the water in an attempt to increase its temperature, you find that the temperature won't go any higher; instead the water boils all the faster. Faster boiling does not cook food any faster. Only the temperature affects the cooking rate. Since the maximum temperature of liquid water is less at the lower pressure of high altitudes, cooking takes longer. Pressure cookers can of course be used at high altitudes to reduce cooking time (but backpackers beware!).

5. HEAT ENGINES AND THE SECOND LAW OF THERMO-DYNAMICS

The first law of thermodynamics says that energy is conserved. Yet we can think of many processes that would conserve energy but that do not actually occur in nature. For example, when a hot object is placed in contact with a cold object, heat always flows from the hot to the cold object. The hot object never gets hotter and the cold object colder; yet this would not violate the principle of conservation of energy. The energy that left the cold body would enter the hot body, so energy would be conserved. Why

Figure 9–16

(A) A rock falls to the ground and its kinetic energy is transformed into thermal energy upon hitting the ground. Energy is conserved. (B) If the rock were suddenly to cool and the thermal energy transformed into kinetic energy, the rock would rise. Energy would be conserved. Does this ever happen?

is it that heat never flows spontaneously from a cold to a hot object? Consider another example: When you drop a heavy rock, its potential energy changes to kinetic energy as it falls; upon striking the ground it comes to rest. Its kinetic energy all goes into thermal energy. The molecules in the rock, and in the ground where it hit, speed up slightly. Energy is conserved in this process. And energy would be conserved in the reverse process as well. But have you ever seen the reverse happen? Can you imagine a rock resting on the ground [Fig. 9–16(B)] that suddenly rose up into the air because the thermal energy of the molecules was changed into kinetic energy of the rock as a whole?

Could it be that the impossibility that these two events could happen suggests a new principle? The answer is yes. This new principle is the **second law of thermodynamics**. Scientists hypothesized this law in order to "explain" why the above processes, and others like them, do not actually happen. Of course this new law, like all physical laws, is a unifying principle that relates all the forbidden processes; it does not really "explain."

One way of stating the second law of thermodynamics is this: *the natural flow of heat is always from an object at a high temperature to one at a low temperature;* or, *heat never flows spontaneously from a cold object to a hot object.* This is not a very general statement, however. For instance, it is not at all apparent how to relate it to the example of Fig. 9–16. A broader statement is needed, one that will include the variety of phenomena that do not violate conservation of energy but that are not observed to happen. This is no easy task, for the second law of thermodynamics is a strange law, perhaps the most abstract in all of physics. Yet it is one of the most interesting, and it has stimulated a great deal of philosophical discussion.

We now investigate how a general unifying statement of the second law was developed, and we will examine some of the ramifications of that law.

Heat Engines

The development of the second law of thermodynamics was prompted by studies on the changing of mechanical energy into heat, and the reverse, the changing of heat or thermal energy into mechanical energy. Changing mechanical energy into heat is easily accomplished, as was shown for instance in Fig. 9–16(A) or by simply rubbing your hands together. In any process in which friction is present, mechanical energy is transformed into thermal energy. But what about the reverse process? Is it possible to change thermal energy into mechanical energy that can be used to do mechanical work? We know that doesn't happen spontaneously, as in Fig. 9–16(B), but is there some other way it can be done?

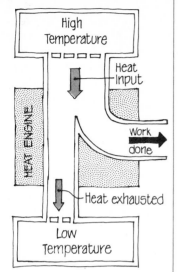

Figure 9–17

General principle behind any heat engine. When heat is allowed to flow from a substance at a high temperature to one at a low temperature, part of this heat can be turned into work.

It *is* possible to get mechanical energy from thermal energy, and the devices that do so are called **heat engines**. Common examples are the steam engine and the internal combustion engine.

Early in the nineteenth century the French engineer and physicist Sadi Carnot (1796–1832) investigated the process of transforming heat into mechanical energy and recognized that there are natural limitations on the work that can be done by a heat engine. He realized that mechanical energy can be obtained from heat only when heat is allowed to flow from a substance at high temperature to one at a lower temperature; some of this heat can then be diverted into mechanical work, as diagrammed in Fig. 9–17. The high and the low temperatures are called the *operating temperatures* of the engine.

Let us see why a *difference* in temperature is necessary. A useful heat engine will produce mechanical energy from heat in a continuous fashion, and we consider an idealized method of doing this in Fig. 9–18. This ideal heat engine makes use of a gas expanding against a movable piston; the energy given to the piston can be used to turn a wheel. The gas is heated to a high temperature by contact with a high-temperature furnace, Fig. 9–18(A). As the heated gas expands (remember, an increase in temperature gives rise to increased pressure and/or volume), it pushes against the piston that, through the linkage, turns the wheel. Now, in order to obtain mechanical energy continuously from this engine, the gas must be recompressed so that it can again expand and do work; thus the process must be a cyclic one—that is, a process that repeats itself. After the expansion, the wheel continues around and the linkage pushes the piston in, which compresses the gas; but the work required to compress the gas in this part of the cycle is the same as that obtained from the expansion in the first place; so the wheel comes to rest. Thus, under these conditions, no net work or mechanical energy can be obtained in the total process. However, if

Figure 9–18

Diagram of an idealized cyclic heat engine. (A) Heat enters the gas, causing it to expand against the piston, which pushes the wheel around. (B) Gas is recompressed at a low temperature. Heat flows out of the gas.

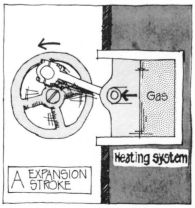

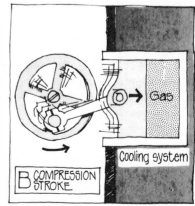

Figure 9–19
The steam engine
(reciprocating type).

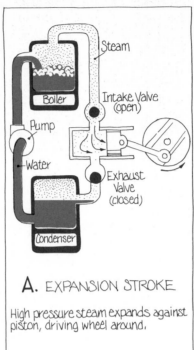

A. EXPANSION STROKE

High pressure steam expands against piston, driving wheel around.

B. EXHAUST STROKE

Steam, now at low pressure, is pushed out so that new steam can enter cylinder as in A. Expanded steam is cooled in the condenser and pumped back to boiler to be heated and compressed.

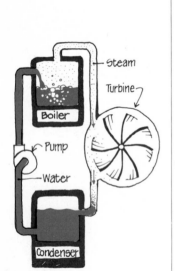

Figure 9–20
A steam turbine.

the recompression of the gas is carried out at a low temperature, the pressure of the gas will then be lower, Fig. 9–18(B). Therefore less work will be required to compress it and a net amount of mechanical energy will be obtained in each cycle.

In summary, if the gas expands and contracts at the same temperature, no mechanical energy can be obtained and no net work can be done. But if the gas is allowed to expand at a high temperature, and then is compressed at a low temperature, mechanical energy *is* obtained, and the wheel will continue to turn.

It is impractical, however, to construct an engine in which the gas is moved back and forth between two substances, one at a high temperature and the other at a low temperature. In most real engines the gas itself is continually exchanged through the use of valves. The practical case of a steam engine is illustrated in Figs. 9–19 and 9–20. There are two main types of steam engine, the "reciprocating piston" type shown in Fig. 9–19 and the steam *turbine*, which differs only in that the moving piston is replaced by a rotating turbine, Fig. 9–20. The working of an internal combustion

Figure 9-21

Four-cycle internal combustion engine: (A) The fuel-air mixture rushes into the cylinder; (B) the piston moving upward compresses the gas; (C) the spark plug fires, igniting the fuel-air mixture and raising it to a very high temperature; (D) the high-temperature, high-pressure gases expand against the piston; this is the power stroke; (E) the burned gases are pushed out to the exhaust pipe. Finally, the intake valve opens and the whole cycle repeats.

engine is shown in Fig. 9-21. In this case, the high temperature is obtained by the burning of the fuel-air mixture; the heat input comes from the molecular potential energy that is released in the burning process. At this high temperature the molecules have a great deal of kinetic energy and they force the piston down. The burnt gases are exhausted to the atmosphere during the exhaust stroke and a fresh fuel-air mixture is then brought in and compressed at a low temperature. Note that if the whole engine were at the same high temperature as the burning gases, no mechanical work would be done, as discussed above. Thus the engine must be cooled. The engine must be at a lower temperature during the compression of the gases than it is during the expansion of the gases.

In any heat engine, heat must flow out to the atmosphere; this is the heat released at the lower temperature, as shown in Fig. 9-17. In the internal combustion engine the heat is released in two ways: it is carried away in the exhaust by the burned gases, and it is carried away by the engine coolant, which then releases it to the atmosphere through the radiator. The schematic drawing of Fig.

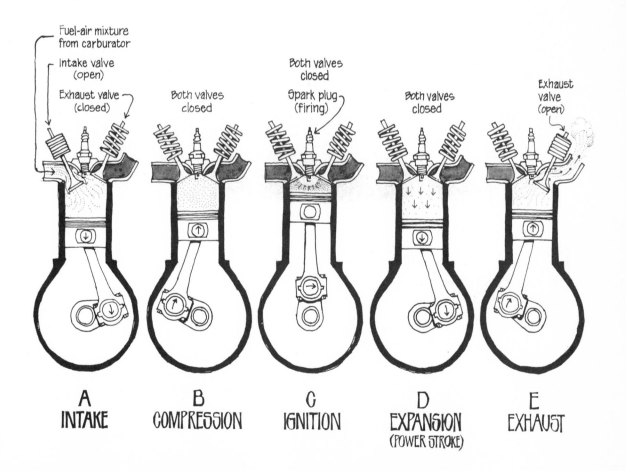

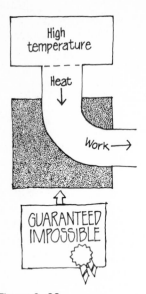

Figure 9–22

Schematic diagram of a hypothetical perfect heat engine in which all the heat intake is transformed into work. It is not possible to construct such a perfect heat engine.

9–17 applies to all heat engines; thermal energy is taken in at a high temperature, some of it is transformed into mechanical energy, and the rest is exhausted at the lower temperature.

You might wonder if it is possible to construct an engine so that all the heat intake at the high temperature is transformed into work, and none is exhausted at the lower temperature, as in Fig. 9–22. Carnot found that this was not possible. If it were possible, it would mean that a low-temperature substance would not be needed to absorb the exhaust heat. Unfortunately, nature insists that some heat be exhausted at a lower temperature, as we have seen. Carnot's discovery is an alternative way of stating the second law of thermodynamics: *it is impossible to construct an engine that transforms a given amount of thermal energy completely into mechanical energy, without any heat being exhausted in the process.* Another way to state this is to say that there is no perfect engine such as the one diagrammed in Fig. 9–22.

Efficiency of Engines

The efficiency of a heat engine is defined as the ratio of the work done to the heat input (see Fig. 9–17). That is, the more work an engine can do for a given amount of heat input, the greater its efficiency. Carnot showed that the maximum efficiency of any heat engine is related to the operating temperatures of the engine. The relation he derived was: maximum efficiency = (work output)/(heat input) = $1 - T_L/T_H$ where T_H is the Kelvin temperature of the high-temperature source of heat and T_L is the low temperature (Kelvin) of the exhaust heat. Clearly, the higher the input temperature and the lower the exhaust temperature, the greater the efficiency. For example, a steam engine at a power plant may take in steam at 400°C and exhaust it at 200°C; to find the maximum possible efficiency we must first change these to Kelvin: $T_H = 400 + 273 = 673$ K and $T_L = 200 + 273 = 473$ K. Then the efficiency is $1 - \frac{473\,K}{673\,K} = 1 - 0.70 = 0.30$; or, as a percentage (we multiply by 100), it is 30 percent efficient. The efficiency given by this formula is the *maximum* that is consistent with the laws of thermodynamics. Well-designed real engines reach 60 to 80 percent of this ideal value. Thus, the engine in our example would have an actual efficiency of perhaps 18 to 25 percent.

Refrigerators and Air Conditioners

The operating principles of refrigerators and air conditioners are just the opposite of those of heat engines. As shown in Fig. 9–23, by doing work W an amount of heat Q_L is taken from a low-temperature

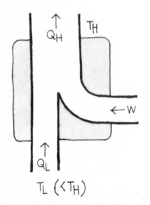

Figure 9–23

Schematic diagram of the operation of a refrigerator or air conditioner.

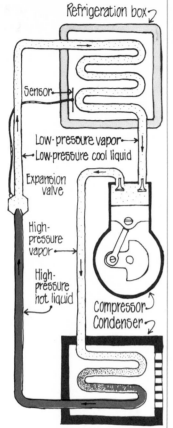

Refrigeration box

Sensor

Low-pressure vapor→
←Low-pressure cool liquid

Expansion valve

High-pressure vapor→

High-pressure hot liquid

Compressor

Condenser

Figure 9–24
Typical refrigeration system.

area, T_L (the inside of a refrigerator), and heat Q_H is exhausted at a high temperature, T_H (the room). You can feel this heat blowing out at the bottom of a refrigerator. The work W is usually done by an electric motor driving a compressor that compresses a fluid, as illustrated in Fig. 9–24. A perfect refrigerator—one in which no work is required to take heat from the low-temperature region to the high-temperature region—is not possible. This was implicit in our first statement of the second law of thermodynamics: that heat does not flow spontaneously from a cold object to a hot object. To accomplish such a task, work must be done.

Thermal Pollution

The fact that all heat engines require a coolant at a low temperature to absorb the exhaust heat means that all heat engines create **thermal pollution**. All power plants at the present time, whether they burn fossil fuels or use nuclear energy, employ heat engines to transform the energy obtained from the fuel into usable energy. Most power plants use water from a nearby lake, stream, or sometimes from the ocean, as the coolant. This naturally causes the temperature of the water in the vicinity to rise, and that temperature rise reduces the amount of dissolved oxygen in the water. This may adversely affect fish and other organisms that normally exist there and encourage the growth of other (perhaps alien) organisms such as algae. The ecology of an area can thus be seriously disrupted.

Order to Disorder:
The Second Law of Thermodynamics

We have mentioned several different applications of the second law of thermodynamics. But we have still not found the general principle behind these examples. It was not until the end of the nineteenth century that a general statement of the **second law of thermodynamics** was finally arrived at. It can be stated as follows:

> Natural processes tend to proceed toward a state of greater disorder.

The statements of the second law given earlier were examples of this more general statement. Let us look at those earlier statements, and some additional ones as well, and see how they satisfy this general statement.

When a hot object is placed in contact with a cold object, the hot one cools and the cold one warms until both reach the same intermediate temperature. Before this process occurs, we can dis-

245

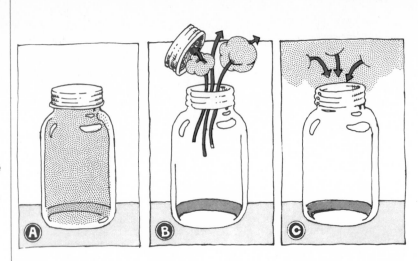

Figure 9–25

A gas compressed in a jar (A) will rush out in a disorderly fashion (B), but the inverse process of the gas running back into an open jar (C) is not observed to happen.

tinguish two classes of molecules: those that have a high average kinetic energy and those that have a low average kinetic energy. After the process has run to completion, when the two objects are at the same temperature, the molecules of both objects have the same average kinetic energy and we no longer have the more ordered arrangement of molecules into two classes. They are now all in one jumbled category. Order has gone to disorder; or, disorder has increased. We can see how order has gone to disorder in this example from another point of view. The separate hot and cold objects could, in principle, serve as the high- and low-temperature regions for a heat engine and thus useful mechanical energy could be obtained from them. But once they are mixed and come to the same temperature they are no longer capable of running a heat engine. This is another indication that order has gone to disorder.

When a falling rock strikes the ground, its kinetic energy is transformed into thermal energy. So long as the rock had kinetic energy, it was in a relatively ordered state. The molecules were all moving in the same direction, downward, and the rock was capable of doing some work—for example, driving a stake into the ground. After the rock hits the ground, its energy goes into increasing the random disordered motion of molecules and the energy is no longer available to do work. Again, disorder increases. Both examples suggest that heat or thermal energy can be considered "disordered" energy, not only because the molecules are moving in random disordered motion but because this energy cannot as readily be transformed into useful work.

Let us consider some other simple processes and see whether disorder increases, as the second law tells us it must. When you remove the lid of a bottle containing a gas, the gas molecules rush out into the room in a disorderly fashion [Fig. 9–25(B)]. Order goes to disorder. You would not expect the molecules to all run back into the bottle [Fig. 9–25(C)] and thus return to a more orderly state. Similarly, you would not expect all the molecules of air in your bedroom to suddenly order themselves into one small corner of the room. Processes in which disorder goes to order are simply not observed to happen. If you put a layer of salt and then a layer of pepper into a single jar and shake it, the salt and pepper become mixed together. No matter how hard and long you shake, the salt and pepper will not return to the more orderly state of two distinct layers.

Processes in which the net effect is an increase in order are not observed to occur in nature. That is what the second law of thermodynamics is all about.

Entropy: A Measure of Disorder

The quantitative measure of disorder is known as **entropy**. This new term enables us to state the second law of thermodynamics in yet another way:

The entropy of the universe continually increases.

Entropy Is Time's Arrow

The second law of thermodynamics is also related to the direction in which time flows. Suppose we made a motion picture of any of the naturally occurring processes we have just discussed. If we ran the film backward, we would certainly sense that it wasn't natural. When we saw a rock resting on the ground suddenly leap into a person's hand, or the mixed salt and pepper separate into two distinct layers, we would say that the film must be running backward because processes like those are not observed in real life. Yet the other laws of physics would not have been violated. It is the second law of thermodynamics that indicates in which direction natural processes will proceed. We have all seen a teacup drop to the floor and break. But who has seen the disorderly array of broken pieces spontaneously come back together and form that ordered object, a teacup? In a movie running backward you may have seen this happen, but you knew that time was being simulated to run backwards. Time only goes forward, of course, and the direction in which time naturally flows could be defined as that direction in which disorder increases. It has been said that entropy is "time's arrow."

The process of biological evolution is sometimes thought to be a violation of the second law of thermodynamics. But it isn't. It is certainly true that evolution has resulted in more and more complex and highly ordered organisms, the most ordered of which is *Homo sapiens*. Yet a careful examination of all the biochemical processes involved in evolution, or in just the metabolism of a single organism, shows that the waste products produced by biological organisms have such large amounts of disorder that the total disorder, organism plus waste products, increases in time. In most natural processes some matter becomes more ordered while other matter becomes less ordered; but the decrease in order of the one is always greater than the increase in the other, so the total amount of order decreases.

Heat Death

Philosophers often discuss one of the important predictions of the second law of thermodynamics, the so-called "heat death" of the universe. According to the second law, as natural processes continue, the order in the universe will continually decrease until eventually everything will reach a state of maximum disorder. Matter will approach a uniform mixture. Heat will have flowed from high-temperature regions to low-temperature regions until the whole universe is at one uniform temperature. All the energy in the universe will have degenerated to disordered heat energy. All change will have ceased. Thus will the heat death of the universe be reached.

This final state of the universe appears to be inevitable—although, to be sure, very far in the future. Yet it is based on the assumption that the universe is finite, an assumption that cosmologists are not really certain of. Furthermore, cosmologists are beginning to question whether the second law of thermodynamics, as we know it, actually applies throughout the vast reaches of the universe. Some surprises may be in store for us.

REVIEW QUESTIONS

1. The energy of molecular motion is manifested as
   (a) potential energy   (c) chemical energy
   (b) thermal energy   (d) friction

2. Temperature is a measure of the
   (a) total kinetic energy of molecules
   (b) average kinetic energy of molecules
   (c) potential energy of molecules
   (d) amount of the heated substance

3. Which has the most thermal energy?
   (a) an ice cube 1 cubic inch in volume
   (b) a tea kettle of 1 quart volume full of boiling water
   (c) a bathtub filled with warm bath water (20 gallons)

4. Heat is
   (a) a fluid called caloric
   (b) the average kinetic energy of molecules
   (c) a transfer of energy because of a difference in temperature
   (d) a nonmaterial substance

5. Heat is most closely related to
   (a) temperature   (c) momentum
   (b) energy   (d) force

6. A casserole and a paper clip have been in a 325°F oven. If you take them out with your bare hands the casserole will burn you more severely than the paper clip because
   (a) the casserole has a higher temperature
   (b) the casserole can transfer more heat
   (c) the paper clip is metal
   (d) of conservation of energy

7. We obtain energy from the sun by
   (a) conduction   (c) radiation
   (b) convection   (d) long-distance molecular travel

8. The direction in which heat is conducted between two bodies depends on
   (a) their temperatures
   (b) their pressures
   (c) their thermal energies
   (d) whether they are solid, liquid, or gas

9. A heated part of a fluid moves by natural convection because
   (a) of molecular collisions
   (b) its density is less than that of the surrounding fluid
   (c) its molecular motions become parallel
   (d) its total thermal energy is greater than that of the surrounding fluid

10. Thermal radiation is emitted
    (a) only by hot bodies such as the sun
    (b) only by bodies whose temperature is greater than that of their surroundings
    (c) only by bodies that have greater thermal energies than their surroundings
    (d) by all bodies

11. When 1 kg of ice at 0°C absorbs 80 kilocalories of heat, the ice undergoes
    (a) a change of state   (c) a rise in temperature
    (b) a loss of energy   (d) an increase in volume

249

12. A drop of alcohol feels cool on your skin because heat is
    (a) absorbed from your skin
    (b) removed from the alcohol
    (c) transformed into temperature
    (d) none of these

13. According to the second law of thermodynamics, which of the following is possible?
    (a) a 100-percent efficient engine
    (b) a rock rising in the air as heat energy turns into kinetic energy
    (c) a mixture of salt and pepper separating into separate layers when shaken
    (d) none of the above

14. A heat engine is a device
    (a) to heat houses
    (b) to change heat into mechanical work
    (c) to change heat energy completely into work without any wasted heat being produced
    (d) to avoid thermal pollution

15. The second law of thermodynamics has to do with
    (a) temperature
    (b) order going to disorder
    (c) heat as energy
    (d) thermal energy

EXERCISES

1. What happens to the work done when a jar of orange juice is shaken vigorously?
2. What does the expression "heat flow" mean?
3. From the point of view of kinetic theory, describe how a hot oven heats food.
4. Why would you expect the water temperature at the bottom of a waterfall to be slightly higher than at the top?
5. Down parkas and sleeping bags are often specified as so many centimeters or inches of "loft," which refers to the actual thickness of the garment when fluffed up. Why is this term used?
6. Why does a down parka become useless if it gets soaked?
7. Why are air temperature readings always taken with the thermometer in the shade?
8. Why do Bedouins wear several layers of clothes in the desert even when the temperature reaches 50°C (122°F)?
9. Explain why cold drafts are so uncomfortable.
10. Why does a tile floor seem so much colder to your bare feet than a carpet at the same temperature?
11. Sea breezes often occur at the shore of a large body of water. Explain, using the fact that the temperature of the land rises more rapidly than that of the water as a result of the sun's radiation.
12. A premature baby in an incubator can be dangerously cooled even when the air inside is warm. How can this be?

13. A house can be kept cooler on a hot sunny day by closing the windows and curtains. Explain.

14. Heat loss occurs through windows (a) by drafts through cracks around the edges, (b) through the frame, particularly if it is metal, (c) through the glass panes. For each of these, which mechanisms are involved: conduction, convection, and/or radiation? Which of these will heavy curtains reduce?

15. Why does the earth cool off at night more quickly when the sky is clear than when it is cloudy?

16. The amount of heat needed to heat a room whose windows face north is much greater than is needed to heat a room whose windows face south. Explain.

17. Can convection occur in solids, or only in liquids and gases?

18. Why does dirty snow melt faster than clean snow?

19. A thermos bottle consists of a bottle with two glass walls with shiny silvered surfaces separated by a vacuum. Explain how this construction reduces conduction, convection, and radiation.

20. If you hold a Kleenex tissue over a hot plate, why does the tissue wave up and down? (Try it!)

21. Is it possible to heat a material and not change its temperature? If so, give an example.

22. Why does water condense on the inside of windows on a cold day?

23. Why do you feel cold when you step out of a swimming pool into a warm breeze?

24. You can determine the wind direction by wetting your finger and holding it up in the air. Explain.

25. When your body overheats, it perspires. What is the value of this?

26. Describe briefly what happens at the atomic level as alcohol at room temperature is heated to its boiling point of 78°C.

27. Describe briefly what happens at the atomic level as a block of silver melts at 960°C.

28. Why does water in a canteen stay cool if the cloth jacket surrounding the canteen is kept moist?

29. Alcohol evaporates more quickly than water at room temperature. What can you infer about the molecular properties of alcohol relative to those of water?

30. What happens to the thermal energy of water vapor that condenses on the outside of a cold glass of water? Is work done or heat exchanged?

31. Is it possible to cool down a room by leaving the refrigerator door open? Explain.

32. The oceans in the tropics have a temperature at the surface of about 25°C, whereas deep below the surface the temperature is about 5°C. The possibility of building heat engines to generate electricity using this temperature difference is being considered. What would be the maximum efficiency of such an engine? Why might such an engine be

practical in spite of its low efficiency? What adverse environmental effects might occur?

33. A refrigerator takes heat from its cold interior and transfers it to the warmer exterior. Why is this not a violation of the second law of thermodynamics?

34. Give three examples from everyday life of the degradation of usable energy into heat.

35. Give several examples, other than those in the text, of naturally occurring processes in which order goes to disorder.

10

WAVE
MOTION
AND
SOUND

1. WAVES

When you throw a pebble into a lake, circular **waves** form and move outward from the point of disturbance, Fig. 10–1. A rope, a flexible rubber tube, or a Slinky stretched out straight on a table will exhibit wave motion when one end is moved back and forth in a vibratory (or "oscillatory") fashion, Fig. 10–2. Water waves and waves traveling along a rope are two common examples of wave motion. There are many other examples as well—sound is a wave motion, and in Chapters 11 and 12 we will see that light behaves like a wave, as do radio and TV signals.

Wave Motion

Have you ever stood on a cliff and watched ocean waves move toward the shore? Or have you ever watched the motion of waves formed by a fast-moving boat? Clearly the waves move, and they move with a recognizable velocity. But does the water itself move toward the shore? It might seem so,[1] but in fact the water does not move in with the waves. The water remains essentially in place, merely moving up and down as the waves move past. This can be demonstrated by casting a rock into a still pond on which a few leaves are floating. The leaves are not carried forward by the spreading waves. They merely move up and down, remaining in the same place, because the water itself is only moving up and down. A wave set up on the rope in Fig. 10–2 moves to the right; yet each

Figure 10–1

A rock thrown into a lake gives rise to circular waves, which move outward from the center.

[1]Do not be confused by the breaking of ocean waves. When an ocean wave breaks it has interacted with the ground and is no longer a simple wave.

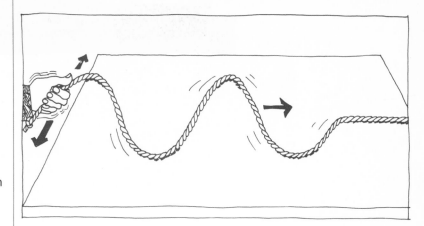

Figure 10–2

A hand moves back and forth, giving rise to waves on the rope that move to the right.

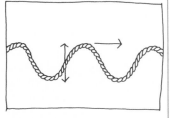

Figure 10–3

Each point on the rope moves up and down, but the wave shape moves to the right.

point of the rope moves in a very limited way, up and down, staying more or less in place, Fig. 10–3. This is a general property of waves: waves can move over long distances, but the matter, or the "medium" (the water or the rope), itself moves only slightly. It is the wave *shape*, not the matter, that travels large distances.

Waves Transport Energy

Although waves do not transport matter, they do transport energy. The energy that is transferred to water by a thrown rock or by wind far out at sea is transported to the shore by waves. The energy carried by a water wave is clearly evident when it breaks against the shore.

EXPERIMENT

Fill a tub with water. Float a cork at each end. Wait until the water is smooth and then move *one* of the corks up and down. Does this produce traveling waves? Does the second cork respond? Is work done on the second cork? If so, how was the energy transferred?

All forms of wave motion transport energy. The hand that vibrates the rope in Fig. 10–2 transmits energy to the rope, and the resulting waves carry that energy down the length of the rope. If a bird happened to be sitting on the rope, it would be jostled by the passing wave.

We thus conclude that energy can be transported from one place to another by either of two methods: by means of material objects or particles and by means of waves. Let us clearly distinguish the two.

First, let's review what we know about material particles as carriers of energy. When a boy throws a baseball he does work on the ball, giving it kinetic energy. The baseball flies through the air, strikes a window, and breaks it. The ball gives up all or part of its kinetic energy to the window. The energy was transported by the ball, that is, by matter, by a material object. Another example: consider water flowing down a river. As the water molecules lose potential energy in their downward flow, they gain kinetic energy. If the water falls over a dam, the kinetic energy gained by the water molecules can be put to useful work turning a turbine that produces electrical energy. The flight of a baseball, a bullet, or a rocket, or the flow of water—whether in a river, over a dam, or in ocean currents—constitutes a transport of energy from one point to another by means of material particles.

When energy is transported by waves, however, it travels from one place to another without the transfer of matter. In our example of waves on a rope, energy is being carried in a particular direction (Fig. 10–3) even though the material particles of the rope are not moving in that direction.

Wave Pulses and Continuous Waves

How are waves formed and how do they travel? What makes them move? To answer these questions we must look into the nature of waves. We begin by examining a single wave **pulse**. A single pulse can be formed on a rope, as shown in Fig. 10–4, by a quick up-and-down motion of the hand. The hand has snapped up one end of the rope and, because the end piece of rope is connected to adjacent pieces, the adjacent pieces also feel an upward force and they too begin to rise. As each succeeding piece of rope moves upward, the wave "front" moves outward along the rope. Now the hand returns it to its normal position, Fig. 10–4(C). Then, as each succeeding piece of rope reaches its peak, it too is pulled back down; thus each succeeding piece of rope returns to its normal position.

In summary, a disturbance (the movement of the hand) gives rise to a wave pulse, and the cohesive forces between adjacent pieces of the rope cause the wave pulse to travel outward. Wave pulses in other media are likewise created by disturbances and travel outward by a similar process.

A **continuous wave**, such as the one shown in Fig. 10–3, is

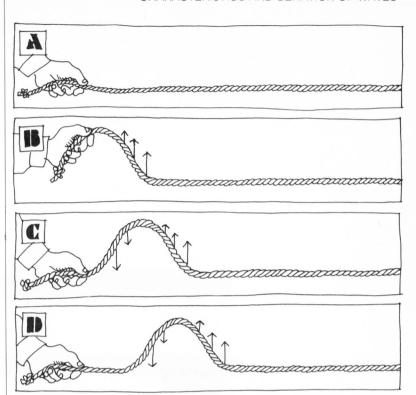

Figure 10–4

How a wave pulse travels along a rope (see text). The arrows indicate the speed of each tiny piece of rope.

created if the disturbance is continuous and oscillatory. For example, the continuous wave of Fig. 10–2 was produced by the sustained up and down motion of the hand. Continuous waves in general are produced by the **vibrational motion** of some object. For example, water waves may be produced by any vibrating object at the surface of the water, such as a moving hand, or by the up-and-down vibration of the water itself when the wind blows against it or when a rock strikes it. Sound waves in air, as we will see shortly, are produced by vibrating strings, membranes, or columns of air; and light waves are produced by vibrating charged particles.

2. CHARACTERISTICS AND BEHAVIOR OF WAVES

Properties of Waves

Some of the important terms used to describe a simple continuous wave are given in Fig. 10–5. The highest points of the wave are called peaks or crests. The lowest points are called troughs. The **amplitude** of the wave is the distance from the midpoint of the wave to the crest, or from the midpoint to the trough. The distance between successive

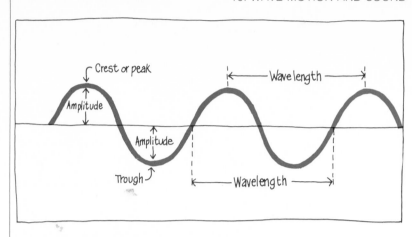

Figure 10–5

Terms used to describe a wave.

crests is called the **wavelength** of the wave. The distance between two successive troughs is also one wavelength, as is the distance between any two successive identical points on the wave. For example, if the distance between two successive crests of an ocean wave is 30 meters, then the wavelength is 30 meters. The **frequency** of a wave is the number of crests that pass a given point per unit time interval. For example, if three wave crests pass a small rock in a pond every second, the frequency of the wave is three wave crests per second, or in more common language, three "cycles per second." By a "cycle" we mean one up-and-down movement. If ocean waves pass a pier every 10 seconds, the frequency is 1/10 of a cycle per second. Since a wave is caused by a vibrating source of some kind, the frequency of any wave is equal to the frequency of the source that produces it. The frequency of the source is simply the number of up-and-down oscillations it makes per second.

Wave velocity is defined as the velocity at which the wave crest or wave shape appears to move. The velocity is related to the wavelength and the frequency:

$$wave\ velocity = wavelength \times frequency.$$

An example will show why this is true. Suppose a wave has a wavelength of 5 meters and a frequency of 3 cycles per second. This means that three wave crests pass by a given point every second, and those crests are 5 meters apart. Thus, 15 meters of wave must pass a given point each second; or, at the end of one second, the crest that passed the point at the beginning of that second will be 15 meters away. Therefore the wave must be moving with a velocity of 15 meters per second.

EXPERIMENT–PROJECT

Throw a ball into a still pond or swimming pool and observe the waves created. Note the following:

1. Is one crest formed? or many? Is the wave continuous, a single pulse, or something in between?
2. What caused the wave? Was it a vibration? of what? To answer this you must observe very carefully.
3. Estimate the wavelength, the frequency, and the velocity of the wave produced. Does the velocity equal frequency times wavelength? To estimate the frequency, count the number of waves that pass a stick or a small rock within a particular time interval as measured on your watch.
4. By using different-sized rocks or by throwing a rock into the water in different places, or by some other means, can you change the amplitude, wavelength, or velocity of the resulting waves?
5. What happens when the wave strikes a barrier, such as a rock or the shore?

Two Kinds of Waves: Transverse and Longitudinal

In a water wave or a wave on a rope, the oscillatory motion of the water or rope particles is mainly *perpendicular* to the motion of the wave itself (see Fig. 10–3). Such waves are called **transverse waves**. In another kind of wave motion the particles of the medium oscillate along the same direction as that in which the wave itself is moving. Such waves are called **longitudinal waves**. The nature of a longitudinal wave can be illustrated with a stretched spring. If one end of the spring is alternately compressed and stretched, as shown in Fig. 10–6(A), a series of compressions and expansions moves along the spring. The compressions are those regions in which the coils of the spring are momentarily close together, and the expansions, or "rarefactions," are those regions in which the coils are relatively far apart. A spring can also be made to carry a transverse wave, as shown in Fig. 10–6(B).

Because the compressions and expansions travel along the spring with a distinct velocity, and because the individual coils of the spring move only slightly, longitudinal wave motion closely resembles transverse wave motion. Wavelength, frequency, and velocity have meaning for a longitudinal wave just as they do for a

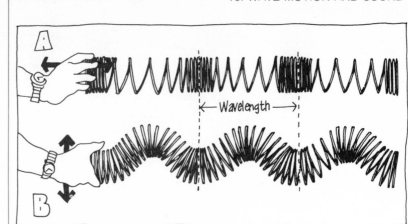

Figure 10–6

(A) A longitudinal wave and (B) a transverse wave, both produced on a coil spring. Notice movement of hand to produce each kind of wave.

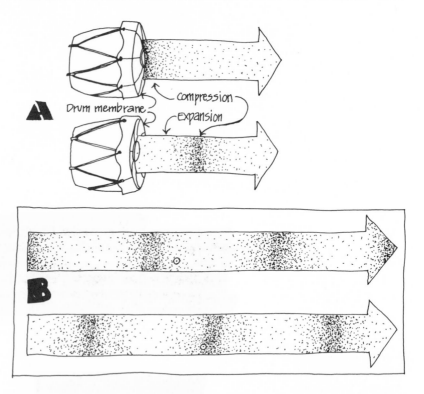

Figure 10–7

(A) When a drum membrane is struck, it vibrates back and forth, alternately compressing and expanding the air. (B) The resulting sound wave in air at two different moments, showing the density of air molecules as the wave moves. Each air molecule (e.g., the one circled) moves only slightly, although the wave pattern moves large distances. It is the wave pattern that travels to your ear, not the molecules themselves.

transverse wave. The wavelength of a longitudinal wave is the distance between successive compressions or between successive expansions; the frequency is the number of compressions that pass a given point per second; and the wave velocity, defined as the velocity at which each compression or expansion appears to move, is equal to the product of the frequency and wavelength.

The most common example of a longitudinal wave is a sound wave traveling through air. A vibrating drum membrane, for example, alternately compresses and rarefies the air adjacent to it, Fig. 10–7. This produces a longitudinal wave traveling outward in the air. In compressions, the air molecules are closer together than normal, and hence the density is higher. In expansions, the air molecules are farther apart than normal and the density is lower.

A graphical representation of a longitudinal wave can be made by graphing the density of air molecules versus position, Fig. 10–8. Notice that the graphical representation looks just like a transverse wave.

Waves Reflect Off Obstacles

When a wave strikes an obstacle, or comes to the end of the medium in which it is traveling, part of the wave is **reflected**. Water waves striking a rock are reflected, and an echo is the result of a sound wave reflecting off a mountain or a building. A wave produced on a rope is reflected when it reaches the end of the rope. Figure 10–9 shows how the amplitude of a wave changes as the wave is reflected. Notice that if the end of the rope is fixed, the reflected pulse is upside down; if the end of the rope is free, the reflected pulse is upright.

Reflected waves tend to have a smaller amplitude than the incident wave. (By the **incident** wave we mean the wave before it strikes the reflecting obstacle.) This happens because part of the energy of the wave is transmitted to the obstacle or barrier. Waves also lose energy through friction; you may have noticed that the

Figure 10–8

Graphical representation of a longitudinal wave.

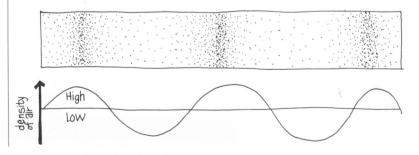

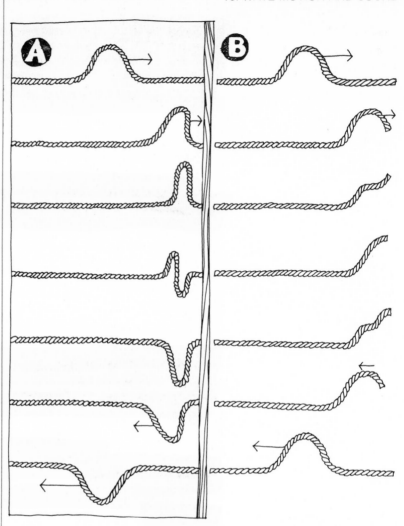

Figure 10–9

Reflection of a wave pulse traveling along a rope that is (A) fixed at the end, (B) free at the end.

amplitude of a wave pulse decreases as it travels down a Slinky or a rope.

Waves Diffract Around Obstacles

If you stand next to a building [Fig. 10–10(A)] and someone tries to throw a ball to you from around the corner of the building, the ball will not bend around the corner and reach you. But if the person shouts at you, you can hear him, Fig. 10–10(B). This is because waves, including sound waves, bend around obstacles. This bending of waves around corners is known as **diffraction**. Figure 10–11 shows water waves diffracting around an obstacle. When you did

262

Figure 10–10
Material objects or particles do not bend around corners (A), but waves do (B).

Figure 10–11
Diffraction (bending) of water waves around an obstacle.

263

the wave experiment a few pages back you may have observed that water waves diffract around an obstacle.

Diffraction is another illustration of the difference between energy carried by material particles and energy carried by waves. Particles do not bend around corners; waves do.

Waves Interfere with One Another

A related phenomenon that occurs only when energy is transported by waves is **interference**. Suppose two different waves of equal amplitude approach each other. They could be waves started from opposite ends of a rope, or they could be two sets of circular waves formed by throwing two rocks into a pond. When the two sets of waves meet, they are neither reflected nor absorbed; they pass right through each other. At those points where the waves overlap, the net amplitude of the combined wave will be the sum of the amplitudes of the two separate waves. If at a given point the crests of the two waves arrive at the same time, and the troughs arrive at the same time, then the combined wave is larger than either of the separate waves, Fig. 10–12(A). This is called **constructive interference**. On the other hand, if the crests of one wave arrive at the same time as the troughs of the other, the net amplitude is zero, as shown in Fig. 10–12(B). This is known as **destructive interference**. There is no wave motion at all at this point. When the two waves meet in a manner somewhere between these two extremes, as shown in Fig. 10–12(C), the result is called **partially destructive interference**. If

Figure 10–12

(A) Constructive interference,
(B) destructive interference,
(C) partially destructive interference.

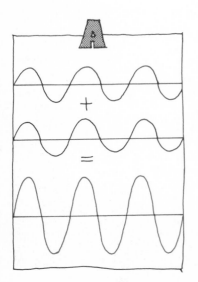
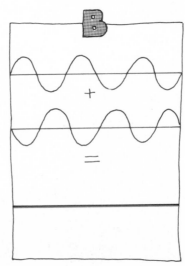
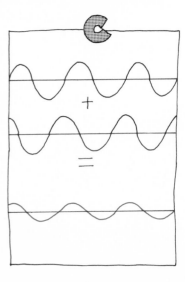

Figure 10–13
Interference of water waves.

the two waves have different amplitudes and/or different wavelengths, again partially destructive interference occurs.

You can demonstrate interference between water waves by throwing two rocks into a still body of water at exactly the same time, as shown in Fig. 10–13. The pattern of the two waves crossing each other is called an "interference pattern." Other types of waves, such as sound waves, also produce interference patterns.

EXPERIMENT–PROJECT

Simultaneously throw two rocks or tennis balls into a still pond a few feet apart. Locate regions of constructive and of destructive interference. The regions of constructive interference are those where the water moves up and down with the greatest amplitude, and the regions of destructive interference are those where the water remains almost still. These regions are easier to see if you float a few leaves on the water's surface. Diagram the waves produced, indicating the regions of constructive and destructive interference. Repeat the experiment with either a larger or a smaller distance between the rocks or tennis balls. Again diagram the waves and note differences in the interference pattern. You may have to perform this experiment several times before you are able to discern the regions of constructive and destructive interference.

Once again, the fact that interference and diffraction occur only for energy carried by waves and not for energy carried by material particles is an important distinction between these two types of energy transport. This distinction will be useful later when we discuss the nature of light and the properties of atoms.

Standing Waves

If you shake one end of a rope up and down, keeping the other end fixed, you produce waves that travel down to the other end and are reflected back. As you continue to vibrate the rope, waves traveling in both directions develop; quite a jumble results as the waves traveling in opposite directions interfere with each other. However, if you shake the rope with exactly the right frequency, the regions of constructive and of destructive interference will occur at fixed positions, and a **standing wave** will be produced, Fig. 10–14. This wave is called a "standing" wave because it doesn't seem to move along the rope. The points of complete destructive interference are called **nodes**, and the points of maximum constructive interference are called **antinodes**. The lowest frequency of vibration that will give rise to a standing wave produces the pattern shown in Fig. 10–14(A). Standing waves will also be produced at twice this frequency

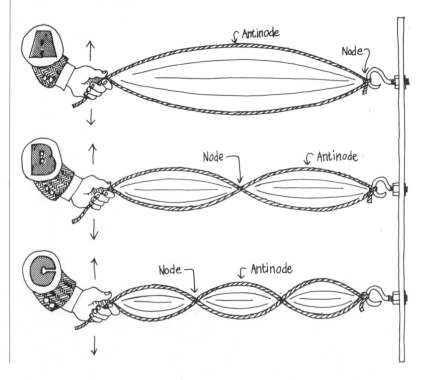

Figure 10–14
Standing waves produced by shaking a rope. The frequency at which the hand shakes the rope is twice as large for (B) as it is for (A), and three times as large in (C) as in (A). The points on the rope that do not move at all are called *nodes;* the points that vibrate up and down with greatest amplitude are called *antinodes*. Note that the higher the frequency, the smaller the wavelength.

[Fig. 10–14(B)], three times this frequency [Fig. 10–14(C)], and at other whole-number multiples of the lowest frequency. But standing waves will *not* be produced at other (intermediate) frequencies. The particular frequencies at which standing waves are produced are known as **resonant frequencies**.

An interesting feature of standing waves is that we do not see the waves traveling back and forth along the rope. The rope merely seems to vibrate up and down. When the rope is vibrated at a resonant frequency, it takes very little effort to obtain a large amplitude. When the rope is vibrated at a frequency in between the resonant frequencies, traveling waves can be recognized, but they are quite jumbled since the interference pattern is not stationary; furthermore, the amplitude is normally quite small.

Suppose now that a string stretched between two supports [as in Fig. 10–15(A)] is plucked like a guitar string; waves of a great variety of frequencies will begin traveling along the string, will be reflected at the ends, and will travel back in the opposite direction. Most of these waves will destructively interfere with each other and will quickly die away. But those waves whose frequencies correspond to standing waves will persist; the ends of the string, since they are fixed, will be nodes of these standing waves. Thus the string vibrates up and down with a frequency corresponding to any one of the standing wave patterns shown in Fig. 10–15(B), or to a combination of them.

Standing waves like those shown in Fig. 10–15 develop when a guitar string is plucked, a violin or cello string is bowed, or a piano string is struck. Standing waves are the basis of operation for all musical instruments, as we shall see later in this chapter. In pipe organs and woodwinds, for example, it is the standing waves of a vibrating air column; and for drums it is the standing waves on the vibrating membrane.

Standing Waves and Resonances

Strings are not the only objects that vibrate at resonant frequencies. All material objects possess natural resonant frequencies and will vibrate at those frequencies when disturbed. Hold one end of a plastic ruler on the edge of a table and leave the other end free to vibrate; when you strike the free end of the ruler, it will vibrate at a certain frequency or combination of frequencies, each of which is a resonant frequency. A window, a drinking goblet, a board, a pail of water, and an air column each possess one or more resonant frequencies.

As we saw in Fig. 10–14, when a rope is vibrated at one of its natural resonant frequencies, standing waves are set up. Very little effort is required to obtain a large amplitude of vibration at a res-

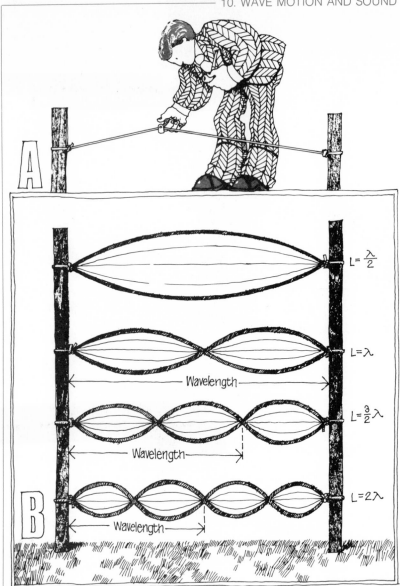

Figure 10–15

(A) A string is plucked.
(B) Only those standing waves that have nodes at the ends persist, and these have specific wavelengths (λ) related to the length (L) of the string.

onant frequency. This phenomenon is known as **resonance**. Resonance occurs when a large amplitude of vibration is induced in an object by applying an external vibration whose frequency equals a resonant frequency of the object. It is only at a resonant frequency that an external vibration can produce a large amplitude in a vibrating object. It is said that the great tenor Enrico Caruso could break a water goblet by singing a note of just the right frequency at full voice. This is an example of resonance. When the sound waves sent out into the air by the voice are of the same frequency as a resonant frequency of the goblet, standing waves are set up in the goblet. If the sound wave is strong enough, the goblet will vibrate, or "resonate," with such a large amplitude that the glass will break.

Figure 10–16
Resonance experiment.

EXPERIMENT

Pour water into two identical glasses until they are about a third full, Fig. 10–16. Tap each glass with a pencil and listen to the tone. If necessary, add a little water to one or the other of the glasses until the tones are the same. Then, with the glasses placed about 10 cm apart, tap one of them and place your ear near the second; place your hand on the first to stop its vibrations. Can you hear the second glass vibrating even though you haven't struck it? Do you get the same effect when one of the glasses contains much more water than the other? Why or why not?

Another example of resonance, although one that does not obviously involve standing waves, is what happens when you push a child on a swing. A swing, like any pendulum, swings back and forth with a particular frequency, which is its resonant frequency. If you exert regular pushes at the same frequency as the natural resonant frequency of the swing, the swing goes higher and higher. But if you push on the swing at random intervals, it just bounces around and reaches no great amplitude.

3. SOUND

The phenomenon of sound is associated with our sense of hearing and therefore with the physiology of our ears and with the workings of our brain, which interprets the sensations that reach our ears. Two aspects of any sound, especially a musical sound, are particularly apparent to humans: loudness and pitch. Both of these subjective sensations are related to certain measurable physical properties of sound waves.

Pitch Corresponds to Frequency

Pitch refers to whether a sound is low, like the sound of a string bass, or high, like a high note on a violin or a piccolo. Sound, as we have seen, travels in the form of a longitudinal wave; and *it is* *the frequency of the wave that determines its pitch.* A sound wave whose frequency is low sounds low in pitch, whereas a wave of high frequency sounds high in pitch. It was Galileo who first clearly noted this connection between frequency and pitch.

EXPERIMENT

Hold a card against the spokes of a bicycle wheel and spin the wheel slowly; then increase the speed. What do you notice?

OR

Play a 33 or 45 rpm record at the wrong speed on a phonograph. What happens to the sound? Explain.

Our ears are sensitive only to a certain range of sound wave frequencies. The lowest sounds we can hear have a frequency of about 20 cycles per second and the highest about 20,000 cycles per second, although these limits vary among individuals. The range of frequencies from 20 to 20,000 cycles per second is called the **audible range** for humans. Sound waves outside this range, even when they impinge on our ears with a large amplitude, make no impression on us; we just don't hear them. Dogs, however, can hear sound frequencies as high as 50,000 cycles per second, and bats can hear frequencies as high as 100,000 cycles per second. Sounds whose frequencies are above 20,000 cycles per second, that is, beyond the range of human hearing, are said to be **ultrasonic**.

Loudness Is Related to Amplitude

Loudness, like pitch, is a sensation in the human consciousness. It is therefore a subjective quality, but it, too, is related (though not in a simple mathematical way) to a physically measurable quantity, *intensity*. The **intensity** of a sound refers to the amount of energy the sound wave is carrying and is numerically proportional to the square of the amplitude of the wave. The greater the intensity of a sound, the louder it seems. Thus loudness, as well as intensity, increases with increasing amplitude. The greater the amplitude of a given sound wave, the louder it sounds.

The intensity of a sound is measured by a rather odd unit called a **decibel**. The approximate intensity of various noises measured in decibels (db) is indicated in Table 10–1. The decibel scale is not a linear scale; instead, it is a compressed, or multiplicative, scale known as "logarithmic." The zero point on this scale does not cor-

**Table 10-1
Intensity of Various Sounds**

SOURCE OF THE SOUND	INTENSITY (db)
Jet plane at 100 feet	140
Threshold of pain	120
Loud indoor rock concert	120
Siren at 100 feet	100
Auto interior, moving at 60 mph	75
Busy street traffic	70
Ordinary conversation	65
Quiet radio	40
Whisper	20
Rustle of leaves	10
Threshold of hearing	0

respond to zero amplitude. Rather, the zero point is arbitrarily chosen to be at the threshold of human hearing—at the lowest-intensity sound that the average human ear can detect. On this scale any increase in noise level by 10 db means that the intensity has increased tenfold. For example, a 40 db sound is 10 times more intense than a 30 db sound. An increase by 20 db means the intensity has increased by a factor of $10 \times 10 = 100$. Thus a 60 db noise is not twice as intense as a 30 db noise; it is $10 \times 10 \times 10 = 1000$ times as intense. This kind of scale is used because the range of intensities that a human ear can detect is enormous. A loud sound that just causes pain in the ear is 10^{12} times more intense than the softest sound the ear can detect.

Sound Waves Travel in Air and Other Materials

We are most familiar with sound as a longitudinal wave traveling through the air. The nature of these waves was illustrated in Figs. 10–7 and 10–8. But sound can travel through other materials as well. If you put your ear to the ground you can hear a passing truck. If you put your ear flat on a wooden table and tap the table lightly a few feet away, you will hear the sound. Some time when you are submerged in a swimming pool, have a friend strike two rocks together underwater; you will find that sound waves travel through water too. In fact, sound waves travel through any material: solid, liquid, or gas. In every case, it is the longitudinal oscillation of material particles about their normal positions that constitutes the traveling sound wave. Thus a sound cannot be heard in a vacuum because there are no molecules to oscillate; there is no medium in which the wave can travel. Robert Boyle first established this fact in the seventeenth century. He demonstrated that a ringing bell could not be heard when suspended inside a jar that was evacuated by a vacuum pump. Matter must be present if sound waves are to travel.

The Speed of Sound

The speed of a sound wave in air is about 330 m/sec, or 1100 ft/sec, at normal temperatures. The speed of sound is quite low compared with the speed of light. Consequently, a distant event is seen before it is heard. For example, if you have sat in the balcony of a concert hall, several hundred feet from the musicians, you may have noticed a time lag of a fraction of a second between the time a musician begins to play and the time you hear the sound. The time lag for more distant events is even greater. Thunder is often heard many seconds after the flash of lightning was seen. By measuring the time difference we can determine how far away the lightning flashed. Because the speed of light is so great, we see events on earth almost instantaneously. But sound has a speed of only 1100 ft/sec, and therefore it takes almost 5 seconds for sound to travel a mile—5280 feet. Thus, if thunder is heard 5 seconds after the lightning flash was seen, the lightning must have occurred about a mile away. If the thunder is heard 8 seconds after the flash, the lightning must have occurred about $1^3/_5$ miles away.

The speed of sound is different in different materials. In water, for example, it is about 5000 ft/sec, or four times its speed in air; in iron it is about 17,000 ft/sec.

The Source of Sound Is a Vibrating Object

Sound is a process by which energy is transferred from one place to another by means of waves. We can divide this process into three parts: the source of the sound energy, the transfer of energy away from the source in the form of sound waves, and finally the detection of the sound energy by an ear or an instrument such as a microphone. Let's look for a moment at the sources and detection of sound.

The source of a sound wave, like the source of any other wave, is a vibrating object. We saw in Fig. 10–7 how a drum membrane set into vibration by being struck gives rise to a longitudinal wave in the air. The moving membrane alternately compresses and rarefies the air that is close to it. These compressions and rarefactions travel outward as each set of air molecules exerts forces on adjacent ones. Similarly, when a hammer strikes a nail, both the hammer

272

and the nail vibrate momentarily as a result of the collision, and a brief noise or sound wave is produced.

Vibrations of a confined volume of air can also act as a source of sound. Wind blowing through a small opening between two tree twigs can set the air between the twigs into vibration and can produce a whistling sound. This is what we hear when we say that the wind is "whistling through the trees." The whistle we produce by blowing air through pursed lips is another example of vibrating air as a source of sound.

Many musical instruments make use of vibrating columns of air to produce sound, whereas others use the vibration of solids, such as strings or membranes. We will discuss this subject in the next section.

Detection of Sound: The Human Ear

The human ear is a remarkably sensitive detector of sound. Even the best microphones barely exceed the ear's ability to detect soft sounds.

A human ear is diagrammed in Fig. 10–17. Sound waves enter the ear and strike a diaphragm called the *eardrum*. The waves set the eardrum into motion, causing it to vibrate with the same frequency or frequencies as the sound waves themselves. Three small bones connected to the eardrum transmit the vibrations to a fluid in the inner ear, the *cochlea*. The motion of the fluid is detected by tiny hairs connected to nerves; at this point the vibrations are transformed into electrical signals and are carried by the nerves to the brain, where the sensation of sound is realized. Sound waves can also reach the inner ear by traveling directly through the bones of the skull, as when you tap your head.

Microphones work on essentially the same principle as the ear. Sound waves impinge on a diaphragm inside the microphone, setting it into vibration. The energy of the sound wave is then trans-

Figure 10–17

Diagram of a human ear.

273

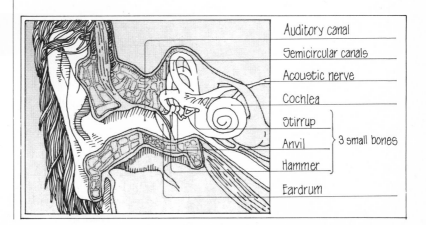

Auditory canal

Semicircular canals

Acoustic nerve

Cochlea

Stirrup ⎫
Anvil ⎬ 3 small bones
Hammer ⎭

Eardrum

formed into an electrical signal of the same frequency, which is amplified and sent to a recorder or to a loudspeaker. The details of this process will be discussed in more detail in Chapters 13 and 14.

4. MUSIC

Musical Instruments Are Simple Vibrating Sources

Musical instruments make use of different kinds of vibrating materials to produce sound. A vibrating membrane produces the sound of a drum; instruments such as xylophones and bells use vibrating pieces of metal. Such instruments as the violin, piano, and guitar make use of vibrating strings; and still others make use of vibrating columns of air. In each case, the source vibrates with a particular frequency or combination of frequencies that are its resonant frequencies. The player, by plucking, bowing, striking, or blowing, causes standing waves to be set up in the source.

We saw in Fig. 10–15 how standing waves of particular frequencies are established along a rope. This is the basis of operation for all stringed instruments. The pitch of a musical sound from a stringed instrument is normally determined by the simplest resonant frequency—when the string vibrates with nodes only at the ends [top diagram of Fig. 10–15(B), and Fig. 10–18]. When the string is shortened by fingering, Fig. 10–18(B), the frequency is higher, since the wavelength is shorter (wavelength × frequency = velocity, and the velocity doesn't change). Since the different strings on a violin or guitar are all the same length, the wavelength is the same for each unfingered string; and yet each string sounds at a different pitch. Why? Because each of the strings differs in size and weight, and the velocity of the waves generated by the heavier strings is less than the velocity of the waves generated by the lighter strings. This makes sense because the heavier strings offer more inertia to a traveling wave and we would therefore expect the wave to travel more slowly.[2] Therefore, since frequency = velocity/wavelength, the frequency or pitch of a heavy string will be lower than that of a light string of equal length.

The wide variety of musical notes that can be played on a violin or guitar thus arises from the different thicknesses of the strings and from the changing of the effective length of each string by fingering. On a piano no fingering is required, because a separate string is used for each note; piano strings vary in length as well as in thickness.

[2]Note that the velocity of the (standing) waves on a string is very different from the velocity of sound waves in air. However, the frequency of the vibrating string is the same as the frequency of the sound wave it produces.

Figure 10–18

The wavelength of a fingered string (B) is shorter than the wavelength of an unfingered string (A). Therefore the frequency of the fingered string is higher (frequency = velocity/wavelength). Only the simplest standing wave is shown.

If stringed instruments relied solely on vibrating strings to produce compressions and rarefactions in the adjacent air, the sounds they produced would not be very loud. That is why nearly all stringed instruments—violin, cello, guitar, piano, and so on—make use of a "sounding board" or a "sound chamber," Fig. 10–19. Because the strings are in contact with the sounding board, usually through a bridge, when the strings are set into vibration the sounding board vibrates as well. And because the area of the sounding board in contact with the air is much greater than the area of the strings themselves, a much stronger sound wave is produced. Thus the sounding board amplifies the sound. A sounding board is less important in an electric guitar, since the string vibrations are amplified electronically.

Figure 10–19

Stringed instruments have a sounding board or sound chamber to amplify the vibrations.

EXPERIMENT

Examine the inside of a piano. Note the different diameters and lengths of the strings. Do you discern a pattern? How do the strings come into contact with the sounding board? Why is it that for some notes the hammer strikes a single string and for others two or three identical strings?

The vocal cords of a human being are another example of a stringed instrument as a source of sound. Instead of strings with different lengths and sizes producing differently pitched sounds, we find only two "strings"—the two vocal chords. The wide range of pitch of which the human voice is capable is produced by tiny muscles that change the shape and the amount of stretch in the two vocal cords. The nose, throat, and mouth play the role of a sounding board or sound chamber.

Many musical instruments, including the woodwinds, the brasses, and the pipe organ, produce sound by means of standing waves excited in the air inside the tube or pipes of the instrument. When a stream of air is directed against one edge of the mouthpiece of a flute, for example, vibrations of the air within the tube are set up. In some woodwinds and in brasses, a vibrating reed or the vibrating lip of the player aids in setting up the vibrations in the air column. The air within the tube vibrates with a wide range of frequencies, but only those waves with a node at the closed end of a tube (where the air is not free to move) and an antinode at the open end (where the air is free to vibrate) will persist. Thus the air within the tube will vibrate in the form of longitudinal standing waves; these are shown graphically in Fig. 10–20 for the simplest resonant frequency in a tube open at both ends and in a tube open at only one end. At the nodes, the air does not move at all, whereas at the antinodes the air is alternately compressed and expanded.

The greater the length of a pipe or tube, the longer its wavelength will be, and therefore the lower its frequency will be. Hence the longest organ pipes produce the lowest-pitched sounds. In woodwind instruments, the length of the vibrating air column, and therefore the pitch of the sound, is changed by covering or uncovering the holes or "stops" along the length of the tube. If all the openings but the end one are covered, the lowest note will sound. The end of the vibrating air column usually occurs at the first uncovered opening, so that the highest note will be produced when every hole is uncovered; in this case the air column is shortest, the wavelength is shortest, and the frequency is highest.

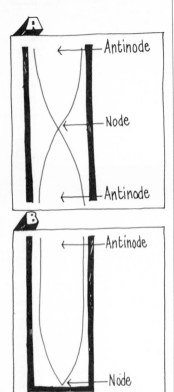

Figure 10–20

Diagram of the standing waves in (A) an organ pipe open at both ends, and (B) an organ pipe closed at one end.

EXPERIMENT

Blow across the top of an empty soft-drink bottle. Now add water to the bottle and blow again. Is the pitch different? Why?

The Quality of a Musical Tone Depends on the Mixture of Frequencies

Three characteristics of musical notes can be distinguished: loudness, pitch, and timbre, or quality. Loudness, as we have seen, corresponds to the amplitude of the sound wave; and pitch corresponds to frequency. Now let's discuss the third characteristic, timbre or quality.

When a note of a particular pitch, say middle C, is played on a piano and then played on an oboe, we notice a marked difference in the two sounds. This is what is meant by **quality**, or **timbre**. Clarinets have a characteristic quality different from pianos, which in turn is different from violins or drums. How does this difference in quality manifest itself physically?

Very few musical tones are pure—that is, consist of a single frequency. Most are a combination or a superposition of many different frequencies. This is because musical instruments possess many resonant frequencies and can vibrate at all those frequencies at once. We have already seen that the resonant frequencies of a string bear a simple relationship to one another. Figure 10–21 shows the various modes of vibration in which a stretched string can vibrate. Each mode corresponds to a different frequency. The lowest frequency is called the **fundamental** frequency. The other

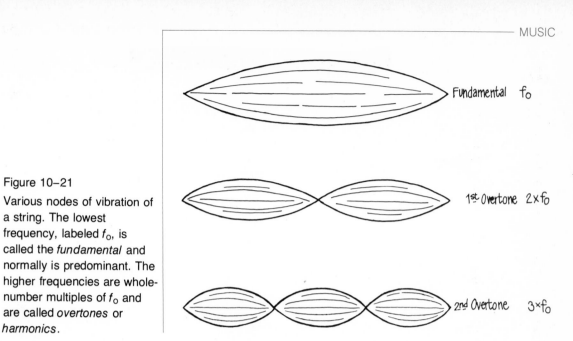

Fundamental f_o

1st Overtone $2 \times f_o$

2nd Overtone $3 \times f_o$

Figure 10–21

Various nodes of vibration of a string. The lowest frequency, labeled f_o, is called the *fundamental* and normally is predominant. The higher frequencies are whole-number multiples of f_o and are called *overtones* or *harmonics*.

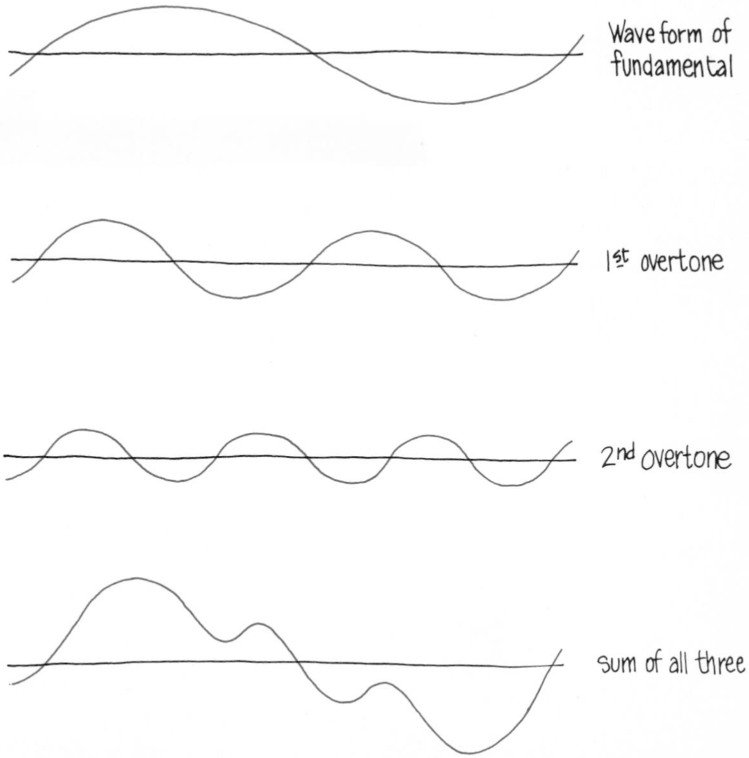

Wave form of fundamental

1st overtone

2nd overtone

sum of all three

Figure 10–22

Diagram showing how the fundamental plus the weaker-amplitude first and second overtones add up to give the composite sound wave.

modes are called **overtones**, or **harmonics**, and the frequency of each is a whole-number multiple of the fundamental frequency. When the string is plucked or stroked with a bow, the fundamental as well as many of the overtones are excited simultaneously. Figure 10–22 illustrates how the fundamental plus the first and second overtones combine to give a composite wave shape. Instruments that use an air column also produce overtones that are whole-number multiples of the fundamental (Fig. 10–20, however, showed the wave shape only for the fundamental).

> ### EXPERIMENT
> Hum into a bottle (Fig. 10–23); start at a low pitch and slowly raise the pitch. At what pitches does the bottle respond? Do you hear "resonance" at different pitches? Explain.

Figure 10–23

Humming into a bottle to detect resonant frequencies.

The overtones on a drum membrane, on the other hand, are not simple multiples of the fundamental; that is why a drum does not sound as harmonious as stringed or wind instruments.

Normally, the fundamental has a greater amplitude than any of the overtones. It is this predominant frequency, that of the fundamental, which we hear as **pitch**. The quality of a musical sound is governed by the mixture of overtones present. Different instruments emphasize different overtones, and this is what gives rise to their different quality. For example, a note played on a piano contains a large number of overtones, whereas a note played on a flute contains very few overtones; on a clarinet the lower overtones are not as strong as some of the higher ones.

The manner in which an instrument is played influences the mixture of overtones. For example, plucking a violin string results in a different sound quality than drawing a bow across it. Even the pitch of a musical note can be changed in some instruments by the manner in which the vibrations are produced. For example, the lips of a flutist can be shaped so that the air is blown more into the hole than directly across it. When this is done the fundamental is excited hardly at all. Thus the first overtone, whose frequency is twice that of the fundamental, predominates; and the pitch, in musician's terms, is an octave higher.

The precise physical shape of an instrument and the kind of material of which it is made influence the particular overtones it will produce. Often small changes in the construction of a musical instrument will have a marked effect on the quality of sound it produces. This is especially true of stringed instruments with sounding boards; not only does the sounding board amplify the sound, but because it has its own resonant frequencies it helps determine which overtones will be emphasized. A good violin maker carefully chooses and shapes the wood that goes into his instrument so that the quality of sound will be pleasing; even the age of the wood is important. To construct a violin with an exceptionally pleasing tone like a Stradivarius is an art.

Music Versus Noise

An ordinary noise, like the sound of a hammer striking a nail, has a definite quality but usually lacks any distinct pitch. Such noises are mixtures of a great variety of frequencies that bear little relation to one another. This is because a complex object like a nail or a hammer, unlike a simple stretched string or a narrow air column in a pipe, has a complicated set of resonant frequencies. No one frequency predominates, and those that are present are not simply related to one another as they are in a musical instrument.

Of course, the distinction between noise and music is not sharp. There exist sounds from a pure tone to a mixture of many random frequencies known as "white noise." Where to draw the line between music and noise is therefore subjective. Moreover, the distinction involves not only the quality of each tone but the succession of tones—that is, whether or not a melody is present.

Some contemporary composers deliberately incorporate into their music sounds normally considered to be noise. But avant-garde composers of past ages—including Beethoven and Brahms—were also accused by their contemporaries of writing "noise." Some people consider rock music, or music from other cultures, to be noise. One person's music is another person's noise. The distinction is subjective.

Noise Pollution

Today we are constantly subjected to noise: the neighbor running his power mower on a Sunday afternoon, or playing his stereo at 6:00 A.M. when you're fast asleep; the noise of garbage disposals and electric dishwashers, of loud parties and jet planes, of freeways and factories. Noise affects us in various ways, particularly psychologically. Sometimes it is merely annoying, but at other times noise can make us irritable or anxious.

Loud noise can also cause loss of hearing; this is a serious problem in factories and other industrial works where the noise persists at a high level for long periods. Rock musicians, too, suffer from hearing loss, for levels as high as 120 db are commonly produced.

Regardless of source, hearing loss caused by noise is particularly common in the frequency range from about 2,000 to 5,000 cycles per second, the very range in which speech and music usually occur.

Research on noise pollution is getting under way, but we still don't know a great deal about it. Standards for industrial noise have been set, with limits varying with the length of time one is subjected to a given noise level. For example, 85 db experienced over a period of 6 hours per day is regarded as a maximum by the federal government. But people subjected to this noise level for a period of 20 years suffer a hearing loss that is perhaps 20 percent greater than that experienced by the general population.

The problem of noise control is a difficult one. Isolating the source of noise by setting up physical barriers is sometimes attempted, but such methods are expensive and not always feasible. Attacking the source of the sound is often the best remedy. One effective method is to reduce the area of vibration of machinery, for the greater the area of vibration the more air that can be "pushed" and the louder the sound. Reducing the amplitude of vibration by encasing the machinery in a rigid housing, or coating the surface of the machinery with energy-absorbing material, also helps. Industrial engineers are careful not to place machinery on a floor, a wall, or another object that may tend to resonate in response to the machine's vibration, thereby increasing the amplitude. They also insist on careful maintenance, because poor lubrication, loose bolts and worn parts permit vibrations.

Fortunately, the human organism is amazingly adaptable. People who have grown used to the urban roar are sometimes uncomfortable in a truly quiet environment. In the design of office buildings, architects and engineers often introduce background noise—perhaps by making the heating and ventilation system somewhat noisy —so that workers will feel more comfortable. It is even said that although a number of quiet home appliances have been invented, people won't buy them. Apparently a vacuum cleaner has to make noise if it is to clean properly!

Musical Scales

Nearly all music is made up of notes or tones that have a definite pitch relationship to one another. A basic set of notes that are normally played is called a musical **scale**. Many different scales have been known throughout the ages and in different cultures. The simplest one in Western music is the **diatonic major scale**. This is the scale most of us learned in school as "do-re-mi-fa-sol-la-ti-do." Each of the notes on the diatonic scale is named by a letter from A to G and each corresponds to a definite frequency. The difference in pitch between each note on the diatonic scale is called a **whole interval** except for the intervals between "mi" and "fa" and between "ti" and "do," which have only about half the difference in pitch that the others have. These are called **half intervals**. The pitch of the first "do" can be chosen arbitrarily, but then the pitches of the subsequent notes conform to a regular pattern. The diatonic scale beginning with middle C as "do," taken to be 264 cycles per second, is shown in Table 10-2, along with the frequencies of each note and the intervals between them. This is the C major scale.

The interval from middle C to C above middle C (or C′) is called an **octave**, from the Latin word for "eight," since there are eight notes in that interval (counting both C's). Notice that C′ has twice the frequency of middle C. This is always the case as we go up the scale. Each C has twice the frequency of the preceding C, each A has twice the frequency of the preceding A, and so on. Thus the frequency of each note doubles with each octave.

If the diatonic major scale begins on D, let us say, the intervals between successive notes must maintain the same relationships of whole and half intervals. Some of the notes must therefore be raised or lowered a half interval, as shown in Table 10-3; when a note is raised by a half interval, it is said to be a "sharp" (♯) and if it is lowered by a half interval, it is a "flat" (♭). Sharps and flats are also necessary when one of the **minor** scales is used, in which the intervals between notes are slightly different than for the major scale.

Table 10-2
Diatonic C Major Scale*

NOTE	LETTER NAME	FRE-QUENCY (cycles/sec)	FRE-QUENCY RATIO	INTER-VAL
do	C	264		
re	D	297	9/8	whole
mi	E	330	10/9	whole
fa	F	352	16/15	half
sol	G	396	9/8	whole
la	A	440	10/9	whole
ti	B	495	9/8	whole
do′	C′	528	16/15	half

*Only one octave is included.

Table 10-3
Diatonic D Major Scale*

NOTE	LETTER NAME	FRE-QUENCY (cycles/sec)	FRE-QUENCY RATIO	INTER-VAL
do	D	297		
re	E	334	9/8	whole
mi	F♯	371	10/9	whole
fa	G	396	16/15	half
sol	A	445	9/8	whole
la	B	495	10/9	whole
ti	C♯	557	9/8	whole
do	D′	594	16/15	half

*Only one octave is included.

Table 10-4
The Equally Tempered Chromatic Scale*

NOTE	FREQUENCY (cycles/sec)	FREQUENCY RATIO	INTERVAL
C	262		
		1.06	half
C♯ or D♭	277		
		1.06	half
D	294		
		1.06	half
D♯ or E♭	311		
		1.06	half
E	330		
		1.06	half
F	349		
		1.06	half
F♯ or G♭	370		
		1.06	half
G	392		
		1.06	half
G♯ or A♭	415		
		1.06	half
A	440		
		1.06	half
A♯ or B♭	466		
		1.06	half
B	494		
		1.06	half
C'	524		

*Only one octave is included.

A scale that includes all the sharps and flats is called a "chromatic scale." Table 10–4 illustrates the "equally tempered chromatic scale," the most common scale in use today for Western music. On this scale C♯ and D♭, for example, are taken to have exactly the same frequency. Originally the frequencies of C♯ and D♭ were slightly different, and solo violin and other string players often play them that way still. The equally tempered scale has a great advantage for fixed-note instruments like the piano and flute, since the number of keys or stops necessary is reduced; that is, the same key is used to play both C♯ and D♭. If C♯ and D♭ and other similar pairs were different, there would have to be many more keys on a piano. The interval between each note and the next on the equally tempered chromatic scale is a half interval, and each succeeding note has a frequency that is about 1.06 times larger than that of the preceding note.

Many other musical scales have existed in past centuries and in other cultures. A great variety of scales existed during the Middle Ages in the West, and it was out of those scales that our present diatonic and chromatic scales evolved. In Africa and the Orient other scales are often used; one of them is the pentatonic scale, which consists of just five notes.

Tuning an Instrument

In order for an orchestra or a band to sound pleasing, all the instruments must play the same frequency for any given note. For example, if one instrument plays middle C at 262 cycles per second and a second instrument at 257 cycles per second, the two sound waves would interfere with each other and result in an unpleasant sound. Thus, all instruments must be tuned to the same frequencies. Since the relative values of the notes on a scale follow a fixed pattern, however, the frequency of only one note has to be chosen; the frequencies of the other notes then follow according to pattern. Usually the standard is taken to be A above middle C at 440 cycles per second.

Tuning woodwinds is accomplished by changing the length of the tube in order to alter the length of the air column; this changes the wavelength of the resonant modes and therefore the frequency as well. Stringed instruments must be tuned often. However, the length of the strings and therefore the wavelength of the resonant modes are not changed; instead, the velocity of the standing waves on the strings is changed by adjusting the tension in each string. An increase in tension increases the velocity of the waves and therefore raises the pitch. (Frequency is proportional to velocity when wavelength is unchanged, frequency = velocity/wavelength.) Reducing tension reduces the pitch. The membranes of a drum are tuned in a similar way by increasing or decreasing the tension.

The tuning of an instrument should perhaps be called "fine tuning," since only slight changes in frequency are accomplished. The "gross" tuning was done by the manufacturer of the instrument who chose the length and diameter of each string or the general length and position of holes on a woodwind.

5. THE DOPPLER EFFECT AND THE SONIC BOOM

The Doppler Effect: Pitch Changes If Source or Listener Is Moving

You have probably noticed that the pitch of a siren on a rapidly moving fire engine drops sharply as it passes you. When a speeding race car passes you, the sound of its engine changes pitch as well—aaaaaaahhr-rooooooomm. The pitch of a moving source of sound is higher than normal when it is approaching a listener and lower than normal when it is moving away. This is known as the Doppler effect.

EXPERIMENT

Find a small round whistle that will fit into a funnel, as shown in Fig. 10–24. Tie a string as shown and have someone rotate the funnel-whistle in a horizontal circle. Notice how the pitch changes as the whistle approaches and retreats from you. (Try to distinguish between loudness and pitch changes.)

What causes the Doppler effect? The siren of a fire engine at rest emits sound waves in all directions at a particular frequency, Fig. 10–25(A). When the fire engine is moving rapidly toward an observer [Fig. 10–25(B)], the sound waves ahead of it are closer together than normal. This is because as the fire engine moves forward, it is "catching up" with the previously emitted wave crests. To put it another way, since the fire engine is moving forward, the same number of waves must fit into a shorter distance. Hence the distance between subsequent crests is less. An observer standing on the sidewalk will detect more wave crests per second, and hence the frequency and the pitch will be higher. On the other hand, the waves emitted by the fire engine in the backward direction [Fig. 10–25(B)] will be farther apart than normal, since the fire engine is racing away from the wave crests and is consequently leaving more space between them. In the latter case, an observer will detect fewer wave crests per second and the frequency will be lower.

The Doppler effect also occurs when the listener moves toward or away from a stationary source of sound. If he moves toward the source, the wave crests will pass him at a faster rate, so the pitch

Figure 10–24

A funnel-whistle used to detect the Doppler effect.

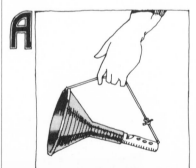

Figure 10–25

(A) Both observers on the sidewalk hear the same frequency from the fire engine at rest. (B) Doppler effect: the observer toward which the fire engine moves hears a higher-frequency sound, and the observer behind the fire engine hears a lower frequency.

is higher. If he moves away from the source, the wave crests take a little longer to catch up with him; therefore the frequency with which they pass him will be less, and the pitch will be lower.

Shock Waves and the Sonic Boom

When an airplane exceeds the speed of sound, it is said to have a **supersonic** speed.[3] Supersonic speeds are sometimes rated in "Mach numbers," named after the physicist-philosopher Ernst Mach. The Mach number is defined as the ratio of the speed of the object to the speed of sound in air. Thus "Mach 2" means the object is moving at twice the speed of sound, and "Mach 0.5" means it is moving at half the speed of sound.

We have seen that a moving source of sound gives rise to changes in pitch known as the Doppler effect. In a sense, the source is chasing the sound waves moving out in front of it. If the moving source exceeds the speed of sound, far more dramatic ef-

[3]Do not confuse supersonic with *ultrasonic;* the latter refers to a frequency greater than audible frequencies.

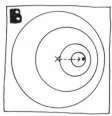

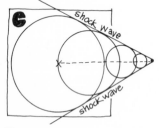

Figure 10–26

Set of circular waves emitted by (A) an object at rest, (B) an object moving at less than the speed of sound [same situation as Fig. 10–25(B)], (C) object moving faster than speed of sound. Straight lines in (C) represent moving shock wave, where wave crests pile up.

Figure 10–27

Bow waves of a boat are analogous to shock waves.

fects occur, since in this case the object is *outrunning* the waves it produces, Fig. 10–26. The wave crests overlap one another and form a single crest (or a small number of crests) of very large amplitude along the lines shown in Fig. 10–26(C). This steep wave front is called a **shock wave**. It is analogous to the bow wave of a boat moving faster than the speed of the water waves it produces, Fig. 10–27.

When the shock wave passes a listener, a very loud "sonic boom" is heard. It sounds much like an explosion and in fact resembles the pressure wave produced by the sudden compression of air caused by an explosion. When we hear an ordinary sound, we receive each wave crest separately, one at a time. When we hear a sonic boom, we receive all the wave crests at once. Consequently, a shock wave contains a tremendous amount of energy. Shock waves produced by supersonic aircraft can break windows and can cause severe damage to fragile structures.

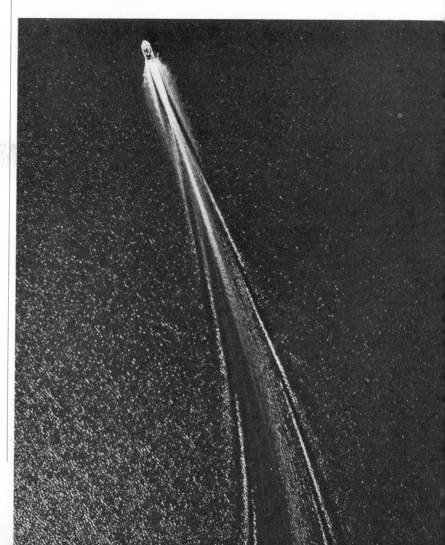

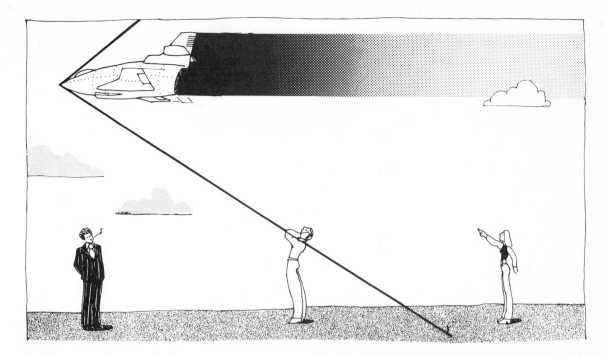

Figure 10–28

Sonic boom has already been heard by person on right, is just being heard by person in center, and will shortly be heard by person on left.

When an aircraft is accelerating up to and beyond the speed of sound, it encounters a barrier of sound waves in front of it just as it reaches the speed of sound. Extra thrust is required to pass through this "sound barrier," but once the speed of sound is exceeded, this barrier no longer impedes the motion. Thus, exceeding the speed of sound is popularly referred to as "breaking the sound barrier."

It is sometimes thought, incorrectly, that a sonic boom is produced only at the moment at which the speed of sound is exceeded. Actually, a shock wave follows behind an aircraft continuously as long as it is moving faster than the speed of sound. A series of observers standing on the ground will each hear a single bang, a sonic boom, as the shock wave passes by, Fig. 10–28.

REVIEW QUESTIONS

1. Which of the following is *not* true of waves in general?
   (a) They transport energy
   (b) they travel outward from the source
   (c) they travel over much greater distances than do the molecules of the medium of travel
   (d) they carry matter with them

287

2. The *origin* of any wave motion is:
   (a) a condensation
   (b) a vibration
   (c) velocity
   (d) the wavelength

3. The distance from crest to crest of a wave is called the
   (a) frequency
   (b) amplitude
   (c) wavelength
   (d) speed

4. The number of wave crests passing a given point per second is called
   (a) frequency
   (b) velocity
   (c) wavelength
   (d) amplitude

5. The velocity of a wave is equal to the wavelength multiplied by
   (a) 2
   (b) frequency
   (c) amplitude
   (d) 1/frequency

6. A sound wave can best be described as a
   (a) longitudinal wave
   (b) transverse wave
   (c) standing wave
   (d) shock wave

7. The difference between a transverse and a longitudinal wave is
   (a) a longitudinal wave carries more energy
   (b) the oscillations are perpendicular to the direction of travel in a transverse wave and parallel in a longitudinal wave
   (c) the reverse of (b)
   (d) a transverse wave has greater amplitude
   (e) none of these

8. The bending of a wave around obstacles is called
   (a) refraction
   (b) interference
   (c) reflection
   (d) diffraction

9. Resonance occurs when an object capable of vibrating is *forced* to vibrate by an external stimulus whose vibrations are:
   (a) violent
   (b) at the natural frequency of the object
   (c) at a high frequency
   (d) at a low frequency
   (e) of a large amplitude

10. What aspect of a sound wave corresponds to its pitch, i.e., its highness or lowness?
    (a) frequency
    (b) speed
    (c) amplitude
    (d) none of these

11. The amplitude of a sound wave determines its
    (a) pitch
    (b) loudness
    (c) quality or timbre
    (d) frequency

12. A sound wave travels a mile in about
    (a) $1/5$ sec
    (b) 1 sec
    (c) 5 sec
    (d) 1100 sec

13. The lowest note that a given string can produce is called
(a) an overtone
(b) the fundamental
(c) the timbre
(d) the amplitude

14. To double the frequency of a given piano string, the string must be made
(a) ½ as long
(b) ¼ as long
(c) twice as long
(d) four times as long

15. The shortening of the length of the air column in a wind instrument primarily
(a) raises the pitch
(b) lowers the intensity
(c) lowers the quality
(d) raises the quality

16. The quality of a sound is determined by
(a) the loudness of the sound
(b) the pitch
(c) the presence of overtones
(d) beats

17. The frequency of sound from an approaching police car heard by a hitchhiker is
(a) lower than the frequency heard by the driver
(b) higher than the frequency heard by the driver
(c) the same as the frequency heard by a policeman driving in the opposite direction
(d) lower than the frequency that will reach the hiker after the car passes

18. The sonic boom is
(a) a very low-frequency sound wave
(b) a high-frequency sound wave
(c) a very large-amplitude sound wave containing one crest or a small number of crests
(d) a large-amplitude electromagnetic wave

EXERCISES

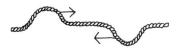

Figure 10–29
Two pulses traveling toward each other.

1. The two pulses shown in Fig. 10–29 are traveling toward each other along a rope. Sketch the shape of the rope at the moment they overlap.
2. Why do you suppose circular water waves decrease in amplitude as the radius increases?
3. What kind of waves will travel along a horizontal rod if you strike it on its end? What if you strike it from above?
4. Is the frequency of a simple continuous wave the same as the frequency of its source?
5. Explain the difference between the speed of a transverse wave traveling down a rope and the speed of a tiny piece of the rope.
6. The wavelength of a wave on water is 20 cm and its frequency is 2 cycles per second. What is the speed of the wave?
7. A water wave has a wavelength of 6 in. and a speed of 2 ft/sec. What is its frequency?

289

8. A wave has a frequency of three cycles per second and a speed of 50 cm/sec. What is its wavelength?

9. Why can you make water slosh back and forth in a pan only if you shake the pan at a certain frequency? (Try it!)

10. Is a car rattle ever a resonance phenomenon? Explain.

11. What do you hear when you put a seashell to your ear? Explain. If you don't have a seashell, try an empty can. Listen to the sounds from a small shell (or can) and from a large one. Why is the sound higher in pitch from the smaller one and lower from the larger one?

12. Why is it difficult for a large group of people to sing in unison?

13. What is an "echo"?

14. Suppose you place two identical tuning forks or two identical glass goblets a foot or two apart on a table. You strike one of the tuning forks, or rub the rim of one of the goblets, and let it vibrate for a moment. If you then grab it with your hand and stop its vibration, you will find that the *other* tuning fork or goblet is now vibrating even though you haven't touched it. Explain.

15. Audible sound waves range from 20 to 20,000 cycles per second, and the speed of sound is about 1100 ft/sec. What, therefore, is the range of wavelengths for audible sounds? Remember, wavelength = velocity/frequency.

16. Suppose a violin string vibrates at 440 cycles per second when unfingered. At what frequency will it vibrate if it is fingered a quarter of the way down from the end?

17. In some concert halls there are locations where it is difficult to hear the performers, yet in adjacent locations it is easy to hear. Explain.

18. What is the evidence that sound is a form of energy?

19. Whistle through your lips and explain how you control the pitch of the sound. Are there any similarities between the interior of your mouth and the length of an organ pipe or an air column in a woodwind instrument?

20. By very lightly touching a guitar string at its midpoint, a very pure tone can be obtained which is one octave above the fundamental for that string. Explain.

21. Why are the frets on a guitar closer together as you move up the neck? Remember that on the equally tempered chromatic scale the ratio of frequencies of any two successive notes is the same all along the scale (see Table 10–4).

22. Why does high C played on the piano sound different from high C played on a violin?

23. The frequency of A above middle C is 440 cycles per second. What is its wavelength in air? How long would an organ pipe, open at one end, have to be in order to produce this sound as its fundamental?

24. What is the frequency of a sound whose wavelength is 0.5 meters?

25. What is the wavelength of middle C, which has a frequency of 262 cycles per second?

26. If a string is vibrating in three segments, are there any points where you can touch it and not disturb the motion?
27. What is the speed of sound in miles per hour? In kilometers per hour?
28. When a sound wave passes into water, where its speed is higher, does its frequency or its wavelength change?
29. A particular string vibrates in three segments at a frequency of 240 cycles per second. What is its fundamental frequency?
30. If two successive overtones of a string are 200 cycles per second and 250 cycles per second, what is the fundamental frequency?
31. A guitar string vibrates at 220 cycles per second as its fundamental frequency. What are the frequencies of its first three overtones?
32. A hiker wants to determine the length of a lake by shouting and listening for the echo from a cliff at the far end. If the echo returns after 2 seconds, how long is the lake?
33. A phonograph record turns at 33 revolutions per minute. What is the speed of the record as it passes the needle 6 inches from the center? The grooves have a wavy shape and will produce a sound of what frequency if their wavelength at one point is 1/8 inch?
34. A sonic boom sounds much like an explosion. Explain the similarity.
35. Will the frequency of a sound heard by a listener at rest with respect to the source be altered if a wind is blowing? Will the velocity or wavelength be changed?

11

LIGHT
AND
LENSES

How do we see? What is it that enters our eyes and arouses the sensation we know as sight? This "something" we call **light**. We will discuss the nature of light here.

1. THE CHARACTERISTICS OF LIGHT: THE RAY MODEL

Children tend to perceive the sense of sight as something that reaches out *from* the eye *to* the object that is seen, much as their hands reach out and touch objects. This view was commonly held in ancient times. Plato believed that the eye sends out "streamers" and that the act of "seeing" an object occurs when those streamers touch the object. Other Greek philosophers, however, held that the eye was a passive organ, like the ear, that merely received the light coming from objects in the field of view. The modern revival of interest in the study of light, which began in the sixteenth and seventeenth centuries, confirms the latter view.

According to the modern view of light, we see an object in one of two ways. First, we can detect light coming directly from a source of light such as an electric-light bulb, a flame, a star, or a bolt of lightning. Second, we can see light reflected off an object, as in Fig. 11–1. Nearly all the objects we perceive are seen by way of reflected light.

What is this mysterious something we call light, which can enter our eyes and produce visual sensations? Let's examine some of its properties.

Have you ever used a magnifying glass to focus the sun's rays to a point on a piece of paper and burned a hole in it? This simple act demonstrates that light is a form of energy and can be transformed into heat. In fact, the light energy from the sun heats the earth and enables living organisms to flourish here. Plants use light energy from the sun directly, in photosynthesis, and animals obtain this energy indirectly by eating plants.

Light travels from one place to another at high speed. Indeed the question of whether the speed of light is infinite or finite occupied thinkers for thousands of years. The idea that light might travel at a finite speed is, at first, an odd idea. After all, when we open our

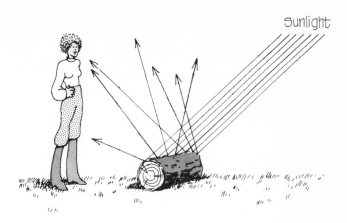

Sunlight

Figure 11–1

We *see* an object when light is reflected off that object into our eyes.

eyes, we see—the light is right there. To have imagined for the first time that light actually moves must have required a great leap of the imagination.

Galileo tried to determine the speed of light by measuring the time it took light to travel a known distance. He stationed himself on one hilltop and had an assistant stand on another a known distance away. Each had a lantern he could flash by removing its cover. Galileo flashed his lamp and instructed his assistant to flash his own as soon as he saw the light from Galileo's. Galileo then measured the time between the flash of his own lamp and the time when he saw the flash from his assistant. The time was so short that Galileo realized it represented nothing more than human reaction time. He concluded that light traveled either with infinite speed or with such a high speed he could not measure it in his experiment.

The first successful measurement of the speed of light was an indirect one performed by the Danish astronomer Olaf Roemer (1644–1710). Roemer had been observing the moons of Jupiter at various seasons of the year. He noticed that when Jupiter was far from the earth its moons' orbits were behind the schedule he had predicted from data obtained when Jupiter was closer to the earth. He concluded that the time delay was due to the extra distance the light had to travel. From his data he was able to make a rough estimate of the speed of light.

Since then, the speed of light has been measured with increasing accuracy by many scientists using many different methods. Among the most important experiments were those performed by the American, Albert Michelson (1857–1931). He began his measurements in 1880 and continued to refine them until the 1920s. He later received a Nobel prize for this work. Michelson's experiments were basically like Galileo's, except that Michelson used mirrors to eliminate human reaction time. He aimed a light source at one face

Table 11-1
Speed of Light in Various Materials

Vacuum	3.0×10^5 km/sec
Air	3.0×10^5 km/sec
Water	2.25×10^5 km/sec
Glass	1.6×10^5 to 2.0×10^5 km/sec
Diamond	1.2×10^5 km/sec

of a rotating eight-sided mirror; light reflected from this mirror traveled to a stationary mirror a large distance away and then back again, as shown in Fig. 11–2. If the rotating mirror was turning at just the right speed—Michelson could vary its speed—the returning beam of light would reflect from a face of the rotating mirror into the eye of an observer. At any other speed of rotation, the returning light would be reflected to the side and the observer would not see it. From the known distance to the fixed mirror and from the speed required of the rotating mirror, the speed of light could be calculated. In the 1920s Michelson made accurate measurements by setting up his rotating mirror on the top of Mt. Wilson in southern California and his stationary mirror on top of Mt. San Antonio (now called Mt. Baldy), 22 miles away.

What Michelson and his predecessors had measured was the speed of light in air. The speed of light in vacuum was found to be only very slightly higher than its speed in air. The accepted value for the speed of light in vacuum is 299,792 km/sec; we usually round this off to 3.0×10^5 km/sec. This corresponds to a speed of 186,000

Figure 11–2

Michelson's speed of light apparatus.

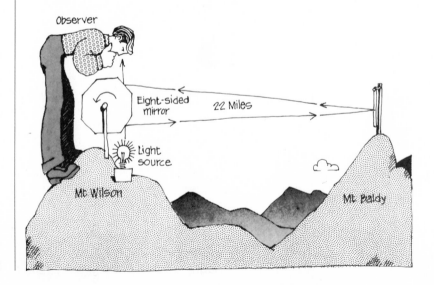

mi/sec. The speed of light in other transparent materials has also been measured and is always found to be less than its speed in vacuum. See Table 11–1.

Light Travels in Straight Lines: The Ray Model

A great deal of evidence suggests that light travels in straight lines. A point source of light, like the sun, casts distinct shadows. Light does not seem to bend around corners; we can hear around walls and doors but we can't see around them. The beam of a flashlight seems to be a straight line. In fact our whole orientation to the physical world is based on the assumption that light does move in straight lines. We infer the positions of objects in our environment by assuming that light moves from the object to our eyes in a straight-line path.

This very reasonable assumption has led to a model for light known as the "ray model": the straight-line paths that light follows are called "light rays." We have already made use of the ray model in Fig. 11–2, which shows a light ray traveling from a source and bouncing off several mirrors before entering the eye of an observer.

Whenever we see an ordinary object, we are aware that the object occupies space and that light comes to us from each point on the object. The object may itself be a source of light or it may be reflecting light from the sun or an electric light. In either case, a bundle of light rays leaves each point of the object and enters our eyes, as shown in Fig. 11–3. Of course a great many other rays leave each point on the object and go off in all directions; but only a few of these rays actually enter our eyes. If we move our head slightly, a different set of rays from a point on the object enters our eyes.

Under certain circumstances, however, light deviates from a straight-line path. For example, when light reflects from a surface, its straight-line path is changed from one direction to another. This is **reflection**. When light travels from one medium to another in which its speed is different, it bends at an angle. This is the phenomenon of **refraction**.

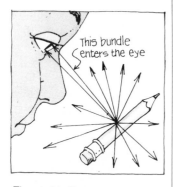

This bundle enters the eye

Figure 11–3

Rays leave each point on an object and go in all directions. A small bundle of these rays enters the observer's eye.

2. REFLECTION OF LIGHT; MIRRORS

Reflection

When light strikes the surface of a material object, some of it is reflected. The rest is either absorbed by the object if it is opaque, or is transmitted through the object if it is transparent (like glass or water). Some

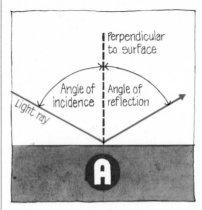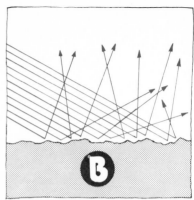

Figure 11–4

(A) When light strikes a smooth surface the angle of incidence equals the angle of reflection. (B) Light striking a rough surface is reflected in many directions.

materials reflect very little of the incident light, whereas others, such as a mirror or a piece of stainless steel, reflect nearly all the incident light.

When a light ray strikes the surface of a material that is very flat, such as the surface of a smooth body of water or a glass mirror, the light is reflected in a regular way: namely, the angle the reflected light ray makes with the perpendicular to the surface of the material always equals the angle the incident ray makes with the perpendicular, as shown in Fig. 11–4(A). This is the law of reflection: *the angle of incidence equals the angle of reflection*. This law was known to the ancient Greeks.

EXPERIMENT

In a darkened room, put a small mirror flat on a table. Shine a flashlight beam on the mirror and notice where the reflected light beam hits the ceiling. Try various angles of incidence and note the angle of reflection. You can see the beams better if you sprinkle a little talcum powder in the air. Do your observations confirm the law of reflection?

AND

Have someone shine a beam of light on the mirror at an angle. Move your head around and note that the reflected light enters your eye only at a certain angle. Now have the person shine the light on a flat piece of white paper. Now the reflected light enters your eye from almost any angle. Why?

Most surfaces are rather rough. Light incident upon a rough surface is reflected in many directions, Fig. 11--4(B), although the law of reflection holds for each ray striking each tiny segment of the surface. Because the surface of most ordinary objects is relatively rough, we can see them from any direction; no matter where our eyes are, so long as there is no obstruction, some of the reflected rays will reach them. But a beam of light—say from a flashlight—will not be reflected from the surface of a small mirror into your eyes unless your eyes are at just the right place. This is what accounts for the unusual properties of mirrors.

Reflecting Mirrors Form Images

When you look into a mirror, you see what looks like yourself, as well as the various objects around and behind you. What you see, of course, is an **image** of all those objects and of yourself. Figure 11–5 illustrates how an image is formed by a flat mirror. Rays from two different points on the leaf are shown. (Of course rays leave each point on the leaf and go in all directions, but the figure shows only those rays that happen to enter the observer's eye.) Note that all the rays are drawn so that the angle of incidence equals the angle of reflection. The diverging rays that enter the observer's eye appear to come from behind the mirror, as shown by the dotted lines. The point from which each bundle of rays seems to emanate is one point of the image. For each point on the leaf there is a corresponding image point. All the image points make up the image. The image is the same distance in back of the mirror as the original object is in front. The image is also the same height as the object.

The image looks just like the object, except that left and right are reversed. Notice that the light rays do not actually pass through the image itself. It merely seems as though the light is coming from the image, since our brain interprets any light entering our eyes as

Figure 11–5

Image formation by a flat mirror. Light from the leaf reflects off the mirror into the observer's eye (rays that do not enter the eye are not shown). The reflected rays seem to come from behind the mirror.

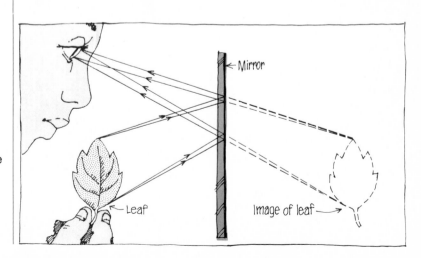

coming from in front of us. Because the light rays do not actually pass through the image itself, there is really nothing material at the position of the image. We could not detect this image with a piece of film placed at the image point, for example. It is therefore called a **virtual** image. Later we will meet **real** images, which can be detected physically; these are the images that are produced by a camera lens, for example, and that can be recorded on film.

EXPERIMENT–PROJECT

Find a mirror long enough for you to see an image of your entire self, from the top of your head to your toes, when you are standing erect several feet away from it. What is the minimum-length mirror required for you to see a full-length image of yourself? To determine this, cover the top and bottom of your mirror with paper if necessary. You will find that the mirror has to be only half as long as you are. Draw a ray diagram, like that of Fig. 11–5, to show why this is so. Does your distance from the mirror influence the size of mirror required? Try various distances and see.

Some mirrors are curved rather than flat. **Convex** mirrors bulge outward in the center and **concave** mirrors are "caved in" in the center (remembering the "cave" in the word "concave" will help you remember which name goes with which mirror).

The images produced by concave and convex mirrors, as shown in Fig. 11–6, differ in size from those produced by a flat mirror. For both Figs. 11–6(A) and 11–6(B) the rays are drawn so that the angle of incidence equals the angle of reflection at the surface of the

Figure 11–6

Image formation by (A) a convex mirror, and (B) a concave mirror.

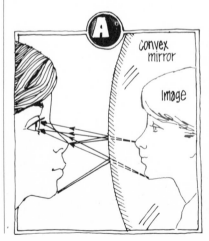
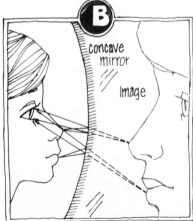

Figure 11–7

Looking into a convex mirror.

mirror. As before, the rays are shown emanating from two different points on the object, and only those rays that enter the observer's eyes are shown.

In a convex mirror the image is smaller than the object, Fig. 11–6(A). Because it makes things appear smaller, a greater area is included within the field of view. Convex mirrors are thus useful to cover a large field of view, as in a store (Fig. 11–7), to aid motorists in seeing around a dangerous corner, and as a rear-view or side mirror on a car or truck.

A concave mirror, on the other hand, magnifies the object—the image is larger, Fig. 11–6(B). Shaving or facial mirrors that magnify are concave. If you are too far from a concave mirror, however, your image is smaller and upside down. You might try to draw a ray diagram to show this.

3. REFRACTION OF LIGHT

When light passes from one transparent material (or "medium") into another, two things happen. Part of the light is reflected at the surface and part of it passes into the new medium. The part that passes into the new medium is bent, and the light changes direction, as shown in Fig. 11–8. This phenomenon is known as **refraction**. The amount of refraction or bending depends on the speed of light in each medium and on the angle at which the light strikes the boundary between the two. It is found experimentally that when light passes from one medium into another medium in which the speed of light is *less,* the light bends *toward* the perpendicular to the surface, Fig. 11–8(A); that is, the angle of refraction is *less than* the angle of incidence. But if the light passes from one medium into another in which its speed is greater, the light bends *away from* the perpendicular, Fig. 11–8(B); the angle of refraction is *greater than* the angle of

Figure 11–8

Refraction of light. (A) Light entering a medium in which its speed is *less* bends *toward* the perpendicular; (B) light entering a medium in which its speed is *greater* bends *away* from the perpendicular.

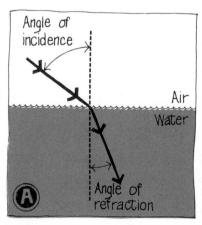
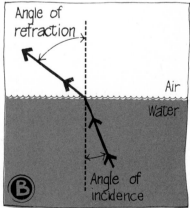

incidence. In general, *when light passes from one transparent medium into another, the light is bent toward the perpendicular if its speed is lowered, and away from the perpendicular if its speed is increased.* The only exception is when the light strikes the second medium perpendicular to its surface; the light then passes straight through without bending at all. Figure 11–8 illustrates another important fact: the reversibility of light rays. If a light ray were directed back along the path by which it had come, it would follow that same path even after refraction.

EXPERIMENT

Put a penny in the bottom of an empty cup. Position your head so the penny is just out of sight, as shown in Fig. 11–9. Now have someone pour water slowly into the cup. Don't move your head. The penny becomes visible and seems to rise. Explain.

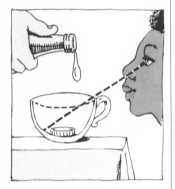

Figure 11–9

The penny *seems* to rise as water is poured into the cup.

Some Optical Illusions Are Due to Refraction of Light

Refraction gives rise to many strange illusions. The apparent shortening of the legs of a person standing in water is due to refraction, as shown in Fig. 11–10. Because the light rays leaving the person's

Figure 11–10 Because of refraction, a person's legs look shorter in water.

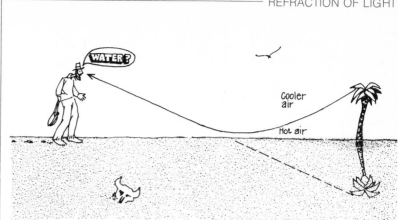

Figure 11–11

Mirages are caused by the bending of light in different densities of air.

foot, for example, are bent upon striking the surface of the water, the light reaching the observer's eye seems to be coming from a higher point [Fig. 11–10], and therefore the leg looks shorter. Similarly, a swimming fish or a rock on the bottom of a lake seems higher than it actually is, and bodies of water do not appear to be as deep as they really are. The apparent bending of a stick placed in a glass of water (try it) is another example of an illusion created by refraction.

Mirages are produced by the bending of light rays in the atmosphere. On hot days, particularly in the desert, a layer of very hot air lies along the ground. Hot air is less dense than cool air, and so the speed of light in this layer is slightly greater than in the cooler air above. Because of refraction, the light rays are turned upward into a distant observer's eye and he "sees" the tree upside down as if reflected in a lake, Fig. 11–11.

Motorists often see a mirage of water on the highway some distance ahead of them, Fig. 11–12. It looks as though the sky were reflected in this body of water; but in fact the motorist is seeing the sky directly, because of refraction.

Figure 11–12

A highway mirage. Vehicles appear to be reflected in water on the road. When you reach that point you find there actually is no water there.

4. TOTAL INTERNAL REFLECTION

In Fig. 11–8(B) we saw that when light passes into a medium where its speed is greater, say from water into air, rays are bent away from the perpendicular. As the angle of incidence increases, a **critical angle** is reached at which the angle of refraction is 90°; that is, the refracted ray passes along the surface of the water, as shown for ray (B) in Fig. 11–13. If the angle of incidence *exceeds* this critical angle, no light passes through the surface—all the light is reflected back into the medium from which it came, as shown for ray (C) in Fig. 11–13. This phenomenon is called **total internal reflection**.

The critical angle depends on the speed of light in each of the two materials. For light traveling in water toward air, the critical angle is 48°.

Total internal reflection can occur at a surface only if the speed of light in the material is less than the speed beyond the surface. For example, light passing from air to water does not undergo total internal reflection; but it will if the light is going from water toward air and the incident angle exceeds 48°. There are many interesting applications of this phenomenon. For example, when you look up from beneath the surface of the water in a lake, you see the outside world compressed into a circle whose edge makes a 48° angle with the perpendicular, Fig. 11–14. And if the water is smooth, you will see reflections from the sides or bottom at angles greater than 48°.

Total internal reflection is the principle behind the transmission

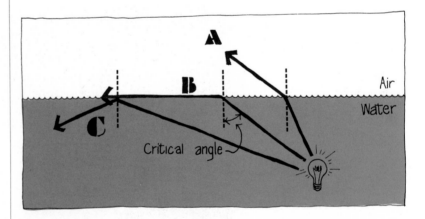

Figure 11–13

Light from an object below the surface of the water. Ray (A) represents ordinary refraction. Ray (B) represents the critical angle where the refracted ray skims along the surface. Ray (C) represents total internal reflection.

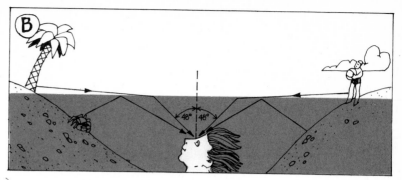

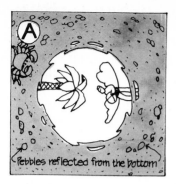

Figure 11–14

View looking upward from beneath the water; the surface of the water is smooth.

of light along "light pipes," or "light fibers," which consist of one or more narrow hollow tubes. As shown in Fig. 11–15, light traveling down such a tube will undergo total internal reflection each time it strikes the edge of the tube (unless a large-angle bend creates a reflection angle that is less than the critical angle). Thus the light is transmitted with almost 100 percent efficiency. Such light pipes or fibers are used in decorative lamps (Fig. 11–16). Bundles of fibers can be used to observe inaccessible places. An instrument employing this principle is used by doctors to examine the interior of patients' stomachs. The end of a tube containing bundles of fibers is inserted into the stomach by way of the mouth. Light travels down to the stomach along one group of fibers; it is reflected from the

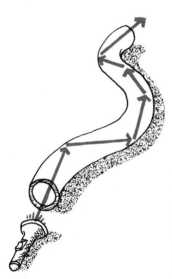

Figure 11–15
A "light pipe."

Figure 11–16
A decorative lamp using fiber optics.

stomach walls and returns along another set of fibers to form an image of the interior. Fiber optics technology is also now beginning to be used to transmit telephone calls and other communications signals. The light traveling along the pipe varies in intensity at the frequency of the signal (say, of people's voices on the telephone).

5. LENSES

The phenomenon of refraction finds a practical use in lenses. The development of optical instruments using lenses really got underway in the sixteenth and seventeenth centuries, although primitive eyeglasses were known as early as the fourteenth century. Today eyeglasses, telescopes, binoculars, cameras, microscopes, and magnifying glasses all use one or more lenses.

Lenses Form Images

Lenses are usually made of glass or plastic. In a simple thin lens the two lens surfaces are ground so that in a cross-sectional view they appear as sections of a circle, Fig. 11–17. (One surface of a lens can be flat, but to act as a lens, the other surface must be curved.) An imaginary line drawn through the centers of the two faces is called the "lens axis." A ray of light that strikes the lens will be refracted at each of the two surfaces according to the law of refraction. That is, at the first surface the ray is bent *toward* the perpendicular (dotted line in Fig. 11–17), since the speed of light is less in glass than in air. At the second surface it is bent *away* from the perpendicular.

Lenses are of two types, **converging** lenses and **diverging** lenses. A converging lens is one that makes rays of light that are parallel when they strike the lens come together (converge) at a point behind the lens. A diverging lens, on the other hand, causes parallel

Figure 11–17

Cross-sectional view of a simple lens, showing how a light ray is bent.

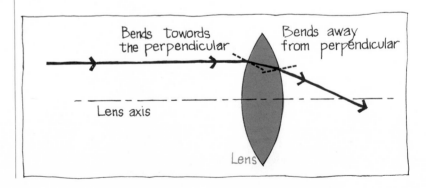

Figure 11–18

(A) A converging lens and (B) a diverging lens, showing focal point and focal length for each.

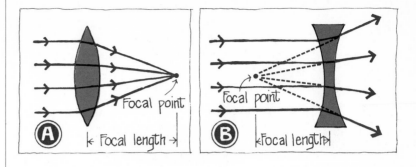

light rays to spread apart (diverge), Fig. 11–18. The point at which parallel rays are brought together by a converging lens is called the **focal point** of the lens. The distance from the center of a lens to the focal point is called the **focal length** of the lens. Focal point and focal length can also be defined for a diverging lens: when parallel rays are made to diverge, the point from which they seem to diverge is the focal point, Fig. 11–18(B). The focal length is a measure of how "strong" a lens is; that is, how much it bends light rays. Thus the focal length is the most important specification of a lens. The focal length is determined by the curvature of the two surfaces and is the same on both sides of a lens; it is thus unchanged if the lens is turned around.

The main purpose of a lens is to form an image of an object. For an object not far from the lens, the rays from each point on the object diverge as they approach the lens (see Fig. 11–19). A converging lens can bend these rays and bring them together at a point behind the lens; this is called the **focus**, or **image point**. The combination of all these points makes up the image of the object. Figure 11–19 shows how several rays leaving one point on an object are refracted by the lens and brought to a focus as the image. (Note that the focal point is the image point for—and only for—an object at infinity; this is because the rays from a point on an object very far away are essentially parallel, Fig. 11–20.)

Figure 11–19

Image formation by a converging lens, showing how rays are refracted by the lens. Notice that the image is upside down.

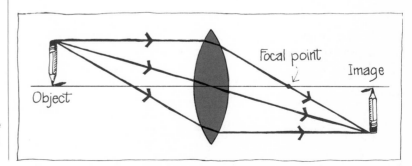

The image formed by this converging lens is called a **real** image, because the light rays actually pass through the image itself. Unlike a virtual image, if a piece of white paper or film is placed at the position of the real image, the image can be seen on the paper or detected by the film. In fact, the ability of a converging lens to form a real image, as in Fig. 11–19, is the operating principle for many optical instruments, as we shall see.

EXPERIMENT—PROJECT

Obtain a simple lens—the lens from a pair of reading glasses, a magnifying glass, or a camera or projector lens. Use a light bulb as the object and project its image onto a piece of white paper. Be sure to position the paper so the image is sharp. Try various lens-to-object distances and notice how the lens-to-image distance changes, as well as how the size of the image changes. Also measure the focal length by finding the image distance for an object very far away. (Note: Do not put your eye where you expect the image to be; put the white paper there and observe the real image on paper.)

AND

Use the lens as a magnifier. Determine where an object must be placed so that the lens produces the largest image. (This time observe the image directly with your eye.) Is it at the focal point, beyond it, or in front of it? Also, determine the magnification of the lens by observing a piece of lined paper directly and through the lens. Compare the spacing between the lines.

Figure 11–20

Rays that enter a camera lens from a distant object are essentially parallel.

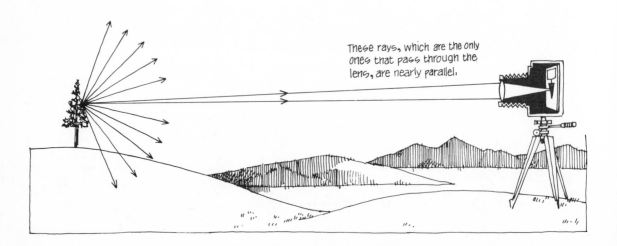

These rays, which are the only ones that pass through the lens, are nearly parallel.

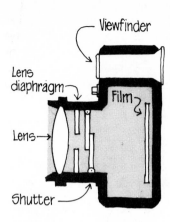

Figure 11–21

Cross-sectional view of a simple camera.

Figure 11–22

The image of a nearby object is farther from the lens than the image of a distant object.

6. THE CAMERA

A camera consists basically of a lens, a light-tight box, a shutter to let light in through the lens, and a place for a strip of film, Fig. 11–21. When the shutter is opened, light enters the camera and is brought to a focus on the film by the lens. That is, the film is located at the position of the image.[1]

Focusing. Unfortunately, objects at all distances are not brought to a focus at the same place. For example, the rays from a distant object are essentially parallel when they reach the camera lens, as shown in Fig. 11–20. Thus they will come to a focus at the focal point behind the lens. (Indeed, the focal point can be defined as the image point for an object infinitely far away.) However, rays from objects that are close to the camera will be brought to a focus at a greater distance behind the lens. This happens because the rays from each point on a nearby object make a steeper angle with the lens, as shown in Fig. 11–22, so that the angle of the rays leaving the lens will not be as great; thus the image will be formed farther back.

Thus, if the film is placed at the focal point, each point on a distant object will make an image point on the film; that is, the image points will be "in focus." But each point on nearby objects will give rise to a circle on the film, as shown in Fig. 11–23(A), and hence the image will be a blur. It will be "out of focus." An out-of-focus picture is a collection of tiny overlapping circles. Similarly, if the lens of the camera is moved farther away from the film, so that the film is placed where nearby objects will be in focus, then distant objects will be out of focus, Fig. 11–23(B).

Good-quality cameras must therefore have a focusing adjustment so that the lens moves with respect to the film. The closer the

[1]A slide or movie projector is just the inverse of a camera, and Fig. 11–19 applies as well; the film plays the role of the object, and the image is projected by the lens onto a screen. Thus the object is much closer to the lens than the image is, just the reverse of the normal situation for a camera.

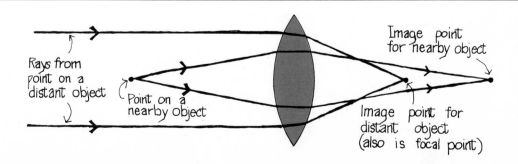

Rays from point on a distant object

Point on a nearby object

Image point for nearby object

Image point for distant object (also is focal point)

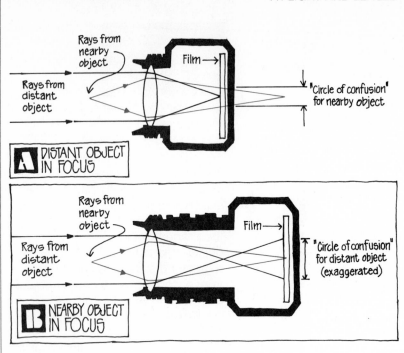

object is, the farther the lens must be from the film. Normally, the range of movement of the lens need not be very large, except for cameras that are used for close-up photography.

A compromise is often made when focusing a camera by adjusting the lens so that it is halfway between the positions for sharp focus of distant and nearby objects; then both nearby and distant objects are only slightly out of focus. Inexpensive cameras often have no focusing adjustment, and reasonably clear pictures are attainable only over a limited range of distances.[2]

Focal length and magnification. The focal length of a camera lens determines the size of the image on the film. A "normal lens" is one that fills the film with a field of view that is approximately the same as the normal field of view of a human being. A "telephoto lens" has a longer focal length and magnifies objects; thus the field of view is smaller. A "wide-angle lens" has a shorter than normal focal length and embraces a wider field of view.

Lens diameter or f-stop. The diameter of a lens is important, since it determines how much light is let in. Ordinary film—a piece of celluloid or plastic coated with light-sensitive chemicals—is sensitive to a remarkably small amount of light. The light necessary to

[2]The sharpness of a picture depends not only on the focusing—the distance of lens to film—but also on the quality of the lens and on the quality or graininess of the film itself.

sensitize the film can be obtained with an exposure of a tiny fraction of a second. On a dark day, when less light is falling on the scene to be photographed, either the exposure time must be lengthened or the amount of light let in by the lens must be increased. Most cameras, particularly those of high quality, have adjustments for changing the shutter speed and also for regulating the amount of light that passes through the lens; the latter is accomplished by an iris diaphragm with an adjustable opening placed next to the lens (see Fig. 11–21). For brightly lighted scenes in broad daylight, the opening is made small; but on dark days, or indoors where the light is dim, the opening is made large. The size of the diaphragm opening is measured by the "f-stop," which is defined as the ratio $\frac{\text{focal length}}{\text{diameter of diaphragm opening}}$. The smaller the maximum f-stop number, the larger the maximum opening for that lens, and the dimmer the light in which pictures can be taken.

Camera lenses are normally specified by two numbers: the focal length and the maximum f-stop they are capable of. Thus a "50 mm f/2 lens" means the focal length is 50 mm and the maximum f-stop is "f/2"; i.e., the maximum diameter of the diaphragm is 25 mm ($\frac{50 \text{ mm}}{25 \text{ mm}} = 2$).

EXPERIMENT

Make a "pinhole camera" by removing one of the ends of a box and covering the opening with a piece of wax paper (or tissue paper). Now make a clean small hole in the opposite end, Fig. 11–24. Stand in a darkened room and point your camera toward a bright scene outdoors. You will see the image of the scene on the wax paper. Draw a ray diagram to show how the image is formed. If you make the hole smaller, the image becomes clearer but dimmer. Why?

wax paper

Figure 11–24
A pinhole camera.

High-Quality Lenses Consist of Several Simple Lenses

It is difficult to design a lens that will have a large opening (small f-stop number) and that will still give a clear picture. For example, a simple thin lens does not focus all the rays from a single point on an object at one point, not even when the rays are parallel rays from a distant object. Rays that strike the edge of the lens are bent more

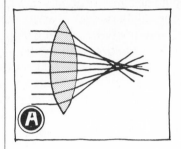
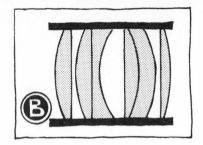

Figure 11–25

(A) Spherical aberration. Single lenses do not bring all the rays to a focus at one point, which leads to a blurry image. (B) A high-quality lens containing six elements.

and more radically, as shown in Fig. 11–25(A). This is called "spherical aberration." Only the central part of a simple lens is usable to produce clear pictures. However, by combining several thin lenses (called "elements"), as in Fig. 11–25(B), we can compensate for this and other aberrations. Today complex lenses that give very sharp images are normally designed with the aid of a computer that traces the rays through a hypothetical lens and adjusts the various elements to minimize aberrations.

7. MAGNIFYING INSTRUMENTS: MICROSCOPES AND TELESCOPES

The Magnifying Glass

A magnifying glass is nothing more than a simple converging lens. For a lens to act as a magnifier, the object to be magnified must be placed slightly closer to the lens than the focal point, as shown in Fig. 11–26. The

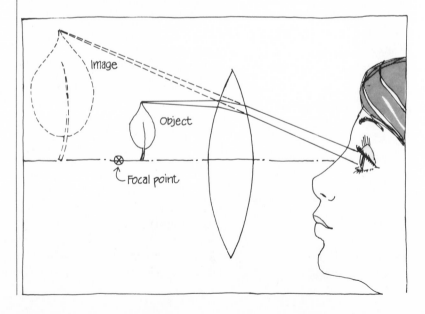

Figure 11–26

Virtual image formed by a magnifying glass (a simple converging lens). Object must be placed within the focal point in order to be magnified.

diverging light rays that reach the eye appear to be coming from a higher point, the tip of the image. The image is a virtual image, since the light rays do not actually pass through it. Notice that the image is not only larger than the object; it subtends a larger angle than the object does. This is what makes the object appear magnified.

The Microscope and the Telescope

Microscopes are used for much larger magnifications than can be attained with a magnifying glass. A microscope uses two lenses, as shown in Fig. 11–27. The **objective** lens, the one closer to the object, has a very short focal length and magnifies the object by forming an enlarged real image. The second lens, the **eyepiece**, then magnifies the image made by the first, forming a virtual image that can be seen by the eye. The eyepiece thus acts like a simple magnifying glass.

A telescope, Fig. 11–28(A), also uses two lenses and forms images much as the microscope does, except that the objective lens has a rather long focal length. This is necessary because a telescope is used to view very distant objects rather than very close objects as a microscope does. The telescope magnifies because, like the simple magnifier, the image subtends a greater angle than the distant object does.

Like the microscope, the telescope forms an upside-down image. This creates no real problem for viewing the heavens with an astronomical telescope, but on the earth an upside-down image is

Figure 11–27

Diagram of a microscope (see text).

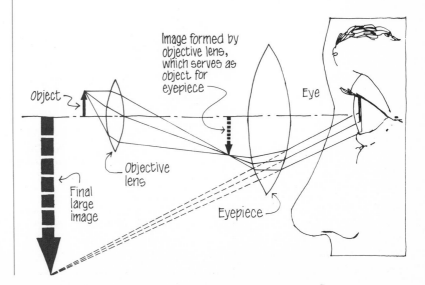

Figure 11–28

(A) An ordinary telescope.
(B) An astronomical telescope using a mirror instead of a lens for the objective.

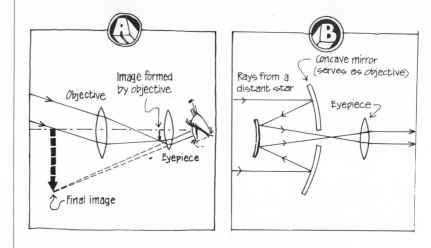

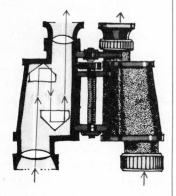

Figure 11–29
Binoculars.

314

a real disadvantage. Therefore terrestrial telescopes use a third, intermediate lens to turn the image right side up. In binoculars, Fig. 11–29, either mirrors or prisms are used to turn the image upright; at the same time they reduce the length of the instrument to a manageable size. Glass prisms are used in all high-quality binoculars because they reflect very nearly 100 percent of the light through total internal reflection. The critical angle for glass is about 42°, so rays striking the prism at 45° as shown are totally internally reflected. Hence prisms in binoculars give a brighter image than do mirrors, since even good mirrors reflect somewhat less than 100 percent of the light that strikes them.

Many astronomical telescopes, particularly the largest ones, use a large concave mirror in place of the objective lens, Fig. 11–28(B).

8. THE HUMAN EYE

The human eye, Fig. 11–30, functions much like a camera. It has a lens, and behind it there is a diaphragm called the iris (the colored part of your eye), which adjusts automatically to control the amount of light entering the eye. You have undoubtedly noticed that the pupil of your eye, which is the opening in the iris, is large in dim light and is small in bright light. The eye does not have the equivalent of a camera's shutter; the corresponding operation is carried out in the nervous system and in the brain. Light from objects in the field of vision is continuously brought to a focus on the retina by the lens. The retina, which plays the role of the film, contains two kinds of light detectors, called rods and cones. They transform the light energy into electrical signals that are carried to the brain by a complex

Figure 11-30
The human eye.

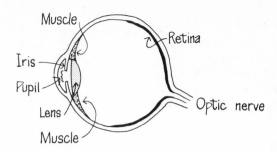

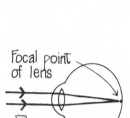

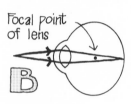

Figure 11-31

(A) The eye relaxed, focused on a distant object. (B) The lens is thickened, shortening the focal length, in order to focus on a nearby object.

array of nerves. The brain somehow analyzes these signals to form images at the rate of about 30 per second.

This can be compared to a movie camera, which operates by taking a rapid series of still photographs at a rate, normally, of 24 frames per second; between each pair of these still photos, a shutter closes over the lens as the film advances. The rapid projection of these still photos on a screen at a rate of 24 pictures per second gives the appearance of motion. The eye works in a similar way, except that the brain distinguishes the separate images and then integrates them into what we perceive as motion.

The eye, like a camera, must adjust to focus on near or distant objects. For example, put your finger a few inches from your eye and focus on it; note that objects far away are blurred. Now relax your eye so that distant objects are clear; your nearby finger will be blurred. The eye does not change focus by changing the position of the lens as a camera does. Instead, the muscles of the eye change the curvature of the lens and therefore its focal length. To focus on distant objects, the eye muscles are relaxed and the lens is fairly thin, Fig. 11-31(A). To focus on a nearby object, the muscles force the lens into a rounder shape [Fig. 11-31(B)] so that the di-

Figure 11-32

(A) A nearsighted eye cannot focus on distant objects; nearsightedness is corrected with a diverging lens. (B) A farsighted eye cannot focus on nearby objects; farsightedness is corrected by a converging lens.

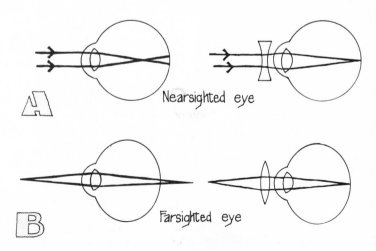

verging rays from a nearby object will be bent sufficiently to be brought to a focus on the retina.

Two common defects of the eye are nearsightedness and farsightedness. Both can be corrected by using supplementary lenses, either eyeglasses or contact lenses. A nearsighted person is one who can see well at near distances but sees distant objects as blurred. Nearsightedness is caused by a lens that is too curved, or by an extra-long eyeball, Fig. 11–32(A); in either case distant objects are focused in front of the retina. A diverging lens corrects this defect. A farsighted person sees distant objects well but sees nearby objects as blurred. Farsightedness is caused by an eyeball that is too short [Fig. 11–32(B)] or by a lens that is too thin; in either case, nearby objects are focused behind the retina. A converging lens corrects this defect.

REVIEW QUESTIONS

1. The ray model of light is based on the observation that light
   (a) travels in straight lines
   (b) bends around corners
   (c) exhibits interference
   (d) has many colors

2. The speed of light is
   (a) infinite
   (b) about 1100 ft/sec
   (c) less than the speed of sound
   (d) greater than the speed of sound

3. The law of reflection says that
   (a) reflection occurs only from smooth polished surfaces
   (b) reflection can occur from rough surfaces in certain circumstances
   (c) the angle of incidence equals the angle of reflection
   (d) a ray bends when it passes into another medium

4. When a ray of light passes obliquely from one medium into a second medium in which its speed is less, the direction of the ray
   (a) is not changed
   (b) is bent toward the perpendicular
   (c) is bent away from the perpendicular
   (d) is perpendicular to the surface

5. When you look into a swimming pool, you are likely to
   (a) overestimate its depth
   (b) underestimate its depth
   (c) see longitudinal waves
   (d) see real images reflected

6. Mirages are explained by
   (a) reflection
   (b) refraction
   (c) diffraction
   (d) interference

7. Total internal reflection refers to
   (a) reflections that occur on the inside of an optical instrument
   (b) complete reflection that occurs when light strikes a surface be-yond which its speed would be greater, but its incident angle is too large
   (c) the high reflective quality of mirrors
   (d) the high reflective quality of any surface

8. Parallel rays of light falling on a converging lens pass through the lens and
   (a) spread out
   (b) are brought to a focus
   (c) remain undeflected
   (d) are reflected back in the original direction

9. The focal length of a lens is
   (a) the diameter of the lens
   (b) the thickness of the lens
   (c) the distance at which an image of any object is formed
   (d) the distance from the lens at which parallel rays come to a focus

10. To use a magnifying glass, the object is placed
    (a) as close to the lens as possible
    (b) just within the lens's focal point
    (c) just beyond the focal point
    (d) some distance beyond the focal point

11. The muscles connected to the lens of the eye contract or relax to ad-just for
    (a) objects at different distances   (c) objects of different sizes
    (b) different amounts of light   (d) different colors

12. The pupil of the eye changes in size to adjust for
    (a) objects at different distances   (c) objects of different sizes
    (b) different amounts of light   (d) different colors

EXERCISES

1. What is the evidence that light is energy?
2. What is a shadow?
3. How far apart would Galileo and his assistant have had to be in order for the light to take one second to travel over and back between them?
4. What is the speed of light in air expressed in miles per hour?
5. How long does it take for light to reach the earth from the sun, which is 150,000,000 km away?
6. How long did it take light to travel the 44 miles in Michelson's exper-iment, Fig. 11–2?

7. Using your answer from the above problem, calculate how many revolutions per second the mirror must rotate so that a new mirror face will just be in position to reflect the light into the observer's eye.

8. How does the size of your image in a flat mirror change with your distance from the mirror?

9. Stand up two pocket mirrors so that they form a right angle, Fig. 11–33. When you look into this double mirror, you see yourself as others see you instead of reversed, as you do in a single mirror. Use a ray diagram to show how this occurs.

10. In Fig. 11–34, is the light entering a medium where its speed is increased or decreased?

Figure 11–33

How is this nonreversed image formed?

11. Draw a ray diagram showing why a stick looks bent in water.

12. When light passes from glass into water, does the beam bend toward or away from the perpendicular?

13. Draw a ray diagram showing how the highway mirage of Fig. 11–12 can occur. (Hint: See Fig. 11–11.)

14. Explain how you "see" a round drop of water on a table in view of the fact that the water is colorless.

15. On certain occasions, sailors have thought they could see a ship or an island of land floating in the sky above the horizon. Explain how this could happen.

16. Stars seem to twinkle when seen from the earth, but not when seen by astronauts beyond the earth. Why?

17. When you look up at an object in the air from under water, does it seem to be the same size as when you look at it from out of the water?

18. The critical angle for total internal reflection in glass is smaller than in water because the speed of light is less in glass than in water. Explain.

19. Diamonds sparkle as a result of total internal reflection. Why do you see different colors? How does the very low speed of light in diamonds (Table 11–1) affect their "brilliance"?

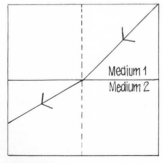

Figure 11–34

An example of refraction.

Figure 11–35

Which of these lenses is converging and which diverging?

20. Draw a few parallel rays impinging on each of the lenses shown in Fig. 11–35 and determine if each is a converging or a diverging lens. Can you make a general rule about whether a lens is converging or diverging on the basis of whether it is thicker at the center or at the edge?

21. In a close-up photograph of a person's face the background is usually out of focus. Explain.

22. Bifocal glasses have lenses that are split so that the lower half has a different focal length than the upper half. Why is this, and what is the advantage?

23. As people grow older, the ability of the eye's muscles to make the lens thicker and rounder is diminished, making it more difficult to focus on nearby objects. For this reason, reading glasses are commonly used by older people. Are reading glasses converging or diverging lenses? See Fig. 11–32.

24. State the difference between a real and a virtual image.

25. Why is a magnifying glass also called a "burning glass"?

26. A camera lens is set to focus on an object 20 m away. How should the lens be moved so that it will focus on an object 4 m away?

27. Suppose you want to take a picture of yourself in a mirror 3 meters away. The camera should be focused for what distance? Can your image and the mirror frame both be in sharp focus?

28. Why can't a prism be used to form an image?

29. The film in a camera plays a similar role to what part of the eye?

12

THE WAVE NATURE OF LIGHT

Light, as we have seen, is a form of energy and travels at very high speeds. How is this energy carried through space? In Chapter 10 we saw that there are two different ways in which energy can be carried from one place to another: by particles or by waves. Does light travel as a stream of particles away from the source? Or does it travel in the form of a wave that spreads out from the source like waves on water? Although the nature of light had been discussed by the ancient Greeks, the question of whether light travels as tiny particles or as tiny waves came to the fore only in the seventeenth century.

The Dutch scientist Christian Huygens (1629–1695), a contemporary of Newton, developed a wave theory of light. Newton himself, however, favored the idea that light consisted of tiny particles. Although Newton was not completely convinced of the particle nature of light, his opinion was very influential and the particle theory of light was to dominate thinking for a century—this despite the fact that the experimental evidence was not very compelling either way.

1. INTERFERENCE AND DIFFRACTION

How can we tell whether light travels by particles or by waves? After all, we can't see individual particles of light nor can we see individual wave crests and valleys of light like the wave crests and valleys on the surface of water. The ultimate nature of light is hidden from everyday experience. Therefore some subtle means must be used to determine if light has a particle or a wave nature. In Chapter 10 we discussed waves and wave motion, and we found certain phenomena that waves exhibit but that particles don't. These were the phenomena of interference and diffraction.

As early as the mid-seventeenth century a Jesuit monk, Francisco Grimaldi (1618–1663), observed what appeared to be diffraction of light. Grimaldi performed an experiment in which sunlight was allowed to pass through a tiny pinhole (Fig. 12–1), and observed that the region of illumination on a white screen behind the pinhole was a circle much larger than the size of the pinhole itself. Further experiments showed that the smaller the pinhole, the larger the illuminated area.

Grimaldi's experiment clearly shows that sunlight is bent into what we might expect to be the shadow region. We would not expect particles to do this, but if light is a wave, this behavior would be normal; it is simply an example of **diffraction**—the bending of waves around obstacles in their path—which we discussed in

Figure 12–1

Grimaldi's experiment showing the diffraction of light.

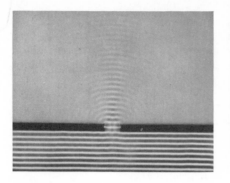

Figure 12–2

Diffraction of water waves.

Chapter 10. The diffraction of *water* waves as they pass through a slit is shown in Fig. 12–2; notice that the bending of the waves is significant only when the size of the slit is close to that of the wavelength, and that the smaller the slit the greater the bending or diffraction. This is just what is observed when light passes through the pinhole of Fig. 12–1. The smaller the hole, the larger the spot on the screen.

Newton, though he was aware of Grimaldi's results, was not convinced that they were due to diffraction; he supposed they were due instead to an interaction of light "particles" with the edge of the slit. Even Huygens failed to see that Grimaldi's experiment supported the wave theory of light.

Young's Double Slit Experiment Discloses Interference

Convincing evidence for the wave nature of light was finally obtained in 1801 by the English physician Thomas Young (1773–1829). Young's famous "double slit" experiment is diagrammed in Fig. 12–3. Light from a small source is allowed to pass through two

Figure 12–3

Young's double slit experiment. (A) We expect to see two bright lines on the screen behind the slits if light is made up of particles. (B) Young observed many lines.

323

very closely spaced slits. Now, if light consisted of particles, we would expect to see two bright lines on the screen behind the slits, Fig. 12–3(A). But Young observed instead a large number of lines, as shown in Fig. 12–3(B). By assuming that light consists of waves, Young was able to show that the series of lines is due to interference between the waves that pass through the two different slits.

To understand Young's explanation in terms of the wave theory, we first must note that the double slit experiment is equivalent to the experiment in which two rocks thrown into a body of water produce an interference pattern (see Fig. 10–13). In place of the two rocks acting as sources of waves, the two slits in the screen serve as two sources of light waves. This is illustrated in Fig. 12–4 in which the light waves are shown as spreading out from each slit just as water waves do. Note that the light *rays*, which we defined as the direction in which the light moves, are perpendicular to the wave fronts or crests. The interference pattern illustrated in Fig. 12–4 is much like that for water waves. However, we cannot directly observe the interference pattern as it spreads through space; rather, we can observe only the illuminations produced by these wave patterns when they strike the screen. The series of bright and dark lines on the screen is also known as an interference pattern. To explain how the interference pattern is produced on the screen, we make use of the diagram of Fig. 12–5. Here the waves that spread out along each ray are shown individually, and the regions on the screen where constructive and destructive interference occur are indicated. We perceive areas of destructive interference—where there is no wave activity—as darkness, and areas of constructive

Figure 12–4

If light is a wave, the light passing through the two slits should interfere.

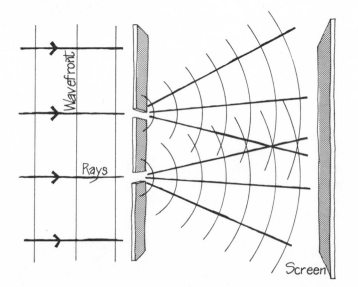

interference as brightness. We see from Fig. 12–5, for example, that a bright area will occur when the paths of the rays from the two slits are the same—for example, at the center of the screen, Fig. 12–5(A). The screen will also be bright where the paths of the two rays differ by exactly one wavelength (or two, or any whole number of wavelengths), since the waves will then interfere constructively, Fig. 12–5(B). For those regions on the screen where the distances from the two slits differ by a half wavelength (or 1½ wavelengths, 2½ wavelengths, and so on), the two waves will destructively interfere, Fig. 12–5(C). Thus the wave theory predicts a pattern of light and dark lines that is precisely in accord with experimental results, Fig. 12–3(B).

Using simple geometry, Young was able to calculate the wavelength of visible light with the data he obtained from his double slit experiment. He found that the wavelength of visible light is very tiny, varying from about 4×10^{-7} meters to 7×10^{-7} meters (1/60,000 of an inch to 1/36,000 of an inch).

EXPERIMENT

Hold a handkerchief at arm's length in front of your eyes and view a small source of light such as a distant street lamp. You will see an array of spots that is larger than the source seen directly. This is an interference pattern produced by the light traveling through the many tiny holes between the threads of the cloth. Observe the interference patterns produced by fabrics of fine weave and of coarse weave. How does the fineness of the weave affect the pattern?

Figure 12–5

Diagram showing how the wave theory explains the pattern of lines observed in the double slit experiment as a result of interference.

Young's experiment was a crucial one, for it had at last settled the argument over whether light has a wave or a particle nature. There was no way to explain Young's results using a particle view of light. Continued experimentation during the nineteenth century

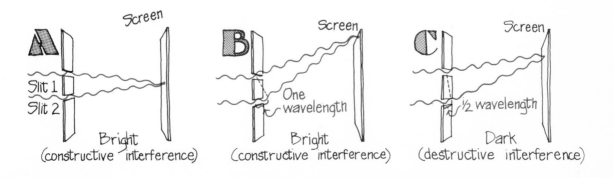

A Screen
Slit 1
Slit 2
Bright
(constructive interference)

B Screen
One wavelength
Bright
(constructive interference)

C Screen
½ wavelength
Dark
(destructive interference)

Figure 12–6

Water waves passing obstacles of various sizes. Notice that the smaller the obstacle compared to the wavelength, the more diffraction or bending takes place into what we might call the "shadow region."

further confirmed the idea that light is made up of waves, and by the end of the century the wave theory of light stood on a firm experimental and theoretical basis.

Diffraction Around Objects

In Fig. 12–2 we saw that water waves passing through a small hole are diffracted significantly only when the size of the hole is not much larger than the wavelength of the waves. If instead of waves passing through a hole, we consider waves diffracting around an object, as in Fig. 12–6, we find a similar result: if the object is as small as or smaller than the wavelength of the waves, there is considerable diffraction around the object; but the larger the object is in comparison to the wavelength of the waves, the smaller will be the effect of diffraction. This can be said another way: *the larger the wavelength compared to the size of the object, the more diffraction will occur.* See especially Figs. 12–6(C) and 12–6(D). This is why light waves, whose wavelengths are very small, do not bend around corners in everyday situations; but soundwaves, whose wavelengths are large, do bend around corners. It is because of its very tiny wavelengths that light seems to travel in straight lines—there is very little diffraction under normal circumstances. It is also for this reason that the ray model we discussed earlier was so successful.

Diffraction Sets a Limit on Magnification

Figure 12–6 illustrates a very important principle when applied to light waves. Objects whose dimensions are about the size of, or smaller than, the wavelength of the light that illuminates them cannot be clearly seen. For example, to view very small objects we usually use a microscope; the light used to illuminate the object is either reflected off the surface of the object or, more commonly, is transmitted through the object from below, Fig. 12–7. In the latter case, we "see" the object because it stops the light from passing through —we see its shadow.

If an object viewed in a microscope is not much larger than the wavelength of the light used, diffraction effects will be very important, either giving rise to a distorted image or, if the object is really small, to no image at all—see the corresponding effects for water waves in Figs. 12–6(A) and (B). Of course, if the object is much larger than the wavelength of light used, diffraction is not important and a clear image can be formed. The wave nature of light thus puts a limit on the size of the tiniest object, or parts of an object, that can be clearly seen. This is summarized in the following general rule:

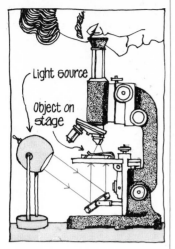

Figure 12–7

Illumination of an object in a microscope.

It is not possible to observe objects clearly whose dimensions are smaller than the wavelength of the light used to illuminate them.

The useful magnification of microscopes using visible light, even if the lenses were perfect, is limited to about 1000 times because of diffraction. Further magnification would merely result in magnifying the distorted patterns caused by diffraction around the edges of tiny objects. However, electron microscopes have been built, which have much greater magnification because, as we shall see in Chapter 15, electrons have very small wavelengths.

The Wave Theory Explains Refraction

The wave theory of light not only accounts quantitatively for the phenomena of interference and diffraction; it also accounts for the refraction of light. The wave theory, but not the particle theory, can explain why light rays bend toward the perpendicular when they enter a region where the velocity of light is less, and away from the perpendicular when the velocity is greater. To see this, we make use of an analogy in which a line of soldiers walking arm in arm represents a wave front or wave crest; see Fig. 12–8(A). At first the

Figure 12–8

(A) A line of soldiers march from firm ground into mud where their marching speed is reduced. Those that reach the mud first are slowed down first. (B) A wave front acts similarly, explaining why a ray bends toward the perpendicular upon entering a medium in which the speed of light is less.

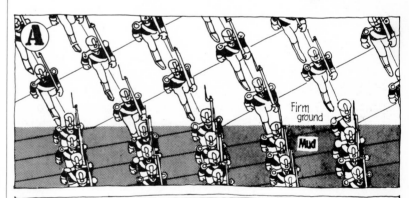

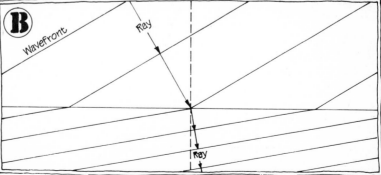

327

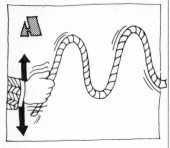

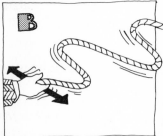

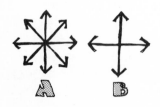

Figure 12–9

A transverse wave polarized
(A) vertically and
(B) horizontally.

Figure 12–10

An unpolarized wave has
vibrations in all directions (A);
these vibrations can be
resolved into two mutually
perpendicular components,
as in (B).

soldiers are marching on hard pavement, but then they approach a muddy region obliquely and they are slowed down. The soldiers who reach the mud first are slowed down first and the line of soldiers bends, Fig. 12–8(A). The same thing happens to a wave front, whether it's a water wave or a light wave. The first part of the wave front to enter the slower medium starts to drag first and the wave front bends. The rays, which represent the direction in which the waves are moving and thus are perpendicular to the wave fronts, are bent toward the perpendicular, as found experimentally. If the light passes into a medium in which its speed is greater, the opposite effect occurs. Thus, the wave theory of light explains the law of refraction.

2. POLARIZATION

Once the evidence had come in that light travels as a wave, a new question arose: are light waves longitudinal or transverse? The properties of light waves we have studied so far would occur for either.

Thomas Young showed that light is transverse because it displays the phenomenon of **polarization**, which occurs only for transverse waves and not for longitudinal waves. To see what is meant by polarization, let us first examine a transverse wave, such as a wave on a rope. A rope may be set into vibration in a vertical plane as shown in Fig. 12–9(A), or in a horizontal plane, Fig. 12–9(B). In either case, the wave is said to be **plane polarized**. A transverse wave can also be **unpolarized**, which means that the source has vibrations in many planes at once, as in Fig. 12–10(A). These various planes of vibration can be resolved into components (like vectors; see Chapter 3) along the horizontal and vertical directions, as shown in Fig. 12–10(B). Thus, an unpolarized wave can be considered as the sum of polarized beams in the horizontal and vertical directions.

Now, if we place a barrier with a slit cut in it in the path of a transverse wave, Fig. 12–11, a plane polarized wave will pass through if it is polarized parallel to the slit (A) but will be blocked if it is perpendicular to the slit (B). Thus if two such barriers with slits are placed one behind the other, one with its slit vertical, the other with its slit horizontal, a plane polarized wave will be stopped by one or the other of them. Even if the waves are unpolarized, they cannot pass through. For, as shown in (C), the first (vertical) slit eliminates all but the vertical component so that the unpolarized wave is turned into a plane polarized wave; and this in turn is eliminated by the second sheet. It is important to note that a longitudinal wave would not be stopped by such a pair of slits.

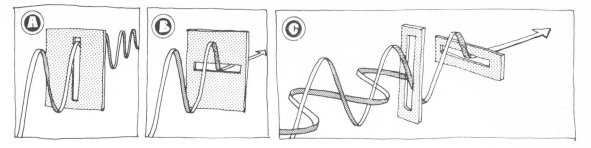

Figure 12–11

A vertically polarized wave can pass through a vertical slit (A) but is blocked by a horizontal slit (B). In (C) an unpolarized wave becomes plane polarized by one slit and is completely stopped by two slits that are perpendicular to each other.

Now you can show for yourself that light is a transverse wave:

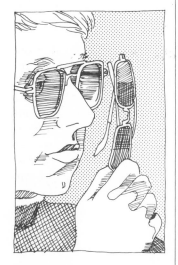

Figure 12–12

Looking through two polarizers. When they are "crossed" very little light passes through.

EXPERIMENT

Obtain two pairs of polarizing sunglasses (Polaroid is one brand). Put on one pair and, keeping only one eye open, place one lens of the other pair in front of your open eye, as shown in Fig. 12–12. Now rotate this lens. You should see that the light coming through the two lenses nearly vanishes at one orientation and is brightest at an orientation perpendicular to this.

The barriers with slits in Fig. 12–11 are called **polarizers**. They are so called because if an unpolarized wave encounters one such barrier, the transmitted wave is plane polarized. Each of the lenses of polarizing sunglasses is a polarizer for light waves. Such polarizing material is generally made of certain long molecules that are all aligned parallel to one another; consequently, they create the equivalent of the slits shown in Fig. 12–11 and permit only one component of polarization to pass through. The light is dimmest when the polarizers are "crossed" (the "slits" crossed) and brightest when their "slits" are parallel.

The light from most natural sources, such as the sun and ordinary light bulbs, is unpolarized. When such light passes through a polarizing sheet, it comes out plane polarized. In this process, 50 percent of the unpolarized incident light is blocked. Thus, polarizing sunglasses cut out much more light than ordinary sunglasses with the same tint.

329

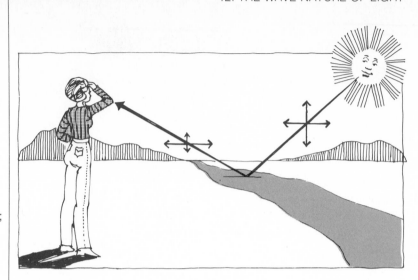

Figure 12-13

Light reflected from most surfaces is partially polarized; the vertical component is absorbed more by the surface.

Figure 12-14

Photograph of the surface of water (A) allowing all light into the lens and (B) using a polarizer that absorbs most of the light reflected from the surface; hence the river bottom can be seen more clearly.

The light reflected from most nonmetallic surfaces is partially polarized. That is, one component is absorbed more than the component perpendicular to it. You can check this by rotating polarizing sunglasses while looking at a lake, a window, or a highway on a bright day. Generally, the light reflected from a surface has less vibration perpendicular to the surface than parallel to it, Fig. 12-13. Since most reflecting surfaces outdoors are horizontal—including highways—the lenses of polarizing sunglasses are usually oriented vertically so as to eliminate the stronger horizontal component. That's why they are so effective in reducing glare. People with fishing experience know that they can see into the water better when they use polarizing sunglasses, Fig. 12-14.

3. COLOR

Before the time of Newton, white light had been considered to be a pure and uniform substance. But Newton performed a simple experiment that showed this wasn't true. He allowed a thin beam of sunlight to impinge on a glass prism and observed that a variety of colors was projected onto the wall on the opposite side of the room. Although some observers believed that the prism somehow *altered* the white light, Newton interpreted the experiment to mean that white light, instead of being pure, is made up of many colors: the colors of the rainbow. To support his argument, Newton tried the reverse experiment. Using a second prism in conjunction with the first, Fig. 12–15(B), he found that the various colors were indeed recombined into white light. This was convincing evidence that white light is a combination of all the colors of the rainbow.

But what is color? What we perceive as color is really a physiological sensation interpreted by our brain. But is there something physical that characterizes the different colors? Some quantity we can measure that is different for different colors? Indeed there is.

Wavelength Corresponds to Color

In his double slit experiment, Thomas Young found that the bright interference lines of Fig. 12–3 are not as far apart for violet light as they are for red light. Young concluded that the different colors of light must correspond to different wavelengths and therefore different frequencies. From the geometry of the double slit experiment, he calculated the wavelength of violet light to be about 4×10^{-7} meters and red light to be about 7×10^{-7} meters. Other colors of the rainbow were found to have wavelengths in between these two. Note that if white light is used in the double slit experiment, each bright interference line, except the central one, is actually a tiny rainbow, a band of colors.

Figure 12–15

(A) White light is broken up into the colors of the rainbow by a prism. (B) The colors are recombined into white light with the use of a second prism.

331

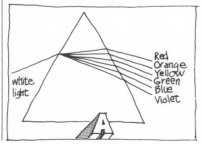

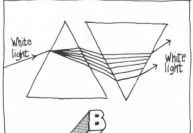

Thus, different frequencies or wavelengths are perceived by our eyes and brain as different colors. This is similar to sound, in which sounds of different frequencies are heard as different pitches. The spectrum of visible-light wavelengths is shown in Fig. 12–16 (see color plate) along with the corresponding colors. Note, however, that the colors flow into one another and that there are many shades in between. The wavelengths are given in nanometers (a nanometer, abbreviated nm, is sometimes also called a "millimicron," meaning millimicrometer, which is abbreviated $m\mu$; 1 nm, or 1 $m\mu$, is defined as 10^{-9} m). White light is not shown in Fig. 12–16 since it does not correspond to a single wavelength. White light can be produced by mixing all wavelengths, or from certain combinations of two or more different wavelengths or colors, as we shall see later.

If you see a particular color, say yellow, it may be a single wavelength, perhaps 570 nm. On the other hand, a mixture of the wavelengths 660 nm and 530 nm, red and green, also produces the sensation of yellow. For some reason, our brains interpret mixtures of wavelengths or colors as other colors. When red and green are mixed we see a yellow that looks just like the pure yellow of 570-nm wavelength. The mixtures of colors or wavelengths also produces colors such as brown that do not correspond to any color in the natural spectrum.

The Speed of Light Is Different for Different Wavelengths

In a vacuum, all wavelengths or colors of light travel at the same speed. This is almost true in air, too. But in most transparent materials, like glass or water, the different wavelengths are slowed down by different amounts. This is why white light is split up into its component colors by a prism. The speed of light in glass is less for violet than it is for red; thus violet is bent toward the perpendicular more than red is, Fig. 12–15(A).

Rainbows are another spectacular example of this phenomenon. You can see a rainbow when you look at rain or a waterfall with the sun at your back. White sunlight hitting droplets of water in the air is broken up into its component colors just as it is in a prism. Figure 12–17 shows how the different colors are bent differently at the surface of a water droplet, reflect off the back surface, and reach the eye of a distant observer. Note in Fig. 12–17(B) that the different colors perceived by the observer originate from different droplets of water. The rays of red light, since they are bent the least, reach the observer's eye from droplets that are higher in the sky, and hence the top of the rainbow is red.

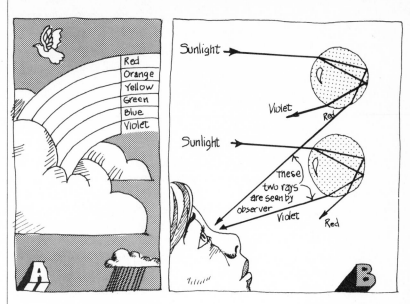

Figure 12–17

(A) A rainbow. (B) A rainbow is produced by the different amounts of bending of the different colors in the water droplets. Notice that a ray of red light enters the eye from a higher angle than a ray of violet light. Thus the red light appears to come from higher in the sky.

Where Do Objects Get Their Color?

Some objects have a certain color because they emit light of particular wavelengths; these emitters or sources of light include light bulbs, neon signs, "red-hot" iron, the sun, and the stars.

But most objects in our environment appear colored not because they are sources of light but because they reflect part of the light that strikes them. Normally we see objects illuminated by white light —from the sun or from electric light bulbs—so all visible wavelengths are incident on the object. Depending on the materials from which the object is made, certain of these wavelengths are absorbed and others are reflected, and it is the reflected light that reaches our eyes. A rose is red because the wavelengths corresponding to red are reflected from its petals and other wavelengths are absorbed. Leaves are green because only green light is strongly reflected. Some clothes are yellow, others red or blue, depending on the wavelengths of light that the material reflects. An object that appears white reflects all wavelengths. A black object reflects almost no light at all; it absorbs essentially all the light that strikes it.

What happens to the light energy that an object absorbs? It is transformed into heat. Thus a black object will be warmer than a white object when subjected to the same illumination. This is why in hot weather people feel more comfortable in light-colored clothing.

The color of the light illuminating an object affects the color of the object. For example, if an object that appears red when illuminated by white light is instead illuminated by red light, it will still look red because it reflects red light. But an object that is green when illuminated by white light will look black when illuminated by red light, because it absorbs red light. Similarly, a red object illuminated by green light will look black, but a green object will still look green.

Why the Sky Is Blue and Sunsets Red

Because the molecules in a gas are far apart, most of the light incident on a gas passes through without being absorbed. However, the oxygen and nitrogen molecules of the air do "scatter" the light (absorb and reemit it) in all directions, and it is the violet and blue colors that they scatter the most. This is what gives air its bluish color and explains why if you look at the sky from any direction, other than directly at the sun, you see blue light, Fig. 12–18. Of course not all the blue light is scattered this way, nor does all the red and orange light pass through unscathed. But much less of the red is scattered than of the blue. Thus the sky looks blue.

The same mechanism gives rise to red sunsets. Late in the day the sun passes through a greater thickness of atmosphere than at midday. A good part of the violet and blue light, and even some of the green, has been scattered by the atmosphere. The light that remains and that passes through to the surface of the earth is thus

Figure 12–18

Air molecules scatter blue light more than other colors; so it is mostly blue light that enters an observer's eye (the observer on the left, for example). At sunset (observer on right) the light reaching the earth has had much of the blue removed and hence appears reddish when reflected off objects such as clouds.

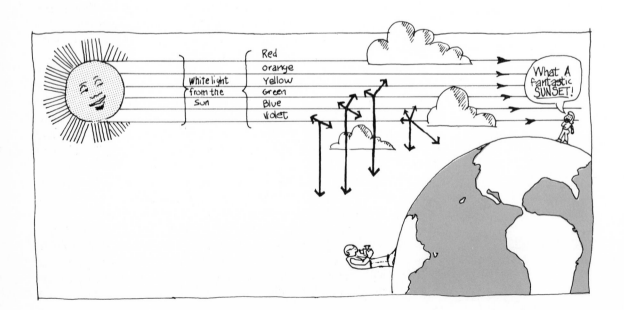

primarily the red, orange, and yellow wavelengths. Thus the light that reaches the earth is reddish-orange, the color we see reflected from clouds and dust near the surface of the earth at sunset.

Mixing Color Pigments— Subtractive Mixing

The mixing of different-colored pigments (or paints) to obtain different colors is part of the painter's art. Artists have long known that nearly all colors can be obtained from mixing only the three so-called **primary colors**, which may be taken as red, yellow, and blue. Actually, there are other three-color sets that can also produce a wide variety of colors when mixed. But the set that seems to produce the widest range, and that is considered today as the standard set of primary colors for mixing pigments, consists of the colors *magenta* (which is reddish-purple), *yellow,* and *cyan* (turquoise).

Each of these colors absorbs part of the white light that falls on it and reflects the rest. The range of frequencies that each reflects is indicated in Fig. 12–19. Let's see how we can produce green by mixing cyan and yellow. When white light falls on a mixture of cyan and yellow, the yellow pigment absorbs light at the blue end of the spectrum, as shown in Fig. 12–19, but reflects other frequencies. The cyan, on the other hand, absorbs all the wavelengths in the red region of the spectrum, whether they are from the original light or reflected by the molecules of the yellow pigment. The only light that won't be absorbed is that which both pigments reflect; this will be the wavelengths in the central part of the spectrum, which are predominantly green. Hence the mixture looks green. Because the result of this mixing depends on what colors are absorbed, or subtracted out, it is called **subtractive** mixing.

The result of mixing any two of the subtractive primary colors in equal amounts is shown in Fig. 12–20 (color plate). We saw how green was obtained from a mixture of cyan and yellow. See if you can make a similar analysis for the mixtures of magenta and yellow, and of cyan and magenta. Other colors are obtained by mixing two or three primary colors in *un*equal amounts. A mixture of all three in equal amounts produces black (center of Fig. 12–20), since light from all parts of the spectrum will be absorbed. Actually, gray or brown is sometimes obtained if the mixture is not exactly equal, or if the pigments are not precisely those indicated in Fig. 12–19.

Complementary colors are any two colors that, when mixed, produce black (or perhaps gray or muddy brown). For example, blue and yellow produce black, as in Fig. 12–20. Since the pigments of complementary colors have no common region of the spectrum in which they overlap, little or no light is reflected.

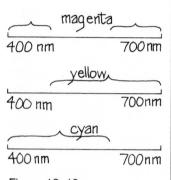

Figure 12–19

The subtractive primaries (for pigments) showing approximately the wavelengths reflected by each. (See also Fig. 12–16.)

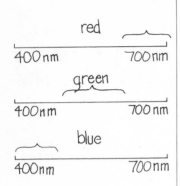

Figure 12–21

The additive primaries, showing approximately the spectrum of light each contains.

Mixing Colored Light— Additive Mixing

If you shine colored spotlights on a screen, new colors will be produced where they overlap. This kind of color mixing is called **additive** mixing, because the color you see is the *sum* of the spectrum of colors contained in each light. Again, nearly all colors can be obtained from three colors, but they are not the same as the subtractive primaries. Rather, the **additive primaries** are taken to be red, green, and blue, and the spectrum of light each contains is shown in Fig. 12–21. The superposition of a red light and a green light on a screen means that all the wavelengths each contains will be reflected by the screen, and so the screen will appear yellow (note the spectrum of yellow in Fig. 12–19). Other combinations of the additive primaries are shown in Fig. 12–22 (color plate). Mixing all three primaries together in equal proportions produces white, since all frequencies are present. And the mixing of two complementary colors also produces white (they are truly complementary only if one contains all the frequencies the other does not). For example, Fig. 12–22 shows that yellow and blue produce white.

Color television works on the principle of color addition. If you look closely at the picture on a color set, you will see that there are tiny dots (or lines) of the three primary colors, red, green, and blue. When you move away from the set, these tiny dots merge in your perception. If all three colors are equally bright, white results. A wide range of colors is achieved by varying the intensity of each of the colored dots. (A description of a TV tube is given in Chapter 13.) To prove to yourself that the eye can mix these colors itself, try the following experiment.

EXPERIMENT
Cut a filing card (or a piece of thin white cardboard) into a circle and paint one side with the three primary additives, as shown in Fig. 12–23(A), and the other side with red and green, as in (B). Make two small holes on either side of center and run a string through them. Hold the string in both hands and rotate the card until the string is very twisted, Fig. 12–23(C). When you release the card and pull gently on the string, the card will spin rapidly. One side will appear white and the other yellow. Explain.

If you look closely at a four-color illustration in a book, or a poster, you will see that it too is made up of many tiny colored dots

Figure 12–23

Experiment to show how the eye mixes color additively.

much like those on a TV screen. Here, however, some colors overlap, so that the subtractive process is involved as well as the additive. Again a wide variety of colors can be obtained by using three primary colors plus black (for emphasis and contrast).

The Impressionist Painters

In general, the process of mixing colors additively tends to produce a brighter color than mixing subtractively with pigments. This fact was used by many painters in the latter part of the nineteenth century, particularly those known as Impressionists and Postimpressionists. An example is Seurat's *Sunday Afternoon on the Island of La Grande Jatte*, Fig. 12–24. These painters tended to use only a small number of paints. They applied the colors with adjacent brush strokes (being sure any underlying paint was dry, so that no mixture of pigments occurred) and left it to the eye of the observer to do the mixing, much as color TV does today. This technique enabled them to produce intense and vibrant colors that seemed to capture the scintillating quality of sunlight and the texture of real objects. The fact that the appearance of the picture changes somewhat as the observer moves away gives the picture a kind of life of its own. Close up, the individual colors from each brushstroke are clear, whereas at a distance they fuse into a single color by additive mixture. When the observer stands at some intermediate point, the eye apparently sees the colors alternately fused and separated, and the painting seems to sparkle or scintillate.

Figure 12–24

Detail from Georges Seurat, *Sunday Afternoon on the Island of La Grande Jatte,* Courtesy of The Art Institute of Chicago.

Color Vision

The retina of our eyes contains two kinds of light receptors, **rods** and **cones**. The rods are very sensitive to light (they respond even to small amounts of light), but they do not distinguish colors. The cones, on the other hand, distinguish the different colors but require

337

brighter light to function. The eyes of most mammals other than humans and primates have only rods and thus do not perceive color.

Near the center of the retina is an area of very closely packed cones called the **fovea**. The sharpest image and the best color discrimination are therefore found in this area. When we view a scene, the eye moves so that the area of greatest interest is focused on the fovea. There are no rods in the fovea, but they are present in the regions surrounding it; in the peripheral regions of the eye, the rods predominate. You can prove this for yourself by having someone bring a brightly colored object from behind you so that you see it only peripherally. You won't be able to tell what color it is! When we see things directly, however, our brain usually remembers their color even when they are off to the side.

In very dim light, objects off to one side seem brighter than those focused at the fovea, because the rods are more sensitive to light. However, we can't perceive colors well in dim light, because the rods don't distinguish colors. Thus, although most stars appear plain white at night, long photographic time exposures reveal that some stars are blue, and others are red or orange.

> **EXPERIMENT**
> Sit in a dark room and let your eyes become adapted to the dark. Open a magazine to a colored picture. Try to tell what colors are present and where. Then turn on a bright light and see if you were right.

A number of theories have been proposed to explain how the cones distinguish different colors. The most likely theory assumes the existence of three kinds of cone corresponding to the three additive primary colors. Although the evidence is by no means conclusive, the fact that only three colors can produce all the colors we perceive gives support to this theory. Studies of colorblindness, Figs. 12–25 and 12–26 (color plates), provide further support. There are several kinds of colorblindness, each of which may be suffered to a greater or lesser extent. About 8 percent of men and 1 percent of women do not have normal color vision. The most common type is red-green colorblindness. A person so afflicted has difficulty distinguishing between red and green or else can't distinguish them at all and sees them both as gray. All the colors they are able to distinguish can be obtained by mixing only two colors. This is consistent with the idea that there are three types of cone and that in colorblindness one type does not function properly. However, scientists have not been able to confirm this directly.

The rods and cones in the eye are connected by way of specialized cells to nerves that carry the information to the brain, which produces the sensation of color and arranges the information from

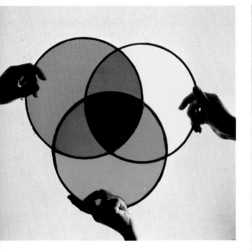

Figure 12-16

The spectrum of visible light, indicating the range of wavelengths for the different colors. A millimicron (mμ) is the same as a nanometer (10^{-9} m).

Figure 12-20

Mixing of the subtractive primaries. In this case, light is being passed through colored filters; each subtracts a portion of the spectrum and when they overlap, additional wavelengths are subtracted out. The result is subtractive mixing as for pigments.

Figure 12-22

Mixing of the additive primaries by shining colored lights on a screen.

400 mμ 500 mμ 600 mμ 700 mμ

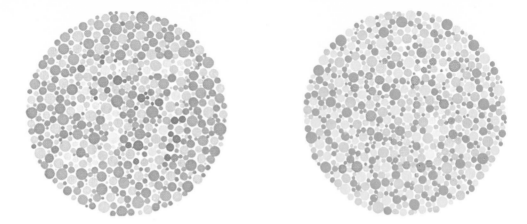

Figure 12-25

A test for colorblindness. Left, certain people who are red-green colorblind see only the number 5, others see only the number 7, and still others see no number at all. Right, a person with red-green colorblindness sees no number; a person with normal vision sees the number 15.

Figure 12-26

A person with normal color vision sees this Persian miniature painting as it appears in the upper left. Upper right: The same painting as it would look to someone with red-green colorblindness. A person with yellow-blue blindness sees the painting in the lower left, and a totally colorblind person sees it as it appears in the lower right.

the different receptors into a coherent whole, the "picture" we see. Actually, some of the receptors (rods and cones) are interconnected within the eye itself, so that some of the "thinking" or organizing is actually done there.

REVIEW QUESTIONS

1. The spreading of a light beam after it passes through a small hole is called
   (a) refraction
   (b) diffraction
   (c) defocusing
   (d) focusing

2. Young's double slit experiment demonstrated that
   (a) light travels as rays
   (b) light has a wave nature
   (c) light has a finite speed
   (d) different colors correspond to different speeds

3. Interference is a phenomenon that may occur in the case of
   (a) sound waves
   (b) water waves
   (c) light waves
   (d) all of the above

4. We can hear sound around corners but we cannot see around corners because
   (a) sound travels slowly
   (b) light has a shorter wavelength
   (c) the amplitudes are different
   (d) the speed of light is greater

5. In "plane polarized light" the light
   (a) vibrates in one plane only
   (b) vibrates in several planes
   (c) travels in a plane path
   (d) reflects only from plane surfaces

6. The color of light depends on its
   (a) amplitude
   (b) intensity
   (c) wavelength
   (d) velocity

7. An object that reflects all colors equally when illuminated by sunlight appears to be
   (a) red
   (b) yellow
   (c) black
   (d) white

8. A book appears blue in sunlight because the blue light is
   (a) diffracted
   (b) reflected
   (c) absorbed
   (d) transmitted

9. Mixing of pigments is called subtractive because
   (a) you have to subtract one wavelength from another to find their resultant
   (b) some of the wavelengths are subtracted out
   (c) all wavelengths are present and none is subtracted
   (d) those wavelengths are subtracted that correspond to the color of the object

10. We see a color picture on a TV set because of
    (a) subtractive mixing by pigments on the screen
    (b) subtractive mixing within the set
    (c) additive mixing of colors inside the set
    (d) additive mixing of colors by the eye

11. What are the three additive primary colors?
    (a) red, yellow, and blue   (c) blue, green, and yellow
    (b) red, yellow, and green   (d) blue, green, and red

EXERCISES

1. Why is light sometimes described as rays and sometimes as waves?

2. Our experience indicates that light travels in straight lines and does not bend around corners noticeably; yet sound does bend readily around corners. Explain.

3. Atoms have diameters of about 10^{-8} cm. Visible light wavelengths are about 5×10^{-5} cm. Can visible light be used to "see" an atom? Why or why not?

4. If magnifications greater than $\times 1000$ were used in a light microscope, the interference patterns formed by diffraction around objects in the field of view would be evident. Do you think therefore that greater magnifications, say $\times 5000$, would give a misleading picture of what the object actually looks like?

5. Use an argument, similar to that used for Fig. 12–8, to show why light entering a medium where its speed is faster will bend away from the perpendicular.

6. Is it better to use red light or blue light when photographing a tiny object through a microscope when you want to obtain maximum definition?

7. When a glass lens is placed on a flat glass surface and light is shone from above, Fig. 12–27(A), a series of rings ("Newton's rings") is produced, Fig. 12–27(B), due to interference between rays of light reflected from the two nearly touching surfaces. (Note the different path lengths for the two rays shown.) Give a brief explanation of why rings are produced and why they have a rainbow of colors.

Figure 12–27
Newton's rings.

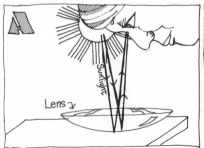
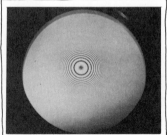

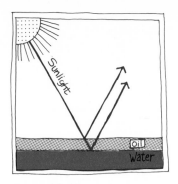

Figure 12–28

A film of oil on top of water reflects various colors.

8. A water puddle with a film of oil on top reflects light with many colors, like a rainbow. Explain, using the fact that light can reflect not only from the top surface of the oil but also from the surface where the oil meets the water (Fig. 12–28).

9. How could you use polarizing sunglasses to tell if light from the sky (not directly from the sun) is polarized? Try it and see.

10. How would you go about determining if a given pair of sunglasses was polarizing or not?

11. A single polarizer ideally transmits 50 percent of the unpolarized light falling on it. How much will be transmitted if two screens are used (a) with their axes parallel, (b) with their axes crossed at 90°.

12. Give an example of a nonspectral color. What is meant by a nonspectral color?

13. Why are you more comfortable on a hot day in light-colored clothes than in dark clothes?

14. What is the physical difference between red and orange light?

15. Describe the physical difference between the light from an object that appears yellow and one that appears green.

16. If the earth's atmosphere were fifty times thicker than it is, would ordinary daylight still seem white, or would it be some other color?

17. A performer wears blue clothes. How could you use spotlights to make them appear black?

18. A red spotlight and a blue spotlight overlap. What color will they produce?

19. When white light passes through a flat piece of window glass, it is not broken down into its spectrum. Why not?

20. An object appears magenta when illuminated by white light (see Fig. 12–19). What color will it appear if it is illuminated by (a) blue light, (b) yellow light, (c) red light?

21. An object appears green in white light. What color will it appear to be if illuminated by (a) magenta light, (b) cyan light, (c) pure blue light?

22. Which colors are complementary to (a) magenta, (b) red, (c) green?

23. Cats do not see in color. How do their eyes differ from ours?

24. Why do objects seen in the moonlight appear so colorless?

25. Impressionist (and other) painters found that shadows are more accurately portrayed if the objects in the shadows are tinted with their complementary color. Explain.

ELECTRICITY

13

Physicists now believe there is a total of four different kinds of force in nature: (1) the gravitational force, (2) the electric force and the related magnetic force, (3) the strong nuclear force, and (4) the weak nuclear force. We have already discussed gravity, and we will discuss the strong and weak nuclear forces in Chapter 16. In this chapter and the next, we will discuss electric and magnetic forces.

Which of these forces do you feel is the most important in our lives? Gravity? The nuclear forces? The electric force? Of course, life would be impossible if any one of these forces ceased to exist. But if we had to designate one as the most important of all, it would undoubtedly be the electric force. It is the electric force that holds atoms and molecules together in liquids and solids. Indeed, we are held together by the electric forces operating between the molecules that make up our bodies. All our bodily metabolic processes occur because of electric forces. Whenever you push or pull an object with your hand, it is the electric forces between the molecules of your hand and those of the object that are responsible for the object's movement.

1. ELECTRIC CHARGES AND FORCES

The word electricity derives from the Greek word *elektron,* meaning amber. Amber is petrified resin and was known by the ancients to have a remarkable property: when rubbed with a piece of fur, a piece of amber would attract small pieces of leaves and dust. This was known as the "amber effect," or—to use the Greek derivative—electricity.

Today we know that many other materials, such as hard rubber, nylon, plastic, and glass, display the amber effect. You can demonstrate the amber effect with readily available materials:

Figure 13–1

Bringing a small plastic ruler up to a few tiny pieces of paper.

EXPERIMENT

Place some tiny pieces of paper (1 cm square or so) on a table. Hold a plastic ruler or comb directly above and close to the bits of paper, Fig. 13–1. Does anything happen? Now rub the plastic ruler or comb with a clean, dry paper towel or piece of wool and again place it directly above and close to the bits of paper. What happens now? Note that the plastic must be rubbed often if it is to maintain its ability to attract.

Today we usually refer to the amber effect as "static electricity." You may have experienced it when combing your hair, or taking a shirt made of synthetic fiber out of a clothes dryer, or touching a door knob after walking across certain kinds of rug. In each case an object becomes "charged" by a rubbing process and is said to possess an **electric charge**.

There Are Two Kinds of Electric Charge

Investigation into the nature of electricity did not really begin until the seventeenth and eighteenth centuries. The first step was the discovery that not only does a rubbed piece of amber attract bits of paper but that two rubbed pieces of amber will *repel each other.* This can be demonstrated by suspending one freshly rubbed plastic ruler from a thread and bringing a second freshly rubbed plastic ruler very close to it, but not touching, Fig. 13–2(A). Similarly, two rods of glass rubbed with a piece of silk will repel each other, Fig. 13–2(B). *But,* if the glass rod is brought near the plastic ruler—each having just been rubbed with the appropriate material—the glass and the plastic will *attract* each other, Fig. 13–2(C).

Clearly the electric charges on the glass and on the plastic must somehow be different. When we test other materials that can be charged by rubbing, we find that they fall into one of two categories: either they are like the plastic and are attracted to rubbed glass and repelled by rubbed plastic, or they are like the glass and are repelled by rubbed glass and attracted to rubbed plastic. Thus there seem to be two distinct kinds of charge. One kind of charge is that found on rubbed glass, the other kind is that found on rubbed plastic. One kind of charge attracts the other kind but repels a charge of the same kind:

Like charges repel. Unlike charges attract.

Benjamin Franklin (1706–1790) introduced the terms **positive** and **negative** to describe the two kinds of electric charge. Which would be called positive and which negative was of course an arbitrary choice. Franklin had to choose, and we still follow his convention today. According to this convention, a glass rod acquires a **positive charge** and an amber or plastic object acquires a **negative charge**, Fig. 13–2.

Figure 13–2

Like charges repel. Unlike charges attract.

Electric Charge Is Conserved

Franklin contributed significantly to the understanding of electricity. He demonstrated the electrical nature of lightning in 1752 with the use of a kite. He was the first to realize that in any process in which a positive charge is produced, an equal amount of negative charge must be produced at the same time. For example, when a glass rod is rubbed with silk, the glass acquires a positive charge and the silk acquires an equal amount of negative charge. The algebraic sum of these two charges is zero. This is an example of the **conservation of electric charge**. It is a conservation law that is as well established experimentally as the laws of conservation of energy and conservation of momentum. It simply tells us that the *net* amount of charge produced in any process is always zero. Electric charge is neither created nor destroyed.

Electric Charge Originates in the Atom

Only during the past century have we learned where electric charge comes from. According to the atomic theory, any piece of matter is made up of a vast number of atoms, and each atom consists of negatively charged electrons orbiting a positively charged nucleus. In a normal atom the positive and negative charges balance each other, so the atom is neutral. When an atom contains too many or too few electrons, it has a net negative or a net positive charge and is called an "ion." In a solid material the nuclei of atoms are rather rigidly fixed with respect to one another, but the electrons are relatively free to move. And the electrons in some materials, such as wool, are held less tightly than those in others, such as plastic. Thus, when wool is rubbed over a plastic ruler, some of the electrons are swept off the wool onto the plastic; the wool, having lost some of its electrons, will have a net positive charge, and the plastic, which now has an excess of electrons, will have a net negative charge.

Since most objects are electrically neutral—that is, they have no net charge—the electric force is ordinarily not obvious. Only when we do something special, like rubbing a plastic ruler with a wool cloth, do we bring about a transfer of electrons and thus a net charge on the plastic and on the cloth. But even in this case, both the rod and the cloth become neutral again after a short time. This happens because water molecules in the air (moisture) carry loosely held electrons that are attracted to the positive rod and eventually neutralize it; and electrons on the cloth will slowly "leak off" onto the water molecules in the air. Thus, static electricity is most noticeable on dry days when there is relatively little water vapor in the air.

Coulomb's Law: A Quantitative Description of the Electric Force

When one charged body attracts or repels a second charged body, as in Fig. 13–2, some kind of force is clearly responsible. We call

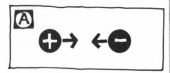

Figure 13–3

(A) Unlike charges attract.
(B) and (C) Like charges
repel. Arrows represent the
force on each charge due to
the other charge.

this force the **electric force**. When a positively charged body attracts a negatively charged body, the electric force is said to be
attractive; each of the bodies exerts an attractive force on the other,
Fig. 13–3(A). If the two bodies have the same charge, each exerts
a **repulsive** force on the other, Figs. 13–3(B) and (C). Electric forces
act even though the two bodies do not touch. As with the gravitational force, the electric force acts "at a distance." The French physicist Charles Coulomb (1736–1806) investigated these electric forces
quantitatively and in 1789 announced the results of his investigations.

Using the apparatus shown schematically in Fig. 13–4, Coulomb
found that the force exerted by either charged ball on the other is
directly proportional to the charge on each of them. Thus, when the
charge on one ball is doubled, the force is doubled. When the
charge on both balls is doubled, the force is four times as great. If
either charge is zero, the force is zero. If we use q_1 to represent the
amount of charge on one body and q_2 to represent the amount of
charge on the second body, we can say that $F \propto q_1q_2$, where, as
usual, "$\propto$" means "is proportional to."

This relationship holds true as long as the distance between the
two bodies is not changed. If the distance is changed, the force
changes as well. To determine just how the force varies with dis-

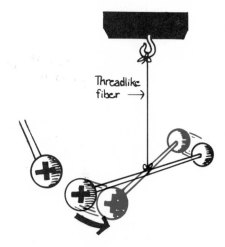

Threadlike
fiber →

Figure 13–4

Schematic diagram of Coulomb's apparatus, which is much like that used
by Cavendish to measure the gravitational force (Chapter 3). When the
charged ball (left) is brought close to the one attached to the suspended
bar, the bar turns slightly. The threadlike fiber resists the twisting motion
so that the angle of twist is proportional to the applied force. Using this
apparatus, Coulomb investigated the strength of the electric force as a
function of the magnitude of the charges and of the distance between
them.

tance, Coulomb measured the strength at various distances while keeping the charge on each ball unchanged. He found that as the distance between the bodies is increased, the force diminishes quite rapidly. Quantitatively, if the distance between the two spheres is doubled, the force is only ¼ as strong. If the distance is tripled, the force is only ⅑ of the original force. Noting that $\frac{1}{2}^2 = \frac{1}{4}$ and $\frac{1}{3}^2 = \frac{1}{9}$, Coulomb concluded that the force between the two charged balls varies inversely as the square of the distance between them. Mathematically, $F \propto \dfrac{1}{d^2}$, where d represents the distance between the two charged bodies.

The dependence of the force on the three variables—the charge on each of the two bodies and the distance between them—can be written as a single equation: $F \propto \dfrac{q_1 q_2}{d^2}$. This will be an equality if we insert a constant of proportionality, call it k:

$$F = k\,\frac{q_1 q_2}{d^2}.$$

This is known as Coulomb's law.

This equation gives the magnitude of the force that one charged object exerts on another. The direction of the force is always along an imaginary line that can be drawn between the two charges; it points away from the other charge if both charges have the same signs, and toward the other charge if they have opposite signs. (See Fig. 13–3.)

The most commonly used unit of measure for electric charge is called the coulomb, in honor of Charles Coulomb. It is always used in conjunction with metric units: meters for distance and newtons for force. The constant of proportionality, k, has the value 9×10^9 newton·meters²/coulomb².

Coulomb's law looks much like the universal law of gravitation, $F = G m_1 m_2 / d^2$. Charge plays the same role for the electric force that mass does for the gravitational force. These two forces, however, are completely independent phenomena. Despite the similarity of the two force laws, there are some very important differences between them. The gravitational force is always an attractive force; the electric force can be either attractive or repulsive. Furthermore, electric forces in general are vastly stronger than gravitational forces. For example, the electric force between two electrons with a charge of 1.6×10^{-19} coulombs is 10^{40} times greater than the gravitational force between them. When dealing with charged objects, the gravitational force can usually be ignored compared to the electric force. Only in the case of a massive body like the earth or the sun is the gravitational force significant enough to be noticed. At the atomic level, the gravitational force is always inconsequential compared to the much stronger electric force.

Charges Can Be Induced

Why does a plastic ruler that has been charged by being rubbed attract a piece of paper that has no charge on it? After all, Coulomb's law tells us that, since one of the objects has zero charge, the force on it must be zero. To see how an attraction can exist between a charged object and a neutral (uncharged) object, we must take into account the atomic nature of matter. In many materials a few of the electrons that normally orbit the nuclei of the atoms are relatively free to move away from their parent atoms. Consequently, when a negatively charged plastic ruler is brought near a small neutral object, the electrons within the neutral object feel a repulsive force and tend to move away from the end of the object nearest to the ruler. That end is left with a slight positive charge; and the opposite end, which now has a slight excess of electrons, will be negatively charged, Fig. 13–5(B). The object still has a net charge of zero, but there has been a *separation* of charge. This separation of charge is said to be "induced" by bringing the charged ruler near the neutral object. Referring again to Fig. 13–5(B), we see that the negatively charged ruler must exert an attractive force on the positive end of the object and a repulsive force on the negative end. Because the positively charged end is closer to the ruler, Coulomb's law tells us that the attractive force must be greater than the repulsive force. Thus the net force on the object is attractive.

Even in those materials in which there are practically no free electrons, the electrons are still free to reorient themselves within each atom. If a negatively charged rod is brought close to such a material, the electrons within each atom will be repelled by the rod and a separation of charge within each atom will occur, as shown in Fig. 13–5(C). The positive part of each atom will be closest to the rod, and again a net attraction will occur.

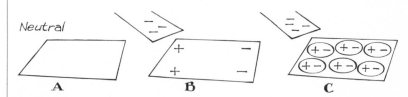

Figure 13–5

(A) A neutral object. (B) A negatively charged body brought close to, but not touching, the neutral object induces a separation of charge in the latter as a few electrons are repelled to the opposite end. (C) Even when there are no free electrons, the electrons in each atom orient themselves as shown.

Conductors, Nonconductors, Semiconductors

We know that electrons can migrate within a solid material. But how many electrons are mobile, and how mobile are they? In nature nearly all materials can be classified into one of two categories according to how free their electrons are to move.

The first category includes materials with a great many electrons that are very free to move. These materials are called "conductors," because they can be used to carry, or conduct, electric charge from one place to another. Conductors include all metals, such as copper, aluminum, iron, and silver.

The second category includes materials with very few electrons that are free to move. These materials are called "nonconductors" and are very poor conductors of electricity. We can use nonconductors to prevent an electric charge from going where it isn't wanted. Hence they are sometimes called "insulators." Some common insulators are glass, rubber, plastic, and air.

Nearly all common materials fall into one or the other of these two disparate categories. There are a few intermediate materials, however, such as carbon, germanium, and silicon, which are called "semiconductors." These materials are used in transistors and other electronic devices, as we shall see.

Electric Field: A New Concept

Most of the forces we are familiar with in everyday life might be called "contact" forces. Contact is required between whatever it is that causes the force and whatever it is that feels the force: as when a man *pushes* (exerts a force on) a lawn mower, or when a tennis racket hits (and exerts a force on) a tennis ball. In other words, when an object is subjected to a force, something has to be "there" to exert the force.

But now we have encountered two forces, the gravity force and the electric force, that act at a distance—that is, they act even when the bodies involved are not touching. The idea of a force "acting at a distance" was difficult for early thinkers to accept, and even today most people find it troublesome.

One way to reduce this difficulty is to introduce the concept of a **field**. We can visualize an electric charge as giving rise to an **electric field** that extends outward from the charge and permeates all of space, as in Fig. 13–6(A). Consequently, the force that one electric charge exerts on a second charge can be described as the interaction between one charge and the field set up by the other. In Fig. 13–6(A) an electric field exists at all points in space due to the charge q_1. A second charge q_2 is placed near it, Fig. 13–6(B);

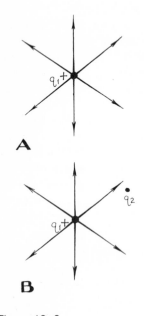

Figure 13–6

(A) Charge q_1 gives rise to an electric field. (B) Charge q_2 feels a force due to the field of charge q_1.

350

the charge q_2 feels a force due to the electric field of q_1. Thus, an electric charge feels a force because an electric field is there to exert the force. Thus the concept of electric field brings us a little closer to the idea of a contact force. Remember, though, that a field is *not* some kind of matter.

Even if we cannot see a charge, we are still aware of its existence because of its electric field. We can detect its electric field at any point by observing whether or not a force is exerted on another charge placed at that point.

Mathematically, the magnitude of the electric field is defined as force per unit charge. In other words, if a charge q feels an electric force F at some point in space, then the electric field at that point in space is defined to be

$$E = \frac{F}{q}.$$

The electric field, like force, has a magnitude and a direction and thus is a vector. It is very useful to represent electric fields pictorially by drawing lines to indicate the direction of the electric field, as was done in Fig. 13–6. The **electric field lines** or **lines of force** are drawn to indicate the direction of the force on a small[1] positive charge placed in the field. The lines then indicate the path that a tiny positive charge would follow if released in this field. For example, the lines of force around a single positive point charge are directed outward, as in Fig. 13–7(A); this is because a positive "test" charge would be repelled from it and move in a straight line directly away. Similarly, the direction of the field around a single negative point charge points inward [Fig. 13–7(B)], since a positive charge would be attracted toward it. Notice in Figs. 13–7(A) and (B) that the lines of force are closest together near the charge. This is just the region where the force on a second charge, and hence the electric field, is strongest. This is a general property of lines of force: the closer together they are, the stronger the electric field.

Figure 13–7

Electric field lines due to different arrangements of electric charges.

[1]We use a *small* electric charge, for "testing," so that the electric field of this charge can be ignored compared to the electric field already there.

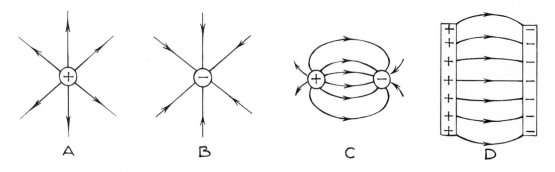

A   B   C   D

Figure 13–8
Gravitational field lines of the earth.

Figure 13–7(C) shows the electric field surrounding two charges of opposite sign, and Fig. 13–7(D) shows the field between two oppositely charged parallel plates. Notice in each case that a small positive test charge would be repelled by the positive charge and attracted by the negative charge. The path taken by a small positive charge would be along one of the lines of force.

The concept of electric field gives us a better sense of how one electric charge attracts another. But note in Figs. 13–6 and 13–7 that only a few of the possible field lines are drawn in each case. Note too that the electric field is not confined to these lines but is continuous and exists between the lines as well.

The concept of a field can also be used for the gravitational force; it too acts "at a distance." A **gravitational field** surrounds every material object. One object attracts another by virtue of its gravitational field, Fig. 13–8. When we drop an object it falls in the gravitational field of the earth. Similarly the earth feels the effect of the gravitational field of the sun.

Electric Potential

As we know, a body placed in a gravitational field feels a force. If it is allowed to fall, the body will gain kinetic energy as it loses potential energy. A particle has potential energy because of its *position* in a gravitational field, and the total energy, kinetic plus potential, is conserved (see Chapter 5). An electrically charged body placed in an electric field also has potential energy by virtue of its position in that field. Take for example the case of two oppositely charged parallel plates, as in Fig. 13–9. A positively charged particle released near the positive plate will, since a force is exerted on it, be accelerated to the right toward the negative plate. As it moves to the right its velocity, and hence its kinetic energy, increases. Conservation of energy tells us that as the kinetic energy increases, the potential energy decreases. A positively charged particle thus has more potential energy when it is near the positive plate than when it comes near the negative plate (see Fig. 13–9). This is in accord with the idea that potential energy is the ability (or "potential") to acquire kinetic energy, or to do work.

A more useful concept than the electric potential energy itself is the **electric potential**; it is defined as the electric potential energy per unit of electric charge and is usually represented by the symbol V. Thus,

$$V = \text{electric potential} = \frac{\text{electric potential energy}}{\text{electric charge}}.$$

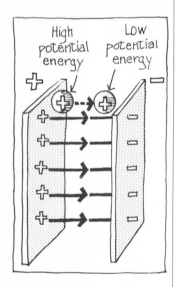

Figure 13–9

A positively charged particle moves away from the positive plate toward the negative plate. It gains kinetic energy but loses potential energy.

The unit of measurement of electric potential is joules/coulomb.

It is called a volt (abbreviated V) in honor of the Italian scientist who invented the battery, Alessandro Volta (1748–1827): 1 volt = 1 joule/coulomb.

Electric potential is a measure of how much energy an electric charge can acquire in a given situation. It can be compared to the height of a cliff: the greater the height difference between the top and bottom of the cliff, the more kinetic energy a falling rock will have when it reaches the bottom. Similarly, the greater the electric potential difference in a given situation, the more kinetic energy a charged object can acquire when released, and the more work it can do.

Electric potential is sometimes referred to as "voltage." For example, a 6-volt battery is one whose electric potential difference between the + and − terminals is 6 volts. A battery of twice the voltage is able to give twice as much energy to electrons that flow in any wire or device connected to the terminals.

Since electric potential is defined as electric potential energy per unit charge (in symbols, $V = PE/q$), to find how much potential energy a certain amount of charge possesses you merely multiply the electric potential by its charge (in symbols, $PE = q \times V$). Thus if twice as much charge moves between the terminals of, say, a 12-volt battery, twice as much work can be done.

2. ELECTRIC CURRENT: CHARGES IN MOTION

Electric Charges Can Flow in Matter

Just as water flows in a river bed, electric charge can flow in a conductor. The "channel" for an electric current is often a metallic wire. In a river a mass of water flows; the more water and the faster it flows, the greater the current. An electric current, on the other hand, is a flow of electric **charge**; the more charge and the faster it flows, the greater the electric current. The magnitude of an electric current is the amount of charge flowing past a given point per second; it is usually represented by the symbol I:

$$I = \text{electric current} = \frac{\text{charge}}{\text{time}}.$$

Electric current is measured in coulombs/second. This unit is called the "ampere" in honor of the French physicist André M. Ampère (1775–1836): 1 ampere = 1 coulomb/second. When we say that one ampere (abbreviated A or amp) flows through a wire, we mean that one coulomb of charge flows by any point of the wire every second. When a current of two amperes flows through a wire, two coulombs of charge pass any point of the wire every second (Fig. 13–10).

In a solid conductor, it is the negatively charged electrons that flow. In a gas or a liquid, both positive and negative ions as well as electrons can flow; together these make up the electric current. It is important to distinguish between charge at rest and charge in motion. When there is a current in a wire, there are charges moving along the wire. Yet the wire has very little or no *net* charge on it. Whenever an electron enters one end of the wire, another electron leaves at the other end. The wire remains electrically neutral even though a current flows through it.

The Electric Cell or Battery

The easiest way to obtain an electric current is to connect two ends of a wire to an electric cell or battery. (Originally "battery" meant "many" cells, but today we often term a single cell a battery.) A battery is a source of electric energy. The chemical reactions that occur inside a battery allow the energy stored in chemical bonds to be transformed into electrical energy. These chemical reactions create a separation of charge. Every battery has two "terminals," or connection points: one positive (+) and the other negative (−), Fig. 13–11. The chemical reactions occurring within the battery supply the negative terminal with an excess of electrons and continuously remove electrons from the positive terminal. When a wire is connected across the terminals, Fig. 13–11(B), the electrons flow

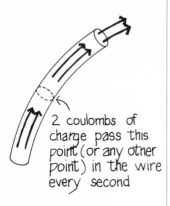

2 coulombs of charge pass this point (or any other point) in the wire every second

Figure 13–10

A section of wire in which a 2-ampere current is flowing.

Figure 13–11

(A) Some ordinary batteries. (B) If a wire is connected across the terminals of a battery, a current flows.

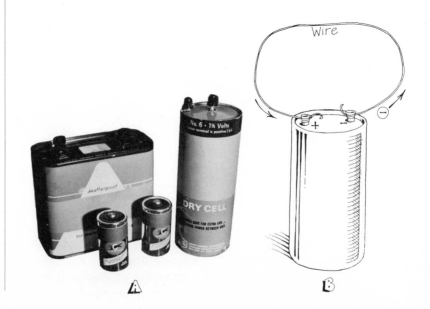

from the negative terminal to the positive terminal. You might expect that electrons arriving at the positive terminal would start piling up (and that is just what they do when the chemicals inside the battery are used up and the battery is "dead"). But when the battery is working properly the chemical reactions continuously remove electrons from the positive terminal and return them to the negative terminal. Because these chemical reactions obey the law of conservation of energy, the potential difference across the terminals of a fresh battery remains essentially constant. When a battery is rated at, say, 1½ volts, or 12 volts, this is the voltage or potential difference that the battery maintains between its terminals.

The electric outlets in homes and offices are maintained at a voltage of about 120 volts.[2] These outlets are connected by electric wires to a source of electric energy, usually a generator at a power plant.

Electric Sparks and Lightning

A dramatic example of a flow of charge is the electric spark formed by lightning. If the potential difference between two close objects—in this case, a cloud and the earth or between two clouds—is great enough, the air, though it is usually a good insulator, breaks down. The strong electric field present pulls electrons off the air molecules, and the air becomes a conductor. This is known as an electric discharge. Electrons flow between the cloud and earth very quickly, eventually reducing the separation of charge and the difference of potential. We see a bright flash because electrical energy is transformed to light energy as electrons recombine with atoms. A streak of lightning (Fig. 13–12) indicates the path of the electron flow from a highly charged cloud to the earth.

[2]Actually, the electric company may vary the voltage somewhat depending on the total energy being used by consumers.

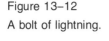

Figure 13–12
A bolt of lightning.

355

Ohm's Law Relates
Current to Voltage

Potential difference is the driving force behind electric current. Generally, the greater the potential difference, the greater the current.

When we put electricity to practical use—in light bulbs, toasters, television sets, and other electric appliances—it is important to know the relationship between current and potential difference.

It was Georg Simon Ohm (1787–1854) who discovered the experimental fact that the current in a wire is directly proportional to the voltage (or potential difference) between its two ends:

$$\text{current} \propto \text{voltage}.$$

This relationship is known as Ohm's law. It tells us that if the voltage between the ends of a wire is doubled, the current flowing in the wire will be doubled. Exactly how much current will flow through a given wire depends on how much resistance the wire offers to the flow of electrons. Electrons frequently collide with the atoms of a wire, and these collisions slow the electrons down. The property of a wire that causes it to resist the flow of current is known simply as "resistance" and is denoted by the symbol R. A high-resistance wire will allow less current to flow than will a low-resistance wire. Current is thus inversely proportional to the resistance. Combining this fact with the above proportionality, we write Ohm's law in its final form as

$$\text{current} = \frac{\text{voltage}}{\text{resistance}}.$$

We can turn this relationship around to read *voltage = current × resistance,* or, in symbols:

$$V = IR.$$

The resistance of a wire depends on the number of collisions electrons make with the atoms of the wire. It thus depends on the arrangement of the atoms in the material—and therefore on the kind of material it is—as well as on the size and shape of the wire. A long wire will offer more resistance than a short one, since there will be more opportunities for collisions to occur in the longer one. A wire with large cross-sectional area will offer less resistance than a wire with a small cross-sectional area, because there is more area for the electrons to flow through.

The unit of electrical resistance is called the ohm. It is defined so that no additional constant of proportionality is required in Ohm's law; thus, since $R = \dfrac{V}{I}$, 1 ohm = 1 volt/ampere. We can use Ohm's law to calculate the current in a wire when we know the resistance of the wire and the voltage across it. For example, if a 6-ohm wire is connected to a 12-volt battery, the current will be $I = \dfrac{V}{R} = \dfrac{12 \text{ volts}}{6 \text{ ohms}}$ = 2 amperes.

Circuits

A continuous conducting path between the terminals of an electrical source of energy such as a battery is called a "circuit," Fig. 13–13(A). If a switch in the circuit is opened, Fig. 13–13(B), or if the wire is cut, the flow of current is stopped and we say that the circuit is "open." If there are no interruptions in the path to stop the current flow, the circuit is "closed."

A great many devices are commonly connected to a source of electrical energy: light bulbs, electric frying pans, toasters, hi-fi equipment, TV sets, and so on. All these permit the flow of current and all offer some resistance. The information plate on most appliances indicates the voltage that the appliance is designed for and the current it will draw. Rarely is the resistance specified, but it can be calculated by using Ohm's law. For example, suppose a toaster is designed to draw 8 amperes when used at 120 volts; its resistance is then $R = \dfrac{V}{I} = \dfrac{120 \text{ volts}}{8 \text{ amperes}}$ = 15 ohms.

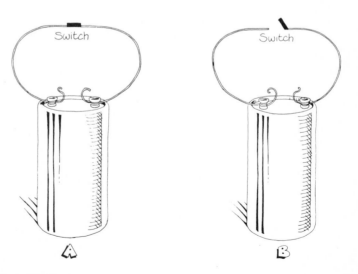

Figure 13–13

(A) A closed circuit. (B) An open circuit.

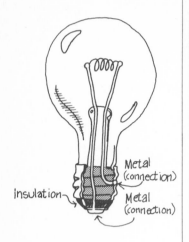

Figure 13–14

Diagram of an ordinary incandescent light bulb.

EXPERIMENT

Take a flashlight apart and figure out how the complete circuit is formed. If it's a metal flashlight, the casing may serve as a conductor. Examine the interior of the bulb, including the filament, with a magnifying glass. Is there a continuous conducting path within the bulb? The two connections to the bulb are made at the conductor on the very bottom, and on the threads (see Fig. 13–14). Connect a circuit consisting of the bulb, one battery, and one or two wires. You can use tape to hold the connections but secure them as tightly as possible. Note how brightly the bulb glows. Reconnect the circuit using two batteries connected end to end; you might lay them in the fold of an open book to keep them in line. Why does the bulb glow more brightly with two batteries than with one?

Series and Parallel Circuits

When two or more light bulbs or electric appliances are connected in one circuit, they can be arranged either in **series** or in **parallel**. When they are arranged in **series**, the bulbs are connected in a single conducting path, Fig. 13–15(A). When they are arranged in **parallel**, each bulb is on a separate path, so that electrons that flow through one do not flow through the other, Fig. 13–15(B).

EXPERIMENT

Obtain two flashlight bulbs, a battery, and several lengths of electrical wire. Connect one bulb to the battery with two pieces of wire. You can make the connections by wrapping the wire around the base of the bulb and securing it with sturdy tape. Next, connect the second bulb into the circuit so that the two bulbs are (a) in series and then (b) in parallel.

Series and parallel circuits have different characteristics. You may have noticed in this experiment, for example, that the two bulbs in the parallel circuit glowed nearly as brightly as when only one bulb was used (see previous experiment); but the two bulbs in the series circuit glowed somewhat less brightly. To see why, let's look at the differences between these two types of circuit.

In the *series* circuit, the same current is flowing in all parts of the circuit (what flows in one end flows out the other). The current through the bulbs in this case is less than it would be if only one bulb were connected to the battery, because each bulb offers resistance to the flow of current. Consequently, the total resistance, which is *the sum of the individual resistances,* is greater. If the two bulbs are identical, the total resistance is double the resistance offered by each. And so the current will be only half what it would be through a single bulb connected to the same battery. This is why the two bulbs in the series circuit glowed less brightly than a single bulb in the circuit. Furthermore, the voltage across the bulbs is divided between them; if they are connected to a 1½-volt battery, the voltage across each bulb is only ¾ volt.

What happens in a *parallel* circuit? Here the current leaving the battery splits into two (or perhaps more) branches. Each of the bulbs in Fig. 13–15(B) has the full voltage of the battery across it, since each is connected directly (by nearly resistanceless wire) to the battery. Thus the current through each bulb will be essentially the same as if only one bulb were in the circuit. Hence bulbs in parallel each burn as brightly as a single bulb would. Since the current flowing from the battery is following two paths, the *total* current is the sum of the individual currents and is thus greater than the current drawn by a single bulb. Thus, arranging the bulbs in parallel actually reduces the net resistance of the circuit. This may seem surprising; but remember that when you add another path in parallel, you give the electrons another channel to flow in; it's like making the wire fatter, so there must be less net resistance. The same is true of a river—if you give it an additional channel to flow through, more water will flow.

A series circuit has one disadvantage that a parallel circuit does not. If one of the lights burns out or comes loose in the socket, the circuit is broken, the current stops flowing, and the other bulb goes out too. Strings of Christmas-tree lights used to be hooked up in series and when one light burned out, they all went out. In a parallel circuit, by contrast, if one bulb goes out, the others are unaffected and continue to glow.

Houses are always wired in parallel to ensure that each device receives the same voltage, usually 120 V.[3] Two lead-in wires are connected to each outlet in parallel, as shown in Fig. 13–16. Thus you can turn one light on or off without affecting the others.

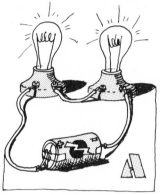

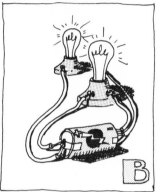

Figure 13–15

Two light bulbs connected (A) in series and (B) in parallel.

[3] In the United States most houses receive 240 V. This full voltage is used for such appliances as electric stoves, but it is split in half for other appliances. In many other countries, 240 V is standard for all appliances.

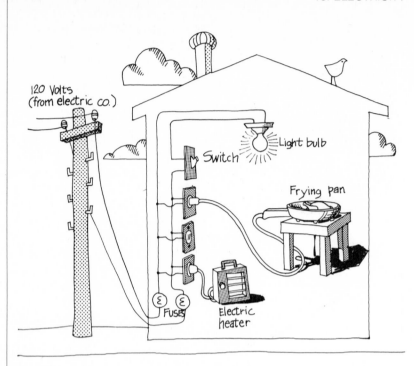

Figure 13–16

In house wiring the electric outlets are connected in parallel.

Electric Current Gives Rise to Heat and Light

Electric energy can be easily transformed into other forms of energy. How is this done? A difference of potential gives rise to an electric current, a flow of charge. The potential energy that each charge loses as it moves through this potential difference is equal to the amount of work it can do. For example, each coulomb of charge that moves through a potential difference of 10 volts is capable of doing 10 joules of work. It can be transformed into heat in a toaster, heater, electric stove, or frying pan, and into light energy in a light bulb. In these appliances the resistance is usually not large, 5 to 500 ohms or so. This means that a fairly large current flows through them. Consequently, many collisions occur between the moving electrons and the atoms of the wire that makes up the heating element of the toaster or frying pan, or the filament of the light bulb. In each collision, a portion of the electron's kinetic energy is transferred to the atom with which it collides. Consequently the atoms move faster. As we saw in Chapter 8, the temperature of a substance is a measure of the average kinetic energy of the atoms or molecules that make it up. So when the atoms in a wire are made to move faster, the temperature of the wire rises—it gets hot! And this heat is conducted to the food in an electric frying pan or toaster, Fig. 13–17; in a hair dryer (Fig. 13–18) and in some electric heaters, a motor turns a fan that blows air across the hot coils of wire. If the wire gets hot enough, the atoms will move so fast that they emit light. You probably have noticed the red glow of the heating element in a toaster or the burner of an electric stove. If the wire happens to be a thin tungsten wire enclosed in a glass bulb from which the oxygen has been removed, it gets very hot and the filament of the light bulb glows white.

We shall see in the next chapter that electric energy can be transformed into mechanical energy by an electric motor and into sound by a loudspeaker. We shall also see, later in this chapter, how electric energy can be used to amplify or modify radio and TV signals so that we can hear and see programmed material.

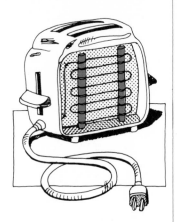

Figure 13–17

Diagram of an electric toaster, showing the heating filament.

Figure 13–18

A hair dryer.

Electric Power

How much energy an electric appliance uses up (and it is energy you pay for on your electric bill) depends on how long you use the appliance and at what rate it uses energy. A radio left on for two hours uses twice as much energy as one left on for only one hour. The characteristic of any given appliance is the *rate* at which it uses energy, which we call *power:*

$$\text{power} = \frac{\text{energy}}{\text{time}}.$$

As we saw in Chapter 5, the unit of power is the watt; 1 watt = 1 joule/second. The total amount of energy used by any appliance is simply the product of its power consumption and the length of time it is on. A 150-watt light bulb uses up 150 joules of energy every second it is on. If it is on for an hour, a total of 3600 seconds, it uses a total of 150 × 3600 = 540,000 joules. Instead of joules, electric companies use a much larger unit for measuring energy—the "kilowatt•hour." One kilowatt is a thousand watts. One kilowatt•hour is the energy consumed at the rate of 1000 watts for one hour.[4] Electric energy costs the consumer a few cents per kilowatt•hour (abbreviated kwh), depending on the locale. It is worthwhile to be able to calculate how much it costs to run an electric appliance. At \$.10 a kilowatt•hour, a 100-watt (100 watts = 0.1 kilowatt) light bulb burning for 20 hours would cost: 0.1 kw × 20 h × \$.10/kwh = \$.20. A 1300-watt electric frying pan used for 2 hours would cost: 1.3 kw × 2 h × \$.10/kwh = \$.26.

361

[4]Therefore 1 kwh = 1000 joules/sec × 3600 sec/h × 1 h = 3,600,000 joules.

EXPERIMENT

Calculate how much energy (in kilowatt·hours) you use in an average day at your residence. You will have to find out the power consumption or "wattage" of each device used; it is usually given somewhere on the appliance. If only the current rating is given, you can use the power relationship below to determine the power. To find the energy used, you must also estimate how long each appliance is on during an average day. Inquire from your electric company how much a kilowatt·hour costs and determine your electricity costs per day and per month. (Does this match your electric bill?) You can also check your calculation by reading the electric meter.

It is useful to know the relationship between power, voltage, and current. In an electric circuit when an amount of charge q moves through a potential difference V, an amount of energy equal to $q \times V$ can be transformed into work or other forms of energy. Then the power transformed must be

$$\text{power} = \frac{\text{energy}}{\text{time}} = \frac{q \times V}{t}.$$

The charge per unit time, q/t, is simply the current I; so

$$\text{power} = I \times V = \text{current} \times \text{voltage}.$$

When the current is in amperes and the voltage is in volts, the power will be in watts. The power relationship can be written in various other ways, using Ohm's law ($V = IR$):

$$P = IV$$

$$= I^2 R \ (\text{since } P = IV = I \times IR = I^2 R)$$

$$= \frac{V^2}{R} \ (\text{since } P = IV = \frac{V}{R} \times V = \frac{V^2}{R}),$$

where P stands for power.

As a simple example, let us calculate the resistance of a 60-watt light bulb. We use $P = V^2/R$ and solve for R, which gives us $R = V^2/P = (120 \text{ volts})^2/(60 \text{ watts}) = 240$ ohms. The current through the bulb can be calculated by using the relation $P = IV$ and solving for

I: $I = P/V = 60$ watts/120 volts = ½ amp; the same result can be obtained by using Ohm's law and the value of R just calculated: $I = V/R = (120 \text{ volts})/(240 \text{ ohms}) = $ ½ amp.

Hazards:
Overloading and Shorts

Being able to calculate power consumption can be a big help in determining the safety of electric appliances and circuits; it also helps when trying to discover why a fuse has blown or a circuit breaker popped open. Wires that carry electricity always have some resistance, though usually very little. When large currents pass through a wire, it heats up. If it gets too hot, it can cause a fire. This is especially dangerous when the wires are inside the walls of a building. When a wire carries more current than is safe, it is said to be "overloaded." To prevent overloading, fuses or circuit breakers are installed in the lines. These safety devices (Fig. 13–19) are designed to break the circuit whenever the current goes above a cer-

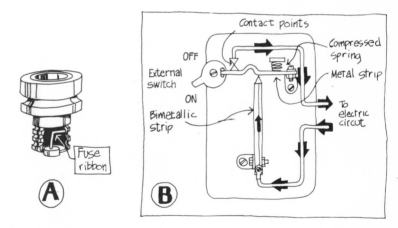

Figure 13–19

(A) A fuse. The ribbon melts when the current exceeds a certain value, thus opening the circuit. The fuse must then be replaced. (B) A circuit breaker. Electric current passes through a bimetallic strip (like that in Fig. 8–7). If the current is large enough, the bimetallic strip is heated and bends so far to the left that the notch in the spring-loaded metal strip drops down over the end of the bimetallic strip. This opens the circuit at the contact points, one of which is attached to the metal strip, and also flips the external switch. When the device cools, it can be reset with the external switch.

tain safe level. For example, a 15 A fuse or circuit breaker will burn out or open whenever the current passing through it exceeds 15 amperes.

When wiring is installed in a building, the fuses are chosen to match the particular wire being used. For example, if the wire can take currents up to 20 amperes safely, a 20 A fuse is used. If the fuse is replaced with a 30 A fuse, overload protection is lost and serious fire danger may result. Therefore, a fuse should never be replaced with one of a higher rating. If one of your fuses or circuit breakers goes out frequently, the first thing to do is to calculate how much current you are drawing in that circuit. Most houses have several different circuits, individually fused, so you will have to determine which electric appliances and lights are on the troubled circuit. This is easily done by noticing which lights and appliances fail to work when the offending fuse is removed or the circuit breaker is off.

Here's an example: suppose you blow a fuse whenever you use a 150-watt overhead light, an electric heater, and an electric frying pan, Fig. 13–16. On each of these appliances you find a plate giving its wattage (although sometimes the current drain is given directly). Suppose the wattage is 1320 watts for the frying pan and 1800 watts for the heater, and they are plugged into a standard 120-volt outlet. How much current is being drawn by each? We use the formula $P = IV$ and solve for I, which gives $I = \dfrac{P}{V}$. The light bulb draws a current $I = \dfrac{P}{V} = \dfrac{150\ \text{watts}}{120\ \text{volts}} = 1.25$ amperes; the frying pan draws $I = \dfrac{1320\ \text{watts}}{120\ \text{volts}} = 11$ amperes; and the heater draws $I = \dfrac{1800\ \text{watts}}{120\ \text{volts}} = 15$ amperes. The total current drawn is $15 + 11 + 1.25 = 27.25$ amperes. You look at the fuse and discover it's a 20 A fuse. No wonder it keeps blowing! The solution is not to use both the heater and the frying pan at the same time, or else to put them on separate circuits with separate fuses.

If you find that you blow a fuse even though the current drain does not exceed the fuse rating—for example, if the light bulb and the frying pan alone caused the 20 A fuse to blow—then something else must be wrong. The most common difficulty is that the insulation around the wires somewhere in the circuit, frequently on an appliance cord near the plug, is badly worn, permitting the two wires to touch. This is known as a "short circuit," since the path taken by the current is shortened. The resistance of the circuit is then very small and very large currents can be drawn. A short circuit should be remedied immediately.

The fuses used in automobiles and electronic equipment such

as hi-fi components serve to prevent fires and to protect delicate electronic circuits.

Electric Shocks Are Caused by Current Through the Body

It is primarily current rather than voltage that causes shocks. A severe electric shock can be damaging to the body or even fatal. Whether or not a shock is damaging depends on the magnitude of the current, how long it acts, and through which part of the body it passes. Current passing between the forefinger and thumb of one hand will be much less damaging than current passing from one hand through the heart to the other hand. Most shocks occur when a person touches a frayed cord or other exposed electrical conductor.

Currents through the body of less than 5 milliamperes (a milliampere, abbreviated mA, is 1/1000 ampere) usually do little damage. If the current is about 15 mA, however, a person may no longer be able to control his muscles well enough to let go. A current of about 100 mA (1/10 ampere) or more, endured for only a second, is usually fatal if it passes through the heart. Such currents cause the heart muscles to contract irregularly so blood is not pumped properly; it is difficult for the heart to recover from this state, which is why death is common. Strange as it may seem, however, if the current is larger than about 1 A, the damage may be less and death less likely. Apparently such large currents bring the heart to a complete standstill; with release of the current, the heart may then be able to return to normal action.

We know from Ohm's law that the current through a person's body depends on the applied voltage and the resistance of the body. The bulk of bodily resistance is in the skin; inner tissue is a fairly good conductor and therefore has low resistance. When the skin is dry, the bodily resistance—say between one hand and the other—is usually on the order of 100,000 ohms or more. But if the skin is wet—water is a fairly good conductor—the resistance drops to a few hundred ohms. When the skin is dry, 10 volts may hardly be felt, since little current will flow through the body. Even 120 volts will not give too violent a shock. But if the skin is wet, 10 volts can give rise to a large current through the body and produce a strong shock. When you are wet, 120 volts may be fatal. This is why you are strongly cautioned against touching anything electrical while in the bathtub or at poolside.

Many appliances are "grounded," which means that a wire runs from the frame of the appliance to a "ground"—often a wire or pipe that leads down into the ground itself. Because of the ground wire, a person standing on the ground will avoid a shock if there should be a short between the electrical wiring and the frame of the appliance. This is because the wire to the ground has less electrical resistance than the person, and therefore the bulk of any current will flow through the wire instead of through the person. Nevertheless, it is a good idea to avoid touching any electrical appliance when you are standing barefoot on the ground or on a cement slab floor. Should you accidentally touch a "hot" wire, a complete circuit may be made through you to the ground, causing shock.

AC and DC

If you look at the information plate on an electric appliance, you may see "110–120 V AC," or "9 V DC," as well as the wattage or current requirements. The "110 V" or "9 V" of course refers to the voltage required. In the former case, you use ordinary house current; you plug it in. In the latter case, a 9-volt battery is needed. But what do "AC" and "DC" mean?

DC stands for "direct current." A direct current is one that stays at a constant value, Fig. 13–20(A). The electrons move in one direction only. Direct current is produced by batteries and by some generators known as DC generators.

AC stands for "alternating current." An alternating current reverses direction many times a second, Fig. 13–20(B). The electrons move first in one direction and then in the other. This is accomplished by applying an alternating voltage produced by an AC generator (which is discussed in Chapter 14). House current in nearly all of the United States and in many other countries is AC. In most cases the current alternates back and forth 60 times per second and is commonly referred to as "60 cycle," or as having a frequency

Figure 13–20

Graphical representation of (A) direct current; (B) alternating current.

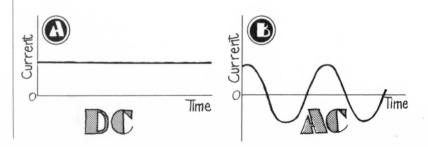

of 60 Hz.[5] The electric company usually maintains the frequency very close to the 60 cycles per second, independent of the voltage or current drawn. In some locales a different frequency is used; for example 50 Hz is used in some countries.

| 3. ELECTRONICS |

We now look at the operation and use of vacuum tubes and semiconductor devices, such as transistors, and some of the instruments that utilize them. This is the field of electronics. Throughout this section we will interpret the operation of these devices in terms of atoms and their electrons.

The first two sections deal with diode and triode tubes that are much less common today than they once were. If desired, the reader may skip these two sections and plunge immediately into the section on semiconductors.

Diode Tubes
Change AC to DC

The vacuum tube "rectifier," or "diode," was discovered by the English physicist J. A. Fleming (1849–1945) in 1904. A diode consists of two separate conductors known as electrodes (hence the name "di-ode") plus a heating element, enclosed in an evacuated glass or metal container; it is shown schematically in Fig. 13–21(A). Wires connected to the heater and to each of the two electrodes emerge from the tube and are connected to pins in the tube base (see Fig. 13–22).

Diodes are used to change AC into DC [Fig. 13–21(B)]. Since nearly all commercial electricity is AC, and most electronic tubes and semiconductors require DC for operation, diodes are found in nearly all electronic instruments, including radios and TV sets.

A diode works in the following way. The heater is connected to a voltage source and the current passing through the heater causes it to become hot. The heater is very close to one of the electrodes —called the cathode—and causes the cathode itself to become very hot. The cathode is made of a special metal or a metallic oxide, such that, when heated, many of the electrons gain enough kinetic energy to escape from it.[6] When the diode is connected to a source of alternating voltage, as in Fig. 13–21(B), the anode of the diode

[5] 1 cycle/second (abbreviated 1 cps) is nowadays called 1 Hertz (abbreviated 1 Hz) in honor of Heinrich Hertz (1857–1894).

[6] The time it takes to warm up a TV set is the time it takes to get the cathode in the tubes hot enough to emit electrons.

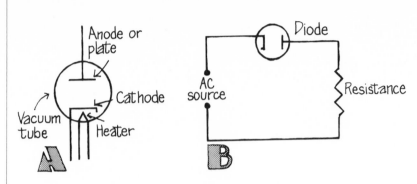

Figure 13–21

(A) Diagram of a diode tube. (B) Diode tube used in a circuit to change AC to DC (heater of the diode is not shown).

will alternately be positive and negative with respect to the cathode. When the anode is positive, the negatively charged electrons emitted by the hot cathode are attracted to it, and a large current flows through the tube. When the anode goes negative, the electrons are repelled by it and remain close to the cathode—no current flows at all. Thus a diode allows the current to flow in only one direction, not both. It changes AC into DC.

Triodes Are Amplifiers

A triode is almost identical to a diode, except that it contains a third electrode called the "control grid," or just "grid." The grid actually looks like a wire grid, Fig. 13–22. Triodes serve different purposes than diodes. One of their main uses is as an **amplifier** of small voltages, such as those that come from a phonograph cartridge, tape recorder head, or over the air as a radio or TV signal.

The grid plays a crucial role in this amplification. Figures 13–23(A) and (B) show a triode with the anode kept at +200 volts DC. Electrons emitted from the cathode are attracted to this positive anode. But if the grid is connected to a voltage source, even a small one, the voltage on the grid can control the flow of the electrons. When the grid is slightly negative, the electrons leaving the cathode are repelled by the grid, and many will be turned back, Fig. 13–23(A). Only the most energetic electrons will have sufficient kinetic energy to make it through the grid and then be attracted to the anode. The greater the negative voltage on the grid, the fewer electrons reach the anode and the smaller the electric current passing through the tube. When the grid is positive, Fig. 13–23(B), the electrons are attracted by it and the current through the tube is large. Thus, a small voltage change on the grid causes a large change in the current through the tube.

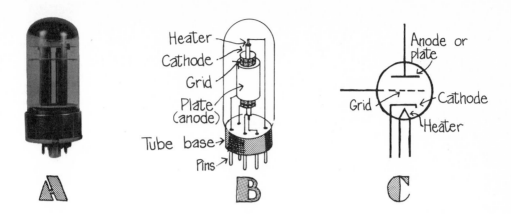

Figure 13–22

(A) Photograph of a tube containing two triodes.
(B) Cutaway drawing of a triode (a diode looks similar except for the absence of the grid). (C) Schematic drawing of a triode.

Figure 13–23(C) shows the triode connected in a circuit as an amplifier. When a large current is passing through the tube, the large current also passes through the "output" resistance R and there is a large voltage drop ($V = IR$) across this resistor. Thus, a small voltage input to the grid will cause a large voltage across R. Hence the triode causes voltage magnification or "amplification." In Fig. 13–23(C) an alternating voltage—it might be 1000 cycles/second, which is in the audible range—is being amplified by the triode.

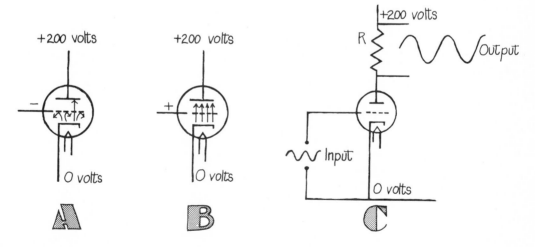

Figure 13–23

Operation of a triode, with 200 volts DC maintained between anode and cathode. (A) When grid is negative, most electrons do not get through. (B) When grid is positive, electrons pass through readily—a large current passes through the tube. (C) Triode used to amplify a small AC signal (see text).

Semiconductors

Vacuum tubes are rarely used today except for specialized purposes. Both the diode tube and the triode tube have been replaced by solid devices made of semiconductor materials. The semiconductor diode performs the same function as a vacuum tube diode, and the transistor performs the same function as a triode. Because semiconductors are solid, devices made from them are often called "solid-state" devices. Semiconductors offer a number of advantages over vacuum tubes; they have a longer life, they are smaller, they need less voltage to operate, and they require less power because no heating elements are needed.

Silicon and germanium are the two main semiconductors used in diodes and transistors. An atom of silicon or germanium has four electrons, which are important in holding the atoms in the regular lattice structure of the crystal, Fig. 13–24(A). Germanium and silicon acquire useful properties for use in diodes and transistors only when a tiny amount of impurity is introduced into their crystal structure. Two kinds of semiconductor are possible, depending on the type of impurity used. If the impurity is a material whose atoms have five outer electrons, such as arsenic, we have the situation shown in Fig. 13–24(B). Only four of arsenic's electrons fit into the crystal structure. The fifth does not fit in and is free to move much like the electrons in a conductor. An arsenic "doped" crystal of silicon is called

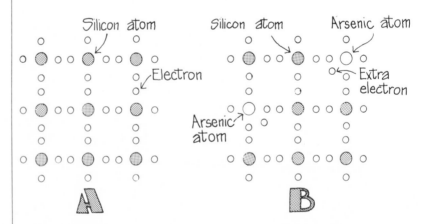

Figure 13–24

A silicon crystal. (A) Each silicon atom is surrounded by four electrons. (B) A few arsenic atoms in a silicon crystal; note the extra electron that does not fit into the crystal lattice. This is known as an *n*-type semiconductor.

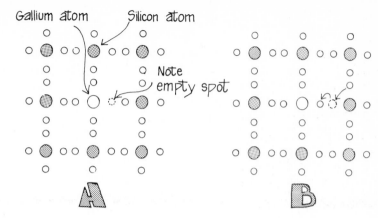

Figure 13–25

(A) Gallium atoms in a silicon crystal as an impurity. Note the empty spot, or "hole," because gallium has only three outer electrons. (B) Electrons can move to fill the hole; the net effect is that the *hole* moves (to the right in the figure). This is a *p*-type semiconductor.

an *n*-type semiconductor, because it is the negatively charged electrons that carry the electric current. On the other hand, if an impurity with three outer electrons (for example, gallium) is added, there is a "hole" in the lattice structure, Fig. 13–25(A), because of the lack of one electron around the gallium atom. Nearby electrons can jump into the hole, filling it; but this leaves a hole where the jumping electron had previously been, Fig. 13–25(B). When the "hole" is near a silicon atom—which it is most of the time, since very few gallium atoms are used—there is a net positive charge in that region. This is so because a silicon atom is neutral only when there are four outer electrons surrounding it. A semiconductor of this type is called a "*p*-type" semiconductor, because it is the positively charged holes that seem to move, carrying the electric current. Note, however, that a *p*-type semiconductor, like an *n*-type, has no *net* charge on it.

Semiconductor Diodes

When an *n*-type semiconductor is joined to a *p*-type semiconductor, a "*p-n* junction diode" is formed. When a voltage is applied across such a diode, current will flow only if it is connected as shown in Fig. 13–26(A). The positive holes in the *p*-type semiconductor are repelled by the positive terminal of the battery, and the electrons in the *n*-type are repelled by the negative terminal of the battery.

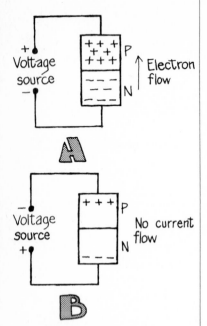

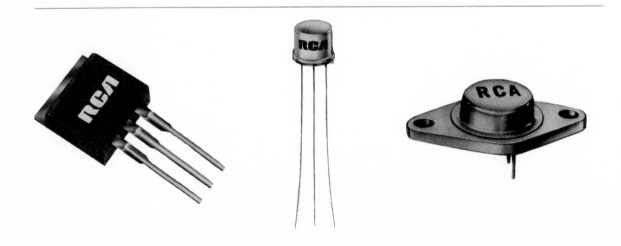

Figure 13–26

Schematic diagram of the operation of a semiconductor diode. Current flows when the voltage is connected as in (A) but not in (B).

Figure 13–27

Transistors.

They meet in the middle, at the "junction," and the electrons cross over and fill the holes. In the meantime, the positive terminal of the battery is continually pulling electrons off the p end, forming new holes, and electrons are being supplied by the negative terminal at the n end. Consequently a large current flows through the diode. But when the voltage is reversed, as in Fig. 13–26(B), the holes in the p end are attracted to the battery's negative terminal and the electrons in the n end are attracted to the positive terminal. The current carriers do not meet near the junction, and no current can flow. Thus a semiconductor diode (like a diode tube) allows current flow in one direction only. It can thus be used to change AC into DC. This is important because the operation of transistors requires a constant DC voltage as we shall see, yet most commercial electricity is AC. Hence, radios and TV sets and other electronic devices that use AC contain diodes to change the AC to DC.

Transistors

The transistor (Fig. 13–27) was developed at the Bell Telephone laboratories in 1948. A transistor consists of a crystal of one type of semiconductor sandwiched between two crystals of the opposite type. Both pnp and npn transistors are made; they are shown schematically in Fig. 13–28(A). The three semiconductors are called collector, base, and emitter. They play the same roles that the anode, grid, and cathode do, respectively, in a triode tube. An npn transistor is shown in Fig. 13–28(B) connected to a 9-volt battery. A transistor, like a triode tube, can act as an amplifier. That is, it can amplify a small electrical signal (an alternating voltage) into a larger one. When the signal applied to the base is positive, the electrons

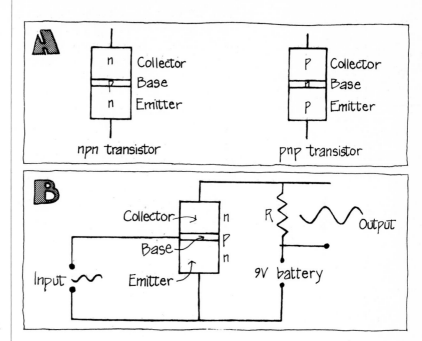

Figure 13–28

(A) Schematic diagram of *npn* and *pnp* transistors.
(B) An *npn* transistor in a circuit acting as an amplifier.

in the emitter are attracted into the base. The base region is very thin, however, and the electrons race right into the collector, which is maintained at a positive voltage by the battery. The large current causes a large voltage drop across the resistor *R*. When the signal to the base is negative, however, electrons in the emitter are not attracted by the base and little or no current can flow. In this way, a small voltage applied to the base is accurately amplified by a transistor. Most amplifiers have several transistors in a row, one after the other, amplifying the output of the preceding transistor. By using many such stages, a voltage of only a few millivolts or less (say that received by a radio or TV antenna) can be amplified to many volts.

Integrated Circuits: The Latest Step in Miniaturization

Transistors were a great advance in the miniaturization of electronic circuits. Although transistors are very small compared to vacuum tubes, they are huge compared to the latest development—integrated circuits, Fig. 13–29. Tiny amounts of impurity (to form *n*- and *p*-type material), only a few molecules thick, can be deposited at different locations in a silicon crystal to form the tiny transistors, diodes, and resistors of a particular circuit. A tiny circuit only a mil-

373

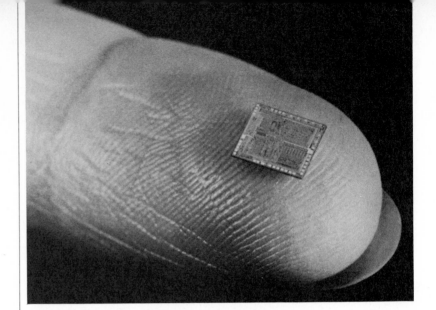

Figure 13–29

An integrated circuit.

limeter on a side may contain over 100 transistors. Integrated circuits are used for many electronic applications, including space instruments and computers.

Radio and TV

Before discussing the essentials of a radio, we must look into the nature of the signal that the antenna of a radio picks up, Fig. 13–30. A broadcasting station takes the audio signal it wants to broad-

Figure 13–30

Signal sent out by the transmitter of a radio station and received by a home radio.

cast (music or words) and adds it to a high-frequency signal known as the "carrier frequency" (more on this shortly). A particular carrier frequency is assigned to each station; it may be anywhere from 550,000 to 1,600,000 Hz (on AM) and from 88,000,000 to 108,000,000 Hz (on FM) (1 Hz = 1 hertz = 1 cycle/second). It is the carrier frequency that allows radio receivers to distinguish one station from another. The combination of carrier and audio signals is then sent out by a transmitter in all directions in the form of an electromagnetic wave (discussed in the next chapter). The rapidly oscillating electromagnetic wave consists of an electric and a magnetic field. The electrons in the antenna of a radio move in response to the oscillating electric fields of all the radio signals that reach it. This tiny current, corresponding to a voltage of only a few millionths of a volt, is amplified in the first stages of a radio, Fig. 13–31. In the "tuning" stages of a radio a special circuit, which makes use of resonance, allows you to choose the particular station you want to listen to by "tuning" to its carrier frequency. For example, "810 on the AM dial" means that the carrier frequency of that station is 810,000 Hz. Only the signal from the tuned station will be allowed to pass into the rest of the radio. All other stations with different carrier frequencies are stopped at this point. The detection stage then separates the carrier frequency from the audio signal, and finally the audio signal is amplified and sent to the speaker where the electric signal is changed into sound. Except for the speaker, which will be discussed in the next chapter, all the stages of a radio use vacuum tubes or transistors and diodes.

Television transmitters and receivers work in a similar way, except that the carrier frequency is mixed not only with an audio signal but with a video signal as well. A TV set, or receiver, has amplification, tuning, and detection stages, just as a radio does. It also has a speaker, and its video counterpart, a picture tube.

Figure 13–31

Stages in a radio (block diagram).

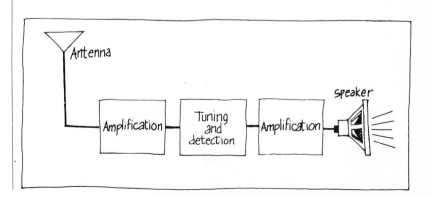

375

Radio and TV Transmission

There are two basic ways in which a radio or TV station can superimpose the program material on the carrier signal: **amplitude modulation** (AM) and **frequency modulation** (FM).

In AM, the frequencies of the program content are simply added to the much higher carrier frequency, as shown in Fig. 13–32. This is called amplitude modulation because the amplitude of the carrier is altered ("modulate" means "to alter"). Commercial AM radio stations have frequencies (assigned by the government) between 550 kHz (1 kilohertz = 1000 cycles/sec) and 1600 kHz.

In FM, the *frequency* of the carrier wave is altered by the programmed frequencies, as shown in Fig. 13–33. This is why it is called frequency modulation. Commercial FM radio stations are assigned much higher carrier frequencies than are AM stations; FM frequencies range from 88 MHz (1 megahertz = 1 million cycles/sec) to 108 MHz. The main difference between an FM and an AM radio is in the method by which they subtract out the carrier signal to obtain the program material.

In FM stereo broadcasting, the carrier wave carries two signals. One signal contains frequencies up to about 17,000 Hz and so includes most audio frequencies; the second signal contains the same range of frequencies but 21,000 Hz is added to all these to distinguish them from the first signal. A stereo receiver must then subtract this 21,000 Hz signal and distribute the appropriate signals to the left and right channels (and speakers). The first signal actually contains the sum of the left and right channels (L + R), so that mono radios will hear all the sound. The second signal contains the difference between left and right (L − R); so the radio must add and subtract these two signals to get pure left and pure right signals to send to each speaker.

Television signals are frequency modulated, and their carrier frequencies are also in the megahertz region; UHF (ultrahigh fre-

Figure 13–32

How the program signal is added to the carrier for AM radio.

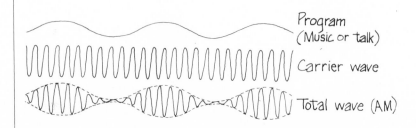

Program
(Music or talk)

Carrier wave

Total wave (AM)

Figure 13–33

How the program signal is
added to the carrier for FM
radio.

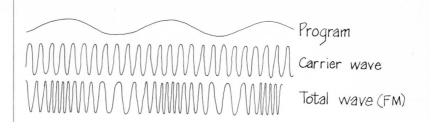

Program

Carrier wave

Total wave (FM)

Figure 13–33

How the program signal is
added to the carrier for FM
radio.

quency) stations have much higher frequencies, in the hundreds of
megahertz region.

Citizens band (CB) receivers are much like ordinary AM radios,
except that the carrier frequencies are around 27 MHz.

The uses to which the various regions of the radio-wave spectrum can be put are assigned by governmental agencies. Besides
those mentioned above, there are regions (or "bands") assigned
for use by ships, airplanes, police, military, amateurs, satellites and
space, and radar.

A TV Picture Tube Is a Cathode Ray Tube

The picture tube of a TV set is often called a cathode ray tube (CRT;
"cathode ray" was an early term for the electron). Cathode ray tubes
are also used in the oscilloscope, a device for visualizing such electrical signals as heart beats (EKG) and brain waves (EEG).

A cathode ray tube is diagrammed in Fig. 13–34. Several electrodes (conductors) are inserted inside an evacuated glass tube.
A heating element is made hot by a current passing through it and
so heats the electrode known as the cathode. When heated in this

Figure 13–34

Cathode ray tube.

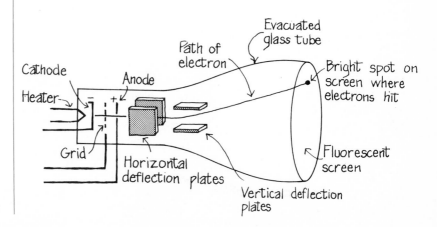

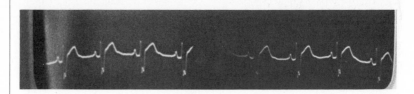

Figure 13–35

An EKG trace.

way, electrons in the cathode acquire sufficient kinetic energy to escape from the metal. These electrons are accelerated by the high positive voltage on the anode (10,000 to 50,000 volts). When the electrons are shot from this "electron gun" they emerge from a small hole in the anode and speed toward the face of the tube. The tube face is coated with a fluorescent substance that glows briefly when struck by electrons. It is this glow that constitutes the visual image. Two vertical and two horizontal plates deflect the beam of negatively charged electrons whenever a voltage is applied across them. The electrons are of course attracted to that plate that at any given moment is positive. In an oscilloscope, the horizontal deflection plates cause the beam of electrons to go from left to right (repeatedly, in most cases); and the signal to be viewed, for example an EKG (Fig. 13–35), is amplified and applied to the horizontal plates.

In a television picture tube, the two sets of plates[7] are used to cause the electron beam to sweep in the manner shown in Fig. 13–36. A complete sweep over the whole screen, a total of 525 horizontal lines, is done 30 times a second; each complete sweep of the whole screen requires 1/30 of a second. The "picture" consists of the varied brightness of the spots on the screen, Fig. 13–37. The brightness at any point on the screen is proportional to the

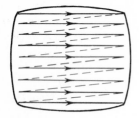

Figure 13–36

Television screen showing how electrons sweep across the screen in a succession of horizontal lines. (The dotted lines represent a very rapid, unnoticeable motion.)

[7]Most modern picture tubes use magnetic deflection coils instead of electric plates, but the effect is the same.

Figure 13–37

Photo of a TV screen showing individual lines.

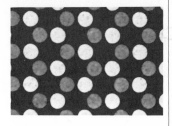

Figure 13–38
Tiny spot on a color TV screen containing three colored phosphors.

intensity of the electron beam when it sweeps past that point. When many electrons strike one spot, that spot glows brightly; if only a few or no electrons strike that spot, it appears dark. The beam intensity is controlled by another electrode, the grid (see Fig. 13–34), whose voltage is determined by the video signal sent out from the TV transmitter. When the grid voltage is strongly negative, the electrons emitted from the cathode are repelled and few can make it past the grid; in this case the screen is dark. When the grid is made positive, the electrons from the cathode easily pass through it, and they are accelerated by the positive anode voltage, so the screen will be bright.

The screen of a color television receiver is more complex than that of a black-and-white receiver; it contains an orderly array of tiny spots, each composed of three phosphor dots (or lines), one for each of the additive primary colors: red, green, and blue. When a spot is struck by the electron beam, it glows either red, green, or blue, or an intermediate color if more than one phosphor in a spot is struck, Fig. 13–38. Except for the more complicated color video signal and the circuitry needed to direct the electron beam to the correct phosphors across the screen, a color TV receiver is basically the same as a black-and-white receiver.

REVIEW QUESTIONS

1. Which of the following are repelled by a negative electric charge?
   (a) other negative charges
   (b) positive charges
   (c) neutral objects
   (d) protons

2. A negatively charged body has
   (a) lost electrons
   (b) gained electrons
   (c) acquired protons
   (d) none of these

3. The magnitude of the electric force between two charges is
   (a) inversely proportional to the distance between them
   (b) directly proportional to the product of the magnitudes of the two charges
   (c) proportional to the charge on only one of them
   (d) proportional to the distance between them cubed

4. If we double the distance between two electrically charged particles, the electric force between them
   (a) decreases by ¼
   (b) decreases by ½
   (c) increases fourfold
   (d) doubles

5. Electrical conductors are materials
   (a) that have an excess of electrons
   (b) that have a deficiency of electrons
   (c) in which electrons move freely
   (d) that repel electrons

6. Ordinary electric currents in a wire consist of the motion of
   (a) ions
   (c) electrons
   (b) protons
   (d) atoms

7. The correct relation expressing Ohm's law is
   (a) $I = VR$
   (c) $V = I/R$
   (b) $I = V/R$
   (d) $R = VI$

8. A 12-volt car battery is connected to a headlight, which has a 3-ohm resistance. The current through the lamp is
   (a) 36 amps
   (d) ¼ amp
   (b) 4 watts
   (e) 36 watts
   (c) 4 amps

9. A light bulb is connected to a battery. When a second light bulb is connected in parallel to the first, the current drawn from the battery is
   (a) doubled
   (c) half as large
   (b) four times as large
   (d) one-fourth as large

10. Repeat the above question for a series circuit.
    c

11. A light bulb glows because
    (a) the electric current glows
    (c) the filament is hot
    (b) oxidation is occurring
    (d) there is a fire inside

12. The current in a 60-watt, 120-volt lamp is
    (a) ½ amp
    (c) 90 amps
    (b) 2 amps
    (d) 7200 amps

13. In the same length of time, which one of the following develops the most heat?
    (a) a 600-watt iron
    (c) six 75-watt lamps
    (b) eight 60-watt lamps
    (d) a ½-kilowatt electric motor

14. A "short" in an electric circuit means
    (a) there is a high resistance
    (b) the current has found a lower resistance path and bypasses the desired circuit
    (c) an electric shock
    (d) a very small current is flowing

15. Ordinary house current is
    (a) AC
    (c) neither
    (b) DC
    (d) zero

16. A battery produces
    (a) AC
    (c) neither
    (b) DC

380 17. Transistors are used primarily for

(a) changing DC to AC   (c) amplification
(b) changing AC to DC   (d) heating

18. An example of a cathode ray tube is
    (a) a TV picture tube   (c) a semiconductor device
    (b) a vacuum tube used for   (d) a transistor
    amplification

EXERCISES

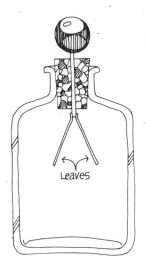

Figure 13–39
An electroscope.

1. An electroscope is a simple instrument consisting of a metal ball attached by a conducting wire to two fine gold leaves that are submerged in a jar, as shown in Fig. 13–39. When the ball is touched by a charged rod, the two leaves, which normally hang straight down, suddenly spring apart. Why? Electroscopes can be used to measure quantity of charge, since the more charge transferred to the ball, the farther the leaves will separate. Explain.

2. How do we know that an amber rod rubbed with a piece of fur acquires a negative charge rather than a positive charge?

3. Why do you sometimes feel a slight shock when you touch a metal railing after walking across a rug?

4. A positively charged rod is brought close to a neutral piece of paper, which it attracts. Draw a diagram showing the separation of charge and explain why attraction occurs.

5. A positively charged object is placed to the left of a negatively charged object, as in Fig. 13–7(C). Consider the electric field close to the positive charge. On which side of the positive charge—top, bottom, left, or right—is the electric field the greatest? the smallest?

6. If a negatively charged particle is placed halfway between two charged parallel plates, one positive and the other negative, as in Fig. 13–9, in what direction will the particle move?

7. Two small objects have the same positive charge. The force between them is 1 N when they are 2 cm apart. What will be the force (a) when they are 6 cm apart, (b) when they are 1 cm apart, (c) when one charge is tripled, (d) when both charges are tripled?

8. Explain, on the basis of the atomic theory, the nature of the force when a hammer strikes a nail.

9. An electron is accelerated by a potential difference of, say, 100 volts. How much greater would its final speed be if it were accelerated by 400 volts?

10. Name three good conductors of electricity and three good insulators. How well do these materials conduct heat? Do you think there might be a connection between the conduction of heat and the conduction of electricity?

11. On dry days, cars and trucks can accumulate considerable charge due to friction between the tires and the road. (You may have felt a

spark upon leaving your car.) Can you guess why trucks carrying flammable materials drag a chain along the ground?

12. Why can lines of force never cross one another?

13. We are not usually aware of the gravitational force or the electrical force between two ordinary objects. What is the reason for this in each case? Give an example of a situation in which we *are* aware of each one and why.

14. A shirt or blouse just pulled from a clothes dryer may cling to your body. Explain.

15. If one body has half the electric potential energy of a second, does the first necessarily have half the electric potential? Explain.

16. Develop an analogy between the flow of electric current and the flow of water in a river. For water, what plays the role of voltage? How does the resistance to water flow depend on the shape of the river's channel?

17. A current of 1 ampere flows in a wire. How many electrons are flowing past any point in the wire per second? The charge on one electron is 1.6×10^{-19} coulombs.

18. Automobile batteries are often rated in "ampere•hours," which is a unit of charge; from the definition of the ampere, determine how many coulombs of charge there are in a new 70 ampere•hour battery.

19. A 9-volt transistor radio battery may be smaller in size than a 1½-volt flashlight battery. What, besides their voltage, do you think might be different between these two kinds of batteries?

20. An ordinary flashlight uses two 1½-volt batteries. What is the voltage applied to the bulb?

21. Only a small percentage of the electrical energy passing through a light bulb is transformed into light. What happens to the rest?

22. Does a wire carrying an electric current necessarily have a net charge on it?

23. What is the difference between a volt and an ampere?

24. Show on a diagram how to make an alarm system that will ring whenever someone opens the door to your room. You have at your disposal a battery, a bell that rings when connected to the battery, a push-button switch (off when the button is "in"), and several lengths of wire.

25. Design a circuit in which two different switches of the type shown in Fig. 13–40 can be used to turn the same light bulb on and off. The switches can be on opposite sides of a room. (Hint: Draw the circuit with only one switch; then "cut" one wire and insert the second switch plus another length of wire—but be sure you get the connection right.)

26. How would the current in a circuit be changed if you doubled both the resistance and the voltage?

27. Why are thick wires preferable to thin wires for carrying current?

28. What is the resistance of a toaster that uses 5 amps on a 120-volt line?

29. If you reduce the voltage to a heater by half, what happens to the current?

30. If you had a set of Christmas-tree lights strung in series and one of the lights burned out, how would you determine which one it was?

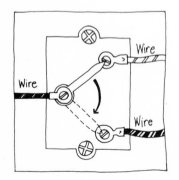

Figure 13–40
A switch.

31. A circuit contains only one lamp. What happens to the current if two more lamps are added in series? What happens if they are added in parallel?

32. By what factor is the current increased when two batteries are used in a flashlight instead of one?

33. A transistor radio uses a 9-volt battery and draws 200 milliamperes of current. What is the net resistance of the radio?

34. Why will a bulb designed for 120 volts burn out if it is placed in a 240-volt circuit?

35. What happens when a light bulb "burns out"?

36. Which draws more current, a 60-watt light bulb or a 75-watt light bulb?

37. How might a transistor be used as a switch?

38. Why is it dangerous to touch an electric switch when you are in the bathtub?

39. A toaster intended for use on a 120-volt circuit has a resistance of 30 ohms. How much current does it draw?

40. Describe the transformations of energy when an electric toaster toasts a piece of bread.

41. In the experiment you performed with the flashlight, why did the bulb glow more brightly when two batteries were used?

42. What is the resistance of a 100-watt light bulb operating at 120 volts?

43. Using the power relationship, $P = V^2/R$, calculate the resistance R of an 1800-watt electric heater operating at a voltage V of 120 volts.

44. Describe how a *pnp* transistor can operate as an amplifier. Remember that in a *pnp* transistor, the positive holes in the emitter play the same role as the electrons in the emitter of an *npn* transistor (Fig. 13–28).

45. A string of Christmas-tree lights contains eight 15-watt bulbs connected in parallel. What are the voltage across each bulb and the resistance of each when connected to its intended voltage of 120 V?

46. A string of Christmas-tree lights contains eight 15-watt bulbs connected in series. What are the voltage across each bulb and the resistance of each when connected to its intended voltage of 120 V?

47. What would happen if a light bulb from a series string of Christmas lights were substituted into a parallel string? How about the reverse? Explain.

14

MAGNETISM
AND
ELECTROMAGNETISM

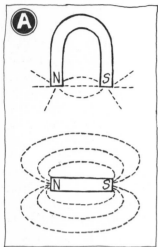

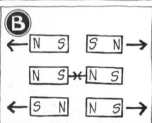

Figure 14–1

(A) A horseshoe magnet and a bar magnet, with north (N) and south (S) poles indicated. (B) Like poles repel; unlike poles attract.

1. MAGNETS AND MAGNETISM

Many centuries ago, people living in the region of Asia Minor known as Magnesia found pieces of rock that attracted each other when placed close together. Today we call such special rocks "magnets," after the place where they were discovered. We know now that the phenomena of magnetism and electricity are intimately related, but that relationship was not discovered until early in the nineteenth century.

Behavior of Magnets: Unlikes Attract, Likes Repel

We are all familiar with horseshoe magnets or bar magnets. A magnet will attract nails, paper clips, and other objects made of iron. Every magnet has two ends or faces, called "poles," where the magnetic force is strongest. About the eleventh or twelfth century A.D. the Chinese discovered that a magnet's poles exhibited a peculiar characteristic: when a magnet is freely suspended, it rotates until one of its poles points toward the north and the other points toward the south. The compass, a great aid to navigation, was born! That pole of a freely suspended magnet that points to the north is called the "north pole" of the magnet; the opposite pole, which points south, is called the "south pole," Fig. 14–1(A).

When two magnets are brought close together, a strong force exists between them. That force may be either attractive or repulsive. If the north poles of two magnets are brought together, the magnets repel each other; the same thing happens if the south poles are brought together. But if the north pole of one magnet is brought close to the south pole of the other, the force between them is attractive, Fig. 14–1(B). In other words: *Like poles repel; unlike poles attract.* Confirm this by performing the following experiment.

EXPERIMENT
Obtain two bar magnets (compass needles may work). Suspend each magnet from a separate thread and determine the north poles of each by noting the pole that points toward the north. Now bring the north poles toward each other and determine the direction of the force. Do the same for the two south poles, and for the north and south poles.

Only iron and a few alloys (mixtures of several metals) exhibit this strong magnetic effect. All other materials show some slight magnetic effect, but it is usually extremely small and can be detected only with very delicate instruments.

Magnetic Fields

Earlier, we found it useful to talk of the electric field surrounding an electric charge. In the same way we can speak of the **magnetic field** surrounding a magnet; we can draw magnetic field lines, or lines of force, just as we drew electric field lines. In Fig. 14–1, the magnetic lines of force surrounding a bar magnet and a horseshoe magnet are shown as dashed lines. The force between two magnets, whether attractive or repulsive, is said to be an interaction between one magnet and the magnetic field of the other.

When iron filings are placed near a magnet, they align themselves in a pattern following the magnetic field lines, Fig. 14–2. Each iron filing in the magnetic field acquires a weak north and south pole in much the same way that an electric field induces a charge separation in an otherwise neutral body (see Fig. 13–5).

If a freely suspended magnet is placed near a stationary magnet, it will rotate until it is parallel to the magnetic field lines of the stationary magnet;[1] the north pole of the suspended magnet will be closest to the south pole of the stationary magnet, as in Fig. 14–3(A). This is how a compass works. The earth acts like a huge but

[1]The direction of magnetic field lines is defined as that direction in which the north pole of a small magnet or compass points when it comes to rest after being placed in that magnetic field; see Fig. 14–3(A).

Figure 14–2

Photograph showing how iron filings line up along magnetic field lines.

Figure 14–3

(A) Suspended magnet rotates until it is parallel to the magnetic field lines of the stationary magnet. (B) The earth acts like a large magnet with its south magnetic pole near the geographic north pole.

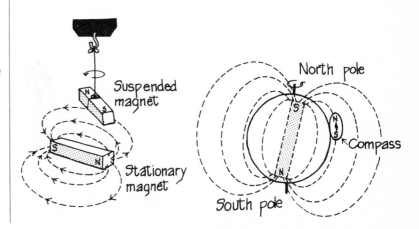

not very strong magnet, Fig. 14–3(B). A compass needle, which is a small magnet, aligns itself in the earth's magnetic field.

The end of a compass needle that points to the north is the north pole of the compass needle. Since the north pole of a magnet is attracted to the south pole of another magnet, we must conclude that it is the earth's "south magnetic pole" that is in the geographic north. And the earth's north magnetic pole is in the south. However, the earth's magnetic poles do not exactly coincide with its geographic poles. (The earth's geographic poles are the ends of the axis on which the earth spins.) The south magnetic pole, for example, is in northern Canada, more than 1000 miles from the geographic north pole. This fact must be allowed for when using a compass. The angular difference between true north and magnetic north (which is the direction in which the compass needle points) is known as the "magnetic declination."

Magnets Are Made Up of Domains

The forces between the like and unlike poles of a magnet are much like the force between positive and negative electric charges; that is, unlikes attract, likes repel. And just as we can isolate positive and negative charges, we might expect that by cutting a magnet in half, we could isolate the N and S poles. But we can't. When we cut a magnet in half we find that we have two new magnets, each with a north and a south pole; the two new poles appear at the cut, Fig. 14–4. If we then cut each of the new magnets in half, we get four magnets, each with a north and south pole. We could repeat this operation many times and end up with more and more magnets, each with a north and south pole. Physicists have long been trying to isolate a single magnetic pole, but it does not seem possible. Apparently single magnetic poles do not exist.

These attempts illustrate an important point, however: an ordinary magnet can be considered to be made up of many microscopic magnets. Indeed, carefully prepared samples of magnetic materials, when viewed under an electron microscope, reveal tiny regions known as "domains." These domains behave like tiny magnets, each with a north and a south pole. In an ordinary "unmagnetized" piece of iron (i.e., a piece of iron that does not act like a magnet), the domains are arranged randomly, as shown schematically in Fig. 14–5(A). Since the effects of each of these tiny magnets cancel one another, this piece of iron will not act as a magnet. However, a piece of iron in which all or nearly all of the domains are aligned will act as a strong magnet, Fig. 14–5(B). It is the combined

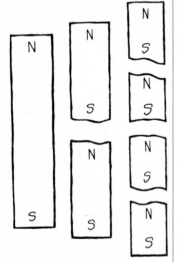

Figure 14–4

When we break a magnet in half, we do not separate the north and south poles; instead, we form two new magnets, each with a north pole and a south pole.

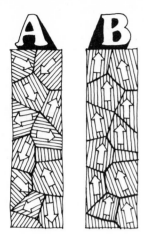

Figure 14–5

Unmagnetized (A) and magnetized (B) pieces of iron, showing orientation of magnetic domains. The tips of the arrows correspond to the north pole and the tails of the arrows correspond to the south pole of each domain.

magnetic effects of each of the tiny magnetic domains that give this piece of iron its strong magnetic power.

Because iron magnets do not lose their magnetism readily, they are often called "permanent" magnets. However, if a magnet is violently struck with, say a hammer, the domains are jarred into randomness, and the magnet loses some or all of its magnetism.

We are now in a position to explain why it is that unmagnetized pieces of iron, such as a nail, can be picked up by a permanent magnet. The magnet's field causes the tiny magnetic domains of the nail to rotate slightly, just as the earth's magnetic field causes a compass needle to rotate. The domains rotate only slightly because of the rigidity of iron. But even this partial alignment of the domains means that the nail has become a temporary magnet with its south pole attracted to the north pole of the permanent magnet (or vice versa). Thus the magnet "attracts" the nail. This is analogous to an electrically charged object attracting uncharged objects by inducing a separation of charge in them.

The domains in the iron nail resist even this slight reorientation and return to their original random positions once the magnet is removed. However, if an unmagnetized piece of iron is placed in a *very strong* magnetic field, the domains will be rotated farther and will remain in their newly aligned positions even after the external magnetic field has been removed. Furthermore, those domains whose magnetization direction is parallel to the external field actually grow larger at the expense of other domains. Both effects are important in making an ordinary magnet. This also is how weak natural magnets, like those found in Magnesia, are made; the external magnetic field that magnetized the rocks was the magnetic field of the earth itself. When we set about making a magnet, the stronger the external magnetic field, the better aligned the domains will be and the stronger the new magnet will be. The easiest way to magnetize a piece of iron is to use the magnetic field of a permanent magnet: we merely stroke the unmagnetized iron repeatedly, in one direction only, with the permanent magnet.

2. MAGNETISM AND ELECTRIC CURRENTS

Electric Currents Produce Magnetism

Magnets are made up of tiny domains. But what are domains made of? Or, to say it another way, what really produces magnetism?

Let's perform an experiment that requires only a compass, a length of wire, and a battery. A battery with screw terminals is best, since the wire must be firmly connected to the battery. However, a

389

good-quality flashlight battery will do, in which case the connection between the wire and the battery can be made by holding them *firmly* together with your fingers. The experiment will work best if a thick wire is used—one that provides low resistance and ample current—although ordinary lamp cord may be satisfactory.

EXPERIMENT

Place a compass on the edge of a table, Fig. 14–6. Attach one end of the wire to one terminal of the battery. Bring a straight vertical section of the wire as close as possible to the compass [Fig. 14–6(A)], keeping the rest of the wire away from the compass. Now touch the free end of the wire to the other terminal of the battery, so that an electric current flows in the wire, Fig. 14–6(B). What do you observe?

Presumably you saw the compass needle move. And what causes a compass needle to move? A magnetic field, of course! The experiment clearly shows that *electric currents produce magnetic fields!*

The experiment you have just performed was first carried out by Hans Christian Oersted in 1820. It clearly demonstrates the intimate connection between electricity and magnetism.

Any electric current is surrounded by a magnetic field. An electric current consists of moving electric charges. Therefore, while a

Figure 14–6

Demonstration that electric currents cause magnetism. Compass needle is deflected by the magnetic field of the electric current.

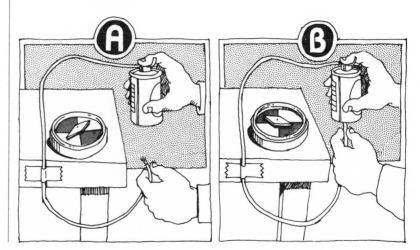

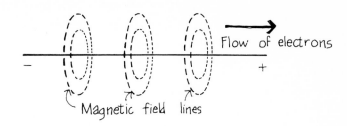

Figure 14–7

Magnetic field lines (dotted) produced by the electric current in a straight wire encircle the wire.

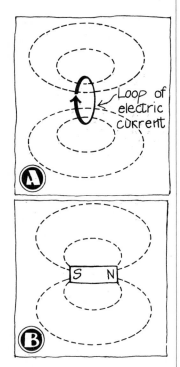

Figure 14–8

Magnetic field due to (A) an electric current in a circular loop of wire, (B) a simple bar magnet. Note the similarities in the two patterns.

391

stationary electric charge fills space with an electric field, a moving charge fills space with both electric *and* magnetic fields.

The orientation of the magnetic field set up by the current in a wire can be easily determined. A compass needle placed in a magnetic field will align itself parallel to the magnetic field lines, with its north pole pointing in the direction of the field lines at that point.[2] In this way it is found that the magnetic field lines encircle the current-carrying wire. For a straight wire, the lines are circles, Fig. 14–7.

The magnetic field lines surrounding a circular loop of current-carrying wire are shown in Fig. 14–8(A). Notice that the pattern of lines is very similar to that of a bar magnet, Fig. 14–8(B).

Now, do the magnetic effects of a current have anything to do with bar magnets and domains? Yes. Everything. The atoms that make up matter contain rapidly moving electrons. Not only do the electrons move in orbits around the nuclei of the atoms; they spin on their own axes as well. Each of these rotational motions constitutes an electric current (remember that electrons are charged). But in most materials the magnetic effects of these atomic currents cancel out because of the random orientations of the many electron orbits. In iron and certain alloys, however, a complicated cooperative mechanism causes all the electrons in a small region (a domain) to spin in the *same direction*. So the tiny magnetic fields of each of these spinning electrons, these currents, add up to a significant magnetic field, the magnetic field of the domain. And, as we have seen, when the domains are aligned, a very strong permanent magnet results.

Since all magnets consist of the many tiny currents of spinning electrons, each just like the current in Fig. 14–8(A), then it is clear why it is impossible to isolate a single magnetic pole. Each spinning electron is itself a tiny magnet, with a north and a south pole. There is no way to divide up the current and obtain a single magnetic pole.

[2]If the current is not large, the magnetic field of the earth will be as large or larger than the field caused by the current, and a compass needle will point in the direction of the resultant *sum* of these two fields.

Physicists today believe that *all* magnetic fields are produced by electric currents.

Magnetic Fields Exert Forces on Moving Charges and Currents

The interrelatedness of electric and magnetic effects is even more profound than we have seen so far.

We just demonstrated that a current-carrying wire exerts a force on a magnet such as a compass needle; this is what convinced us that currents produce magnetism. By Newton's third law, the reverse should be true as well. The magnet should exert an equal and opposite force on the wire carrying the current. This effect too was discovered by Oersted.

A magnetic field exerts a force on any moving electric charge, whether the charge is moving in free space or as an electric current in a metal wire. Strangely enough, this force is a *sideways* force; it is perpendicular to the direction of motion of the charged particles, or the current, and it is also perpendicular to the magnetic field lines, Fig. 14–9. The force is *not* along the direction of the magnetic field lines, as was the case for the electric field; it is not a repulsive or an attractive force toward either pole of the magnet—it is always *perpendicular* to the magnetic field.

The *magnitude* of the force exerted by a magnetic field on a charged particle has been found experimentally to be directly proportional to the charge on the particle, to the velocity of the particle, and to the strength of the magnetic field. Let us use the symbol q for charge, v for velocity, and B for the magnetic field strength.

Figure 14–9

(A) The magnetic field exerts a force upward on the negatively charged particle. (B) If the negative particle comes from the opposite direction, the force due to the magnetic field is downward.

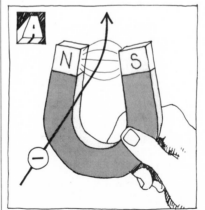
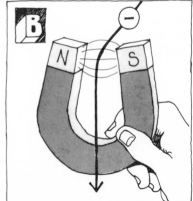

Then, if *v* and *B* are perpendicular to each other, the force is given by $F = qvB$.

The magnetic force is greatest when the charged particles are moving at right angles to the magnetic field. At other angles the magnitude of the force diminishes, and when the charged particles move parallel to the magnetic field lines there is no force at all. The direction of the force is always perpendicular to the magnetic field lines *and* to the direction of motion of the charged particles; this is true even when the particles are not moving perpendicularly to the magnetic field. Experiments show that the force on a negatively charged particle, such as an electron, is always in the same direction relative to the orientation of the magnetic field and the direction of motion of the particle. Thus, in Fig. 14–9(A) the force is upward. If the relative orientation of the field and of the particle's motion is reversed—as when the negative particle comes from the opposite direction, Fig. 14–9(B)—then the force on the particle would be reversed as well; that is, the force would be downward, as shown.

Figure 14–10 shows the situation for positively charged particles—the force is always found to be in the direction opposite to what it would be for a negative particle moving in the same direction.

Since an electric current in a wire consists of moving electrons, a wire placed in a magnetic field will feel a force on it when a current is flowing, Fig. 14–11(A). If the direction of the current is reversed, the direction of the force is reversed as well, Fig. 14–11(B). It would be easy and interesting for you to carry out the experiment illustrated in Fig. 14–11.

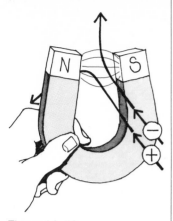

Figure 14–10

Positive and negative charges are bent in opposite directions by a magnetic field.

Figure 14–11

(A) A magnetic field exerts a force on a current-carrying wire. (B) If the direction of the current is reversed (by interchanging the connection to the battery), the direction of the force is reversed.

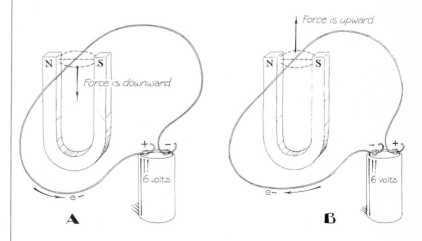

Motors, Galvanometers, Loudspeakers, and Solenoids

A **galvanometer** is a device for measuring electric currents. It consists of a coil of wire suspended by a fine wire in the magnetic field of a permanent magnet, Fig. 14–12. When a current flows through the coil, the magnetic field exerts a force on it, causing the coil and its attached pointer to turn. A small coil spring resists this motion to a certain extent, so the size of the angle through which the pointer turns will be proportional to the magnetic force. The magnetic force, in turn, is proportional to the current in the wire. Thus the deflection of the pointer is a measure of the current flowing through the coil. With the addition of an appropriate amount of resistance, any range of current can be measured; such a device is called an "ammeter." Furthermore, with the addition of a large resistance, the deflection of a galvanometer can be calibrated to measure voltage; such a device is called a "voltmeter." Galvanometers are the essential element in nearly all instruments that measure electric current or voltage.

A **motor** is a device that changes electrical energy into mechanical energy. Motors work on the same principle as a galvanometer, except that the coil of wire is much bigger and is suspended between the poles of a magnet on an axle rather than on a thread of wire. Actually, many turns of wire are wrapped around a piece of iron to form what is called an *armature*. An AC motor is diagrammed in Fig. 14–13. In Fig. 14–13(A) the loop of wire, which in actual motors consists of many turns, is rotating clockwise because of the magnetic force on the current flowing in it. When the loop

Figure 14–12

A galvanometer. The magnetic field exerts a force on the loop of wire, causing it to twist through an angle proportional to the amount of current flowing in the wire.

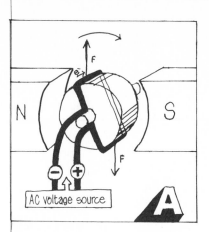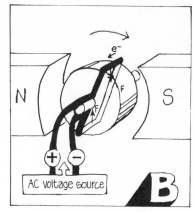

Figure 14–13

An AC motor consists of many loops of wire (only one is shown), which are mounted in a magnetic field so they will rotate when an AC current flows through them. The forces on the wire due to the magnetic field are indicated by arrows and the letter F. In actual motors the lead wires are connected to rings on the main shaft that rub against stationary rings to prevent twisting of the lead wires.

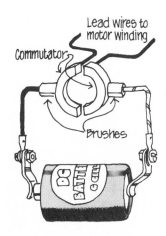

Figure 14–14

A DC motor changes DC to AC using this commutator-brush arrangement. The brushes are stationary; the commutator, with the connected lead wires, is attached to the motor shaft and turns with it.

turns past the vertical position, as it has in Fig. 14–13(B), the magnetic force would ordinarily operate to turn the loop back the other way—the force on the top part of the loop would still be upward, for example. But just as the motor passes through the vertical position, the alternating current is timed to start flowing in the opposite direction. As we saw in Fig. 14–11, the force will then be reversed, and the motor will continue to rotate in the same direction, Fig. 14–13(B). The reversing of the current is crucial to the operation of a motor. Only if the current alternates will the motor turn continuously in one direction.

A DC motor operates from a DC source, such as a battery, and is just like an AC motor except that it has a "commutator" and "brushes" that allow the DC current to reverse every half revolution, Fig. 14–14. In other words, the commutators and brushes change the DC into AC. The commutators are attached to the motor shaft and consist of two curved metal pieces, each connected to one end of the loop of wire. The brushes are connected by wires to the battery and rub against the rotating commutator. The electric current passes through the brushes to the commutator, and at every half revolution each commutator changes its connection to the other brush and causes the current to alternate.

In larger motors an "electromagnet" is used instead of a permanent magnet. An electromagnet is simply a coil of wire wrapped

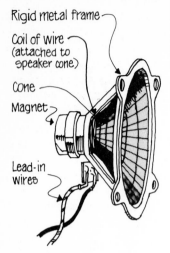

Rigid metal frame

Coil of wire
(attached to
speaker cone)

Cone

Magnet

Lead-in
wires

Figure 14–15

Diagram of a loudspeaker.

around a piece of magnetizable material such as iron. The magnetic field produced by a current in the coil magnetizes the iron. The total magnetic field is due to that of the iron plus that of the current in the coil. Thus the magnetic field of an electromagnet can be made much larger than that of a permanent magnet alone.

A high-fidelity **loudspeaker** also works on the principle that a magnet exerts a force on a current-carrying wire. The output from the radio amplifier is connected to the wire leads of the speaker. The speaker leads are connected internally to a coil of wire, which is itself attached to the speaker cone, Fig. 14–15. The speaker cone is usually made of impregnated cardboard and is mounted so that it can move back and forth freely. A permanent magnet is mounted directly in line with the coil of wire. When the alternating current of an audio signal flows through the wire coil, the coil and the attached speaker cone feel a force due to the magnetic field of the magnet. As the current alternates at the frequency of the audio signal, the speaker cone moves back and forth at the same frequency, causing

Figure 14–16

A solenoid used in a doorbell.

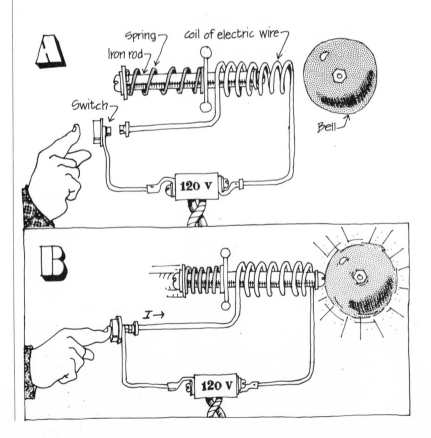

alternate compressions and rarefactions of the adjacent air, and produces sound waves. A speaker thus changes electrical energy into sound energy, and the frequencies of the emitted sound waves are an accurate reproduction of the electrical input.

A **solenoid** is a long coil of wire with a cylindrical piece of iron inserted partially into it. One use of the solenoid is in doorbells, Fig. 14–16. When someone pushes the button of a doorbell, the circuit is closed and the electric current in the coil sets up a magnetic field (like a magnet) and exerts a force on the iron rod. The rod is pulled into the coil and strikes the bell. Solenoids are used for a variety of purposes. The starter motor of a car is not normally in contact with the engine; in order to turn the engine when starting, a solenoid moves a gear on the starter motor shaft so it connects to a gear on the main engine shaft. Thus, when you engage the starter of your car, you not only start the motor turning, but close the circuit on a solenoid, which moves the starter motor gear so it connects to the engine before it starts turning. Solenoids are often used as switches for mechanical devices. Some tape recorders, for example, use solenoid switches—that is, the mechanical parts of the tape transport system are moved by solenoids. Solenoids have the advantage of quick and accurate start-up as compared to moving these parts with a lever as some recorders do.

Van Allen Radiation Belts and the Aurora Borealis

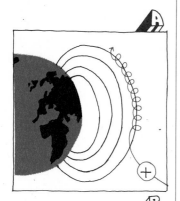

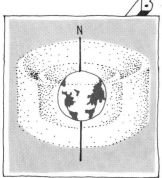

Figure 14–17

(A) Charged particles (cosmic rays) are trapped by the earth's magnetic field. (B) Van Allen radiation belts surround the earth in doughnut-shaped regions.

The magnetic field of the earth exerts a force on charged particles that impinge on the outer atmosphere. Because of their extraterrestrial origin, these charged particles are known as "cosmic rays." They come from the sun, stars, and other regions of the universe. The cosmic rays that come from the sun include hydrogen atoms that have been ionized by the sun's high temperatures. Huge magnetic storms, disturbances in the magnetic field of the sun, cause these charged particles to be expelled. Those that pass near the earth are deflected by the earth's magnetic field, and many are trapped by it, Fig. 14–17. The trapped particles move back and forth between the earth's magnetic poles, following spiral paths around the magnetic field lines of the earth, Fig. 14–17(A). The trapped radiation is largely confined to two doughnut-shaped regions surrounding the earth. These are known as the Van Allen radiation belts [Fig. 14–17(B)], named after James Van Allen, who determined their existence in 1958 from data gathered by the Explorer I satellite.

At the poles, some of the charged particles descend into the earth's atmosphere, ionizing the atoms of the atmosphere and causing them to fluoresce (see Chapter 15); this action produces the beautiful Northern Lights, or Aurora Borealis.

The Earth's Magnetic Field Protects Us from Radiation

Although some cosmic rays penetrate to the earth's surface, most of them are trapped in the radiation belts by the earth's magnetic field. It is fortunate that the earth has a magnetic field to protect us from this onslaught. Fast-moving charged particles cause atoms, including those in biological tissue, to ionize. Hence they can cause biological damage, especially to the genetic material DNA, and may result in mutations or alterations of the genetic makeup of the individual. While some mutations lead to an improvement and better adaptability of organisms, the vast majority of mutations are harmful.

3. ELECTRO-MAGNETIC INDUCTION

A Changing Magnetic Field Gives Rise to an Electric Current

So far in this chapter we have studied two important phenomena that show the relationship between electricity and magnetism. To review briefly:

1. Electric currents give rise to magnetic fields. Even the magnetic field of a permanent magnet is due to the tiny currents produced by spinning electrons.
2. Magnetic fields exert forces on moving electric charges and on electric currents.

Electric currents produce magnetism. Is it possible that the reverse is true? Can magnetism produce an electric current? Will a moving magnetic field exert a force on stationary electric charges and thereby cause a current to flow? Let's see.

EXPERIMENT

Make several small loops with a coil of wire, as shown in Fig. 14–18, and tie the ends of the wire together. Place a compass near a vertical section of the wire. Is the compass needle deflected when you thrust a magnet into the coil of wire, as in (A)? If it is, a current is flowing in the wire. (Be sure the compass is close to the wire but far from the magnet.) Next, let the magnet remain at rest and quickly move the coil toward the magnet, as in (B). Does a current flow in this case?

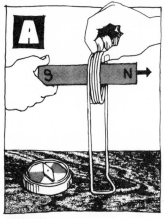

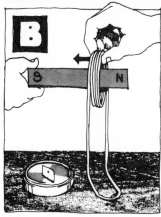

Figure 14–18

A current is induced in a coil of wire when the magnetic field through the loop is changing. This will happen (A) if a magnet is thrust into the coil of wire or pulled away from the coil and (B) if the coil is moved toward or away from the magnet.

The British physicist Michael Faraday (1791–1867) and the American Joseph Henry (1797–1878) investigated this possibility independently in 1831. They discovered that the act of moving a magnet quickly into or out of a coil of wire causes a current to flow in the wire, Fig. 14–18(A). Furthermore, they found that if the magnet is held stationary and the wire is moved [Fig. 14–18(B)], again an electric current is induced to flow in the wire. This phenomenon is known as **electromagnetic induction**.

We know that a current flows in a wire only when a voltage is present. Apparently, the changing magnetic field, whether the change is due to the motion of the magnet or the motion of the wire, causes an "induced voltage." This induced voltage gives rise to a current whose magnitude depends on the resistance of the wire ($V = IR$).

The "principle of electromagnetic induction"—or **Faraday's law**, as it has come to be called—can be stated simply:

> A voltage is induced in a loop of wire when the magnetic field through the loop changes.

Faraday's law tells us that the change in the number of magnetic field lines that pass through a coil of wire gives rise to an induced voltage. More specifically, the induced voltage is *proportional* to the *change* in the number of magnetic field lines passing through the coil. The greater the change in the number of lines per second, the greater the induced voltage. Furthermore, since each loop of the coil receives the same induced voltage, the more loops there are, the greater the total induced voltage.

In Fig. 14–18(B), as the coil is pushed to the left over the magnet, the number of magnetic field lines that pass through the coil increases, and so a current is induced. When the magnet stops moving, the current ceases to flow. If the coil is removed, the number of magnetic field lines that pass through the coil *decreases*. Again a current flows, but it will be in the opposite direction. The faster the coil (or magnet) is moved in either direction, the greater the momentary current through the coil.

The Transformer

Figure 14–19(A) shows two coils of wire; one is called the primary coil, the other the secondary coil. An alternating current is applied to the primary coil. As the primary current changes, the magnetic field it produces changes with it. The changing magnetic field passes through the secondary coil, and, as we know from Faraday's

399

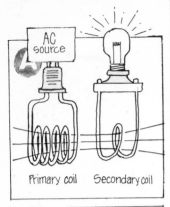

Primary coil   Secondary coil

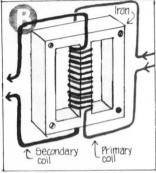

Iron

Secondary coil   Primary coil

Figure 14–19

(A) Schematic representation of a transformer. The changing magnetic field due to the alternating current in the primary coil induces a current in the secondary coil. (B) Diagram of an actual transformer.

law, this changing magnetic field induces a current in the secondary coil.

Such a combination of two coils, in which an alternating current in one coil induces an alternating current in the other, is called a **transformer**. Normally, the two coils are placed very close together, or even intertwined, and a piece of iron is inserted within the loops of the coils, Fig. 14–19(B). The iron guides the magnetic field lines much as a wire guides electric field lines, and thus aids the transfer of energy from the primary to the secondary.

Transformers work only with AC. A constant DC voltage applied to the primary coil gives rise to a constant magnetic field, and no voltage can be induced in the secondary. A voltage can be induced only by a *changing* magnetic field—that is, only by alternating current.

The main use of a transformer is to change the magnitude of an AC voltage, to make it either higher or lower. If the primary has the same number of turns (or loops) as the secondary, the voltage induced in the secondary is the same as that in the primary. But if the secondary coil contains twice as many turns as the primary, the secondary voltage will be twice that in the primary. This is true because the total voltage induced in the secondary is the sum of the voltages induced in each secondary loop. If the number of secondary turns is doubled, the total induced voltage is doubled. In general, the ratio of the induced secondary voltage, V_2, to the applied primary voltage, V_1, is equal to the ratio of the number of turns in the two coils, call them N_1 and N_2, respectively: $V_2/V_1 = N_2/N_1$. If the secondary coil has more turns than the primary, the secondary voltage will be greater than the applied primary voltage. This is called a **step-up** transformer. If the secondary has fewer turns than the primary, the secondary voltage will be less than the primary voltage. This is a **step-down** transformer. Of course we don't get something for nothing with a transformer. True, voltage can be increased with a step-up transformer, but what is gained in voltage is lost in current. The output current of a step-up transformer is smaller by at least the same factor as that by which the voltage is larger. Thus the product voltage × current, which is power, remains the same. (Actually it is reduced, because some of the energy goes into heat.)

Transformers are indispensable for the transmission of electric power. In order to avoid great losses of power during transmission over long distances, high voltage must be used. At a power station the output of the generators is boosted by a step-up transformer to 120,000 volts or more for transmission. Step-down transformers are utilized at the receiving end to reduce the voltage, usually in several stages, to 240 or 120 volts for home and industrial use. The boxes you see on telephone poles contain step-down transformers.

A step-down transformer is the essential component of the converter that allows a battery-operated transistor radio or cassette tape recorder to be used on household voltage. The ignition coil of an automobile is a step-up transformer; it transforms the 12 volts of a car battery to the many thousands of volts necessary to cause the spark plugs to spark and ignite the fuel-air mixture in the cylinders. However, a 12-volt DC car battery alone will not operate a transformer, since alternating current is required. Therefore the automobile battery is connected to a switch, called the ignition "points," which opens and closes the primary circuit. In this way the steady battery voltage alternates between +12 volts and zero volts. This changing voltage can then induce the required high voltage in the secondary of the coil, Fig. 14–20.

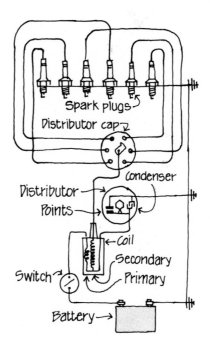

Figure 14–20

Schematic diagram of the ignition system of a car. The ignition "points" (which are actually located inside the distributor) act as a switch; they open and close at a rate determined by the speed of the engine and produce an alternating current in the primary of the "coil" (a transformer). A high voltage is thus induced in the secondary. This secondary voltage is greatest at the moment the points open, and this voltage pulse is sent to one of the spark plugs to ignite the fuel-air mixture in a particular cylinder. The capacitor is a pair of parallel plates (wrapped into a cylinder) that serves to store charge during each cycle and smooth out the voltage.

Generators Work by Induction

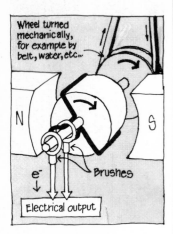

Figure 14–21

Diagram of an electric generator. In principle it is just the opposite of an electric motor. Real generators have many loops of wire, but only one is shown here.

The principle of electromagnetic induction is used in the **dynamo**, or **generator**, which changes mechanical energy to electrical energy. This is the reverse of what a motor does. A generator is essentially a motor run backward. As shown in Fig. 14–21, a generator consists of a coil of wire rotating in a magnetic field, just like a motor (Fig. 14–13). But the axle is not turned by the action of an electric current. Rather, the axle is turned by some mechanical force, and the current induced in the rotating coil is the output. A water wheel—or its modern equivalent, a turbine—is normally attached to the axle. Running water in a river or falling over a dam, or high-pressure steam (produced from burning fossil fuels at a steam plant or from nuclear energy at a nuclear reactor), strikes the turbine blades, forcing them and the attached coil of wire to turn rapidly. Since the rapidly turning coil is situated in a magnetic field, a current is continuously induced in it. At the instant shown in Fig. 14–21, the number of magnetic field lines passing through the coil is increasing and the induced current is as shown. After the coil passes through the vertical position, the number of field lines passing through it will begin to decrease, and the current will flow in the opposite direction. Every half revolution, the current changes direction. It is thus an alternating current and the generator is called an AC generator. AC generators produce nearly all the electrical energy we use.

For certain applications a DC generator is used. A DC generator contains a commutator and brushes just as a DC motor does, and a direct current is obtained as the output.

Had no one discovered electromagnetic induction and its practical application in the generator, life today would be totally different. Technology could not have reached the level it has without them (for good or bad). Even before the Industrial Revolution it had been possible to change the kinetic energy of running water into useful work by means of a water wheel—for example, to turn the wheel of a mill that grinds grain into flour. But the mill had to be situated right on the river. With the development of electric generators and motors, the kinetic energy of moving water could be transformed into electric energy, and the electric energy could then be transported by wires to cities and transformed back into mechanical work. Thus the mill no longer had to be situated at the source of energy, the flowing water. It could be many miles away from the river; it could be built where customers were located. With the building of high dams and large turbines to increase the kinetic energy of the falling water, vast amounts of energy could be produced. Where large dams were unfeasible, steam plants were built to release the chemical energy stored in coal or other fossil fuels by burning, and to use the heat to make the steam that drives the turbines.

4. ENERGY SOURCES AND ELECTRIC GENERATING PLANTS

You often hear it said that we should try to conserve energy because our resources are dwindling. We know, of course, that energy *is* conserved. But what is meant is that we should try to retard the rate at which energy is degraded to less useful forms—namely, low-grade thermal energy.

Nearly all the energy used in the United States comes from the burning of fossil fuels: oil, coal, and natural gas. A small proportion comes from nuclear energy (which uses the energy stored in uranium, which is also a finite resource), and from hydroelectric power (the potential energy of falling water); see Table 14–1. Thus, "conservation of energy" in popular language really means conservation of our natural resources or of fuel.

About 75 percent of the energy produced (or, rather, transformed) comes from the direct burning of fuels to heat houses, offices, and factories; to run cars and other vehicles; and to produce manufactured goods from raw materials (blast furnaces for steel, production of chemicals, refining of petroleum, and so on). The other 25 percent or so is used to produce electrical energy, which is used for some of the same purposes.

We will now examine the various means currently used to produce electrical energy, and some of the possibilities for the future. (See Table 14–2.)

Electricity Produced by Generators

At present, the only practical means used to produce electric power is by an electric generator. As we have seen, an electric generator

Table 14–1 Sources of Energy	
ENERGY SOURCE	PERCENT CONTRIBUTION TO TOTAL ENERGY PRODUCED IN U.S. (APPROXIMATE, 1977)
Coal	18
Oil	48
Natural gas	31
Nuclear power (uranium)	1.5
Hydroelectric power	1.5

Table 14-2
Electric Energy Production

PROCESS	PERCENT OF TOTAL U.S. ELECTRICITY PRODUCTION,* 1970s
I. Generator—Turbine	
A. Fossil fuel steam plants	80
B. Nuclear power	
Fission	5
Fusion	—
C. Geothermal power	Small
D. Tropical seas	—
E. Hydroelectric power	15
F. Tidal power	—
G. Wind power	—
H. Solar power	—
II. Other generators	
A. Solar cells	—
B. Magnetohydrodynamics (MHD)	—

*Fossil fuel and nuclear plants are 30 to 40 percent efficient, whereas hydroelectric plants are nearly 100 percent efficient. So the percentage of energy *consumed* (including waste heat produced) is roughly 90 percent, 5 percent, and 5 percent for fossil fuel, nuclear plants, and hydroelectric plants, respectively.

produces electricity when coils of wire are made to rotate in a magnetic field. A generator is generally connected to a large turbine (Fig. 14–22), which is made to turn by a variety of means.

Fossil Fuel Steam Plants.   In fossil fuel plants, coal, oil, or natural gas is burned to boil water and produce steam that turns a turbine. The basic principles of such steam engines were discussed in Chapter 9. These plants have certain advantages: they are fairly easy to build and are relatively inexpensive to run. But they have real disadvantages as well: The products of combustion create air pollution ("smog"). Like all heat engines, the efficiency of steam plants is limited (30 to 40 percent is typical), and the waste heat they discharge produces thermal pollution. The extraction of the fossil fuels used by these plants can be devastating to the land, particularly in the strip-mining of coal and the recovery of oil from shale. The transportation of the fuels sometimes leads to serious accidents, such as widespread oil spills in the oceans. And finally,

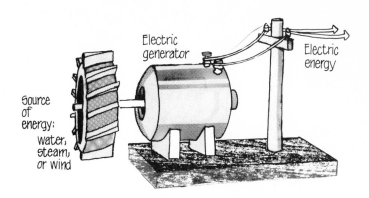

Electric
generator

Electric
energy

Source
of
energy:
water,
steam,
or wind

Figure 14–22

Basic plan of an electric
generating plant.

there isn't a great deal of fossil fuel left—estimates range from a
supply lasting a few centuries to only a few decades.

Nuclear Power Plants. There are two nuclear processes that
release energy: fission and fusion. In fission, the nuclei of uranium
or plutonium atoms are made to split, with the release of vast
amounts of energy. In fusion, energy is released when small nuclei,
such as hydrogen nuclei, combine ("fuse"). These two processes
are described in more detail in Chapter 16.

The fission process is used in all existing nuclear power plants.
Nuclear energy is used to heat steam just as fossil fuels are. Thus
a nuclear power plant is basically a steam engine that uses uranium
as its fuel. Consequently it exhibits the low efficiency that is characteristic of all heat engines and creates the considerable thermal
pollution that inevitably accompanies low efficiency. Although nuclear power plants produce almost no air pollution (so long as no
accidents occur), they do present serious problems: the radioactive
substances they produce are difficult to dispose of; a serious accident could result in the release of radioactive material into the air;
nuclear materials may fall into the hands of terrorists; and the supply
of nuclear fuel is limited. On the other hand, the energy produced
by nuclear power plants per kilogram of fuel is very large, and the
extraction of nuclear fuels is less damaging to the land than is the
extraction of fossil fuels.

The fusion process, compared to fission, would seem to have
fewer disadvantages. Moreover, the fuel supply available for this
process is vast: it consists of the hydrogen in the water molecules
in the oceans! Unfortunately, the fusion process cannot yet be controlled satisfactorily, although it holds genuine promise for the future

405 (see Chapter 16).

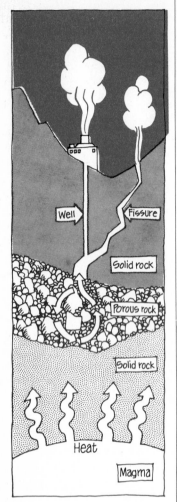

Geothermal Power. Both fossil fuel plants and nuclear plants produce steam to drive turbines. It is possible, however, to obtain natural steam from the earth itself. In many places, underground water in contact with the hot interior of the earth attains high temperature and pressure. It is sometimes expelled at the surface as hot springs, geysers, and steam vents. We can use the hot water or steam that issues through these natural vents, and we can also drill down to steam beds trapped below ground, Fig. 14–23. A large geothermal plant is already in operation at the Geysers in northern California, and a similar plant has been in successful operation in Italy for more than 60 years. Several others are in operation around the world. The number of sites at which water at high temperature and pressure can be tapped is limited. But another possibility is to drill two parallel wells down to hot dry rock in the earth's interior and to pass cold water under pressure down one well so that heated water (or steam) will be forced up the other well, Fig. 14–24.

Geothermal plants appear to be clean in that they create little air pollution, although they produce some (nonsteam) gas emission. Moreover, the spent hot water that must be disposed of leads to thermal pollution. Finally, the mineral content of that water, which is often high, is environmentally polluting and causes serious corrosion to the parts of the apparatus itself. Still, geothermal plants provide a reasonably inexpensive means of production and, with improved technology, may become a more common energy source.

Tropical Seas as Sources of Energy. In tropical seas there is a temperature difference of perhaps 20° C between the surface and the depths. A temperature difference of this magnitude is sufficient to drive heat engines. Steam could not be used in such engines (as in ordinary steam engines); rather a substance with a lower boiling point would have to be used that could drive the turbines. So far, this method has not been tried, since many difficulties have yet to be overcome. The efficiency of the method would be low, because the temperature difference, though adequate, is still small; at best, the efficiency might be only 7 percent. Consequently, losses suffered in the transfer of the heat produced would have to be held to a minimum to keep the efficiency from dropping to zero. The ocean environment itself causes problems, especially corrosion and the fouling of the apparatus by organisms. Moreover, if this method were ever used on a large scale, the ocean currents of the world might be affected. Yet there would be no problem with either air pollution or radioactive disposal, and so the advantages might outweigh the disadvantages.

Hydroelectric Power Plants. Hydroelectric power plants are usually located at the base of high dams where the energy of falling

Figure 14–23
Heat from earth's interior can produce steam to run a turbine generator.

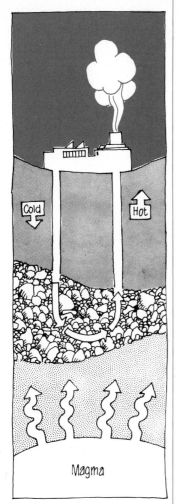

Figure 14–24

A geothermal plant in which cold water is heated to steam by being passed over hot rock in the earth's interior.

water can drive huge turbines. These plants produce practically no air or water pollution. And, since they discharge very little waste heat, they are nearly 100 percent efficient.[3] Unfortunately, however, there are few locations left where dams can be built without inundating land that has great scenic beauty. It is unlikely that many new hydroelectric power plants will be built in the United States.

Tidal Power Plants.   Another promising use of hydroelectric power is the harnessing of the ocean tides. A plant that uses tidal power is in operation in France. Here, as diagrammed in Fig. 14–25, a reservoir behind a dam is filled at high tide; at low tide the water is released to drive turbines. The reservoir is filled again at high tide, and as the water rushes in, it once again turns the turbines. In order for a tidal power plant to work efficiently, it must be built where there is a large difference between high and low tides. Such sites are rare, and even where they do exist huge dams would have to be built across natural or artificial bays. Although the rather abrupt changes of water level might have a harmful effect on wildlife, tidal power plants would otherwise cause little environmental disruption. Unfortunately, however, estimates suggest that tidal power could produce only a small fraction of the world's energy needs. Still, since our future energy needs may have to be satisfied by a variety of modest sources, tidal power may become a significant energy source in the years ahead.

Wind Power.   For centuries windmills served as a major source of energy. Today there is renewed interest in this very practical device, and it is possible that windmills may be used to drive electric generators on a large scale in the future. One proposal calls for the building of 300,000 windmills, each more than 800 feet high with blades 50 feet in diameter, dispersed throughout the Midwest where the winds are strong and steady. Such an array would satisfy a good portion of the energy needs of the United States; of course, it would also be an eyesore and might have some effect on weather patterns. But again, the possible use of wind power should not be overlooked.

Solar Power.   Many kinds of solar energy are already in use: Fossil fuels, after all, are the remains of plant life that grew by the photosynthesis of light from the sun; hydroelectric power depends

[3]Hydroelectric plants change about 80 percent of the energy of the moving water into electricity; the other 20 percent represents mainly kinetic energy of water leaving the turbine. Thus the efficiency, in terms of the original potential energy of the water, is about 80 percent. However, in terms of thermal pollution, hydroelectric plants are close to 100 percent efficient.

on the sun to evaporate the water that later falls as rain; and wind power depends on convection currents produced by the sun's heating of the atmosphere. But the sun's rays can also be used directly to provide energy.

We saw back in Chapter 9 that the absorption of the sun's energy by a black surface can be used to raise the temperature of water to heat a house. The same principle could be used to run electric generators. To do so, we would have to concentrate the sun's rays with large mirrors or lenses onto a small surface in order to create temperatures high enough to produce steam to drive a turbine. Such a system might generate enough electricity for homes and other small buildings, although some sort of backup system would probably be needed for cloudy days. In order to use solar power on a larger scale, we would have to set aside extensive areas of land on which to collect sufficient energy. Indeed, it has been suggested that stretches of Arizona and New Mexico, where the skies are clear nearly all year, be used for this purpose.

The principal disadvantage of relying on solar power is that the sun doesn't shine every day, and not at all at night; thus a highly efficient method of storing energy would be needed. Also, the large areas of land needed for the collectors and concentrators are not available where energy needs are greatest. Although some thermal pollution would result and the climate itself might be affected, there would be essentially no air or water pollution and no radioactivity. Moreover, the technology needed for capturing and using solar energy is fairly simple. Solar energy is one of the most promising future sources of energy.

Figure 14–25

A tidal power plant. Height differences resulting from the rising and falling of the tides are used to produce electricity.

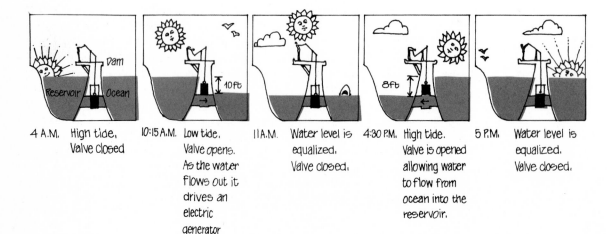

4 A.M. High tide, Valve closed

10:15 A.M. Low tide, Valve opens. As the water flows out it drives an electric generator

11 A.M. Water level is equalized. Valve closed.

4:30 P.M. High tide. Valve is opened allowing water to flow from ocean into the reservoir.

5 P.M. Water level is equalized. Valve closed.

Direct Conversion—
Solar Cells and MHD

Besides the above methods of producing electric energy by driving a generator, there are two other means by which electric energy may be produced in the future without the use of a conventional generator: solar cells and magnetohydrodynamics (MHD).

Solar cells—or, more properly, "photovoltaic cells"—were developed for the United States space program as a means of converting solar energy directly into electricity. The best-designed cells today can make this conversion with about 15 percent efficiency. However, they are very expensive. It is hoped that further research will find ways to reduce the cost and raise the efficiency. If this happens, solar cells will provide a desirable energy source, since they create very little thermal or other pollution. (No heat engine is involved.) For home use the cells could be mounted on the roof. For more extensive use, though, they would require a large land area since the sun's energy is not very concentrated. Another alternative is to place the solar cells (and the accompanying concentrators) in orbit around the earth. The sun's radiation is greater there than it is on the earth's surface because of absorption by the atmosphere, and the orbiting cells would spend little time in the earth's shadow. The electricity produced in this manner might be fed to a microwave generator and transmitted to receivers on earth.

In an **MHD generator**, a gas is heated to such high temperatures (above 2000° C) that electrons are detached from the molecules of the gas, and the gas becomes partly ionized. The resulting collection of positive and negative ions and electrons, called a **plasma**, is then directed between two parallel plates in the presence of a magnetic field. The magnetic field exerts a force on the moving charged particles, impelling positive particles toward one plate, and negative particles toward the other. Thus the two plates become oppositely charged. The potential difference that results can produce an electric current, Fig. 14–26. Although conventional means would have to be used to heat the gas—say, the burning of fossil fuels—the efficiency of an MHD generator can be somewhat higher (50 to 60 percent) than that of ordinary heat engines. Consequently more electricity is produced and less pollution is generated per unit of fuel burned. Also, some of the waste heat might be used to drive an ordinary turbine. Although the idea of MHD generators dates back to the early part of this century, it was not until 1958 that a successful device was built. No large-scale MHD generators have yet been constructed, but they may be in the future. MHD may be the ideal method for obtaining energy from nuclear fusion when and if it can be controlled (see Chapter 16).

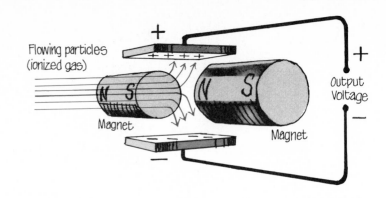

Figure 14–26

Diagram of MHD generator. The magnetic field bends the moving positive and negative ions (and electrons) in opposite directions. When they strike the plates, the plates become charged and an electric current is produced.

The Future and Energy Conservation

Clearly, all forms of energy production have undesirable side effects. Some are worse than others, and not all the problems can be foreseen. We must balance our need for energy against the "price" we must pay to get it. In the days when the demand for electricity was small, the problems of producing an adequate supply were small, too. Now that the demand has grown to such staggering levels, the attendant problems are staggering as well. Society as a whole must decide whether the environmental and health problems posed by energy production are worth the benefits. Although energy production may not decrease significantly in the near future, we can avoid the need for increasing it further by checking the wasteful use of energy. We may find, too, that limitations on future energy production are necessary, and this may require changes in our life styles.

As old reserves of fuel are used up, new forms of energy production will be needed. Many of these were just discussed. No one of these may predominate. Rather, it is likely that many different methods will have to be used in the future to satisfy our energy needs. This would have the added advantage that the problems posed by any one form of energy production might not become so large as to be uncontrollable.

5. ELECTRO-MAGNETIC WAVES

Faraday's law of induction tells us that a changing magnetic field induces an electric current. Now an electric current is a flow of electric charge, and electric charges flow whenever an electric field acts on them. Consequently Faraday's law can be written in a slightly different and more general form:

A changing magnetic field gives rise to an electric field.

It is this *induced* electric field that gives rise to the induced voltage and current in a wire. The electric field is induced, however, even if no wire or other matter is present in the region of space in which the magnetic field is changing. This is the deepest essence of Faraday's discovery. With this insight, Faraday's law takes on even greater power.

A Changing Electric Field Gives Rise to a Magnetic Field

The symmetry of nature is truly remarkable. In 1860 the great physicist James Clerk Maxwell (1831–1879) postulated that the exact inverse of Faraday's law is also true:

A changing electric field gives rise to a magnetic field.

This second induction principle should not come as a surprise. We know that an electric current sets up a magnetic field—this was Oersted's discovery. Now an electric current is simply electric charges in motion. Consider just one electric charge in motion. In Fig. 14–27(A) we see the charge with its electric field. A moment later the charge has moved [Fig. 14–27(B)] and has carried its electric field with it—the electric field at any point in space, say at *P* in Fig. 14–27, is changing! Thus we can reinterpret our statement that moving charges or currents cause magnetic fields and say: a changing electric field induces a magnetic field.

The second induction principle was the last important discovery to relate electricity and magnetism, and it enabled Maxwell to tie all of electricity and magnetism together in a mathematical theory of great beauty. It is one of the great triumphs in all of physics.

We will not go into the mathematics of Maxwell's electromagnetic theory. It is essentially a formulation in mathematical terms of the facts we have already discussed. But we can discuss a startling prediction of Maxwell's theory: *accelerating electric charges or the*

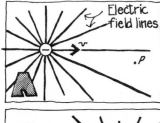

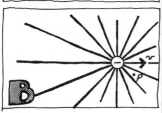

Figure 14–27

Electric field lines of a moving electric charge. Note that the strength of the electric field at point *P* increases as the charge comes closer.

equivalent, changing electric currents, give rise to electric and magnetic fields that move through space. We call these rapidly moving electric and magnetic fields **electromagnetic waves**.

Electromagnetic Waves Are the Result of the Two Induction Principles

Let us see how electromagnetic waves are produced. Figure 14–28 shows a transmitting **antenna** made of wire. The upper and lower arms of the antenna are connected to opposite terminals of an AC source of voltage. At any given instant the electrons in the two arms of the antenna are moving in the same direction, as shown in Fig. 14–28(A). This is true because the electrons are attracted to the positive terminal in one arm and repelled by the negative terminal in the other. A moment later, when the voltage changes sign, the electrons change their direction and start moving in the opposite direction, Fig. 14–28(B). The flow of electrons in the antenna is an electric current that changes direction at the same frequency as the AC source. It thus sets up a magnetic field, which itself is continuously changing. This changing or oscillating magnetic field gives rise to an electric field, as Faraday's law of induction tells us it will. The changing electric field itself produces a changing magnetic field. But these are not two separate processes. They occur together. The oscillating electric and magnetic fields are everywhere linked together.

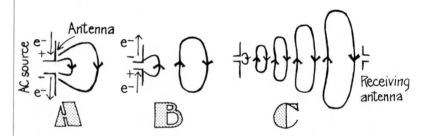

Figure 14–28

Production of electromagnetic waves by application of an AC voltage to a wire antenna. The diagrams show the electric field lines at various moments as the AC voltage on the antenna changes. The magnetic field lines are not shown; they are perpendicular to the electric field lines and therefore point into and out of the page. In (C) a "receiving antenna" (on a radio or TV set) is shown.

412

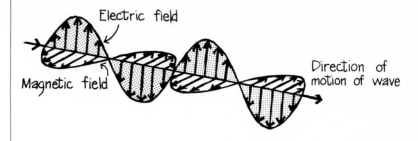

Electric field

Magnetic field

Direction of
motion of wave

Figure 14–29

An electromagnetic wave consists of alternating electric and magnetic fields at right angles to each other. The entire pattern moves in a direction perpendicular to both the electric and magnetic fields, at the speed of light.

The induced electric and magnetic fields produced at any instant do not exist all the way out to infinity. If they did, it would mean that the fields could get from the antenna to distant points in zero time, which is not possible. Instead, the electric and magnetic fields are set up in the vicinity of the antenna and travel outward at a finite velocity. As they move outward, they make room next to the antenna for the oppositely directed electric and magnetic fields that will be set up a moment later when the current changes direction. This is shown in Fig. 14–28(B). As the current in the antenna continues to reverse itself, the newly produced electric and magnetic fields appear next to the antenna, and the "old" fields are "pushed" outward, Fig. 14–28(C). The electric and magnetic fields that travel outward are known as **electromagnetic waves**. They are waves because the direction of the electric and magnetic fields alternates in an undulating fashion. Figure 14–29 is a diagram of the strength of the propagating electric and magnetic fields at a particular instant. The electric and magnetic fields are perpendicular to each other and to the direction of propagation.

Maxwell first predicted the existence of electromagnetic waves on the basis of his theory of electricity and magnetism. It was not until 1885, after Maxwell's death, that Heinrich Hertz (1857–1894) showed experimentally that electromagnetic waves do indeed exist and that they behave in just the fashion that Maxwell had predicted.

Light Is an Electromagnetic Wave

The frequency, or number of oscillations per second, of an electromagnetic wave can be a few cycles per second to hundreds of millions of cycles per second and even higher, depending on the frequency of the AC source.

Maxwell's theory predicted that all electromagnetic waves, regardless of their frequency, would travel at a velocity of 3×10^8 meters/second. This happens to be exactly the speed of light in vacuum. For a long time the nature of light had completely escaped scientists. Light was known to travel at a finite speed, and that speed was known to be 3×10^8 m/sec; but what light "really is" was a great mystery. The only thing known about light, besides its speed, was that it was some sort of wave motion (see Chapter 12); but just what oscillated up and down was not known. Maxwell had at last discovered the solution to the mystery of light. Light was an electromagnetic wave!

Other Kinds of Electromagnetic Wave: The Electromagnetic Spectrum

What distinguishes light from other electromagnetic waves is its frequency or wavelength. Our eyes are sensitive to electromagnetic waves only if the waves have a frequency between about 4×10^{14} and 7.5×10^{14} vibrations per second, which corresponds to wavelengths between 4×10^{-7} and 7×10^{-7} meters (see Chapter 12). Electromagnetic waves in this region are called **visible light**. Electromagnetic waves whose wavelengths are just shorter and just longer than those of visible light are called **ultraviolet** light (UV) and **infrared** light (IR), respectively. Waves with much longer wavelengths that are used for radio and TV transmission are known as **radio waves**. X rays are another form of electromagnetic wave, but

Figure 14–30

The electromagnetic spectrum.

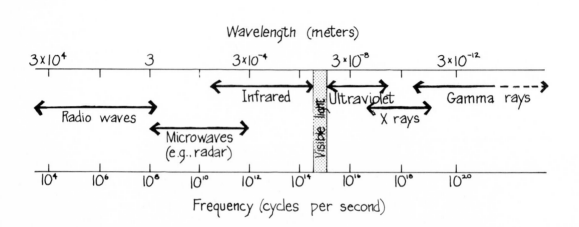

of extremely high frequency. Gamma rays are those with higher frequencies still; they are encountered mainly in nuclear physics (Chapter 16). The classification of electromagnetic waves according to frequency is known as the "electromagnetic spectrum"; it is shown in Fig. 14–30 along with the names given to the various frequency and wavelength regions. Electromagnetic waves of different frequency differ widely in their behavior. For example, as mentioned above, only waves in a certain limited frequency range can be detected by our eyes, whereas only waves of very high frequency are penetrating enough to produce medical x ray photographs. Yet all electromagnetic waves travel through space at the same speed and have the same essential nature. Only the frequency and wavelength are different.

Microwaves

Microwave ovens make use of electromagnetic radiation in the "microwave" region of the spectrum, which can penetrate deeply into materials and heat them rapidly. The term "microwave" applies to electromagnetic radiation whose wavelengths range from less than a millimeter to a few hundred centimeters (see Fig. 14–30). This range includes FM and TV signals, radar, and other uses.

Until recently it was thought that microwaves caused little health hazard and that their main effect was to heat tissue. However, evidence is trickling in that microwaves may have more subtle effects on people, ranging from behavior changes to eye damage that may lead to blindness. Some of these effects may be due to a resonant effect of these waves on the cells of the body. Of course, the level of exposure is important; it is not certain at this time what a safe level is. It might be best not to spend too much time near microwave ovens (which sometimes leak) or near television transmitters.

REVIEW QUESTIONS

1. A magnet
   (a) can be made of any metal
   (b) has two poles
   (c) must be wrapped with wire
   (d) exhibits static electricity

2. "Domains" are
   (a) regions of zero resistance
   (b) regions of large resistance
   (c) regions where the magnetic fields of the atoms are lined up in the same direction
   (d) those parts of an atom that are magnetic

3. The fact that a current-carrying wire deflects a compass needle is evidence that
   (a) the wire is magnetized
   (b) the north pole of the earth has shifted
   (c) the compass needle has an electric charge on it
   (d) the current gives rise to a magnetic field

4. Which of the following does not produce a magnetic field?
   (a) an electromagnet   (c) a stationary electric charge
   (b) a current-carrying wire   (d) a moving electric charge

5. The presence of magnetism in iron magnets is believed to be due to
   (a) the currents of spinning electrons
   (b) tiny magnets within atoms
   (c) electromagnetic induction
   (d) Coulomb's law

6. A stream of electrons passing between the poles of a horseshoe magnet would be
   (a) attracted toward the south pole of the magnet
   (b) attracted to its north pole
   (c) deflected sideways, toward neither pole
   (d) undeflected

7. An electric motor works on the principle of
   (a) induction
   (b) the force between a magnet and an electric current
   (c) the force between two electric charges
   (d) electromagnetic radiation

8. Electromagnetic induction is involved in
   (a) the charging of a storage battery
   (b) the operation of a transistor
   (c) the Aurora Borealis
   (d) the production of a current by relative motion of a magnet and a loop of wire

9. An electric generator works on the principle of
   (a) induction
   (b) the force between a magnet and an electric current
   (c) the force between two electric charges
   (d) electromagnetic radiation

10. A magnetic field can be induced by an electric field that is
    (a) positive   (c) strong
    (b) negative   (d) changing

11. A nuclear power plant
    (a) transforms nuclear energy directly into electricity
    (b) uses kinetic energy of high-speed nuclei to induce electric currents

(c) heats steam to turn a turbine
(d) is at present impossible

12. Which of the following is not being used at present to produce electricity?
(a) tidal power
(b) geothermal power
(c) steam plants
(d) nuclear fusion

13. An electromagnetic wave can be produced by
(a) a positive and a negative charge
(b) a steady current in a wire
(c) changing electric current
(d) a strong magnet

14. Light of shorter wavelength than that to which the eye is sensitive is called
(a) infrared
(b) ultrasonic
(c) microwaves
(d) ultraviolet

15. Radio waves are
(a) electromagnetic waves
(b) sound waves of intermediate frequency
(c) sound waves of very high frequency
(d) shock waves

16. Microwaves are
(a) sound waves
(b) electromagnetic (EM) waves with wavelengths around 1 cm
(c) EM waves with a wavelength of about one-millionth of a meter
(d) EM waves with a wavelength of about a million meters

EXERCISES

1. Why will either pole of a magnet attract an unmagnetized piece of iron?
2. Do you think that heating an iron bar magnet will reduce its magnetism? Explain.
3. Although each iron atom is a tiny magnet, not every piece of iron is a magnet. Why not?
4. Why does dropping a magnet on a hard floor weaken it?
5. Suppose you have three identical iron rods, except that two of them are magnets and the third is not. How would you tell which are the magnets without using any additional objects?
6. Explain the existence of a permanent magnet in terms of tiny atomic currents.
7. Can you set a resting electron into motion with a magnetic field? With an electric field?

8. Two parallel wires each carry an electric current. Is there a force between the two wires? Why?

9. A charged particle is moving in a circle under the influence of a uniform magnetic field. If an electric field is turned on that points along the same direction as the magnetic field, what path will the charged particle take?

10. Bringing a magnet close to a television screen will distort the picture. Why? (Don't do this to a color set; permanent damage may result.)

11. How is a motor similar to a compass?

12. Will a magnet exert a force on an electron at rest? What if it is moving?

13. What kind of field or fields surround a moving electric charge?

14. A current-carrying wire is placed in a magnetic field. How must it be oriented so the force on it is zero? So it is maximum?

15. A loop of wire is suspended between the poles of a magnet with its plane parallel to the pole faces. What happens if a direct current is put through the coil? What if an alternating current is used instead?

16. Why does more cosmic radiation strike residents of Norway than residents of Egypt?

17. Will you find AC or DC passing through (a) the headlights of a car, (b) a flashlight bulb, (c) a light bulb in your house, (d) the secondary of a transformer?

18. What are the similarities between a motor and a generator? What are the differences?

19. Suppose you are holding a loop of wire and suddenly thrust a magnet, south pole first, toward the center of the loop; is a current induced in the wire? Is a current induced when the magnet is held steady within the loop? Is a current induced when you withdraw the magnet?

20. Would a magnet rotating inside a coil of wire induce a voltage in the coil?

21. A model railroad signal is made as shown in Fig. 14–31. Explain how it works.

22. What does a transformer "transform"? Why is an alternating current needed?

23. If a transformer is to change 12 volts to 12,000 volts and there are 5,000 turns in the secondary, how many turns should there be in the primary?

24. A transformer has 30 turns in the primary and 150 turns in the secondary. What kind of transformer is this and by what factor does it change the voltage?

25. A step-down transformer reduces 120 volts to 40 volts. What is the current in the secondary as compared to that in the primary?

26. Does the voltage output of a generator change if its speed of rotation is increased? Explain.

27. A generator produces 100 volts when rotated at a certain speed. If its speed of rotation is doubled, what will be the output voltage?

28. Why is the armature of a motor or generator normally made of iron?

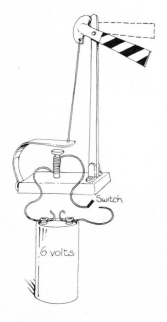

Figure 14–31

29. An astronaut in a fast-moving space vehicle observes that a current is flowing in a closed loop of wire even though there is no battery connected to it. What should the astronaut conclude?

30. In what ways are nuclear and fossil fuel power plants similar? In what ways are they different?

31. Much fuel is used to heat buildings. How might the waste heat from a power plant be used for this purpose? Would it still be called "thermal pollution"?

32. Why is less pollution produced by a more efficient power plant?

33. Water falls 100 cm over a dam at a rate of 12,000 kg/sec. How much electrical power (megawatts) could be produced by a power plant using this energy? (Hint: See Chapter 5.)

34. Solar cells can produce about 40 watts of electricity per square meter of area if they are directly facing the sun. If a house requires about 100 kilowatt·hours of electric energy per day, how large an area is needed to supply all the energy needs through the use of solar cells? (Assume the sun shines for 12 hours a day.) Would this area exceed the roof area on an average house?

35. The bay behind a dam used for tidal power has an area of about 2×10^8 m². The water level between high and low tide averages 1.0 meter. Estimate how much energy could be produced per day. (There are two high and low tides per day; potential energy of the water is given by PE $= mgh$ where m is the mass of the water, g is 9.8 m/sec², and h is the average height change of the water; to get the mass of water, you will need its density $= 1000$ kg/m³.)

36. Discuss the problem of conflicting values between the need for energy and the need to protect the environment.

37. Is sound an electromagnetic wave? If not, what kind of wave is it?

38. Are the wavelengths of radio and TV signals longer or shorter than those detectable by the human eye?

39. How are light and sound alike? How are they different?

40. A concert is broadcast over the radio. Who will hear the opening note of the concert first, a person in the balcony 100 feet away or a person whose ear is glued to his radio 2000 miles away?

41. Why does a television set use an antenna?

42. The carrier frequencies of FM broadcasts are much higher than for AM broadcasts. On the basis of what you learned about diffraction in earlier chapters, explain why AM signals can be detected more readily than FM signals behind low hills or buildings.

43. A useful radio antenna is one whose length is such that standing waves can be set up in it. Explain why an optimum FM antenna is 1½ meters long (5 feet). The carrier frequency for FM stations is around 10^8 cycles/second and the speed of light is 3×10^8 meters/second.

44. A particular FM station broadcasts at 92 MHz. What is the wavelength of this wave?

45. Why, at the beginning of Chapter 13, did we speak of the electromagnetic force as one of the four forces in nature rather than considering the electric force and the magnetic force as two separate forces in nature?

15

QUANTUM THEORY AND ATOMIC STRUCTURE

At the end of the nineteenth century, physicists were feeling rather self-satisfied. Nearly all observed phenomena had been explained through the laws of physics based on Newtonian mechanics and the very successful electromagnetic wave theory of light. A few experimental facts were still unexplained, but it was generally agreed that these would fall into place shortly. This optimism was soon shattered, however. In order to explain those few perplexing experimental results, radical new ideas were necessary, and another great revolution in physics was triggered. This second revolution—the first was that of Galileo and Newton—has had great ramifications for all the sciences and, indeed, for society as a whole.

The first aspect of this second revolution was marked by the emergence of the theory of relativity, which we discussed in Chapter 6. Now we turn to the second aspect—the development of the *quantum theory* and the study of the nature of atoms and molecules.

Physics as it was understood at the end of the nineteenth century is referred to as **classical physics**; relativity and quantum theory, on the other hand, and the even deeper probings into the nucleus of the atom (which we will deal with in Chapter 16) are referred to as **modern physics**. Classical physics dealt very successfully with phenomena of ordinary size. Modern physics came about as a result of attempts to understand the very large (in relativity) and the very small (in the quantum theory).

1. THE BIRTH OF THE QUANTUM THEORY

Emission of Light by Hot Solids Is Explained by Planck's Quantum Hypothesis

One of the phenomena that nineteenth-century physicists could not explain was the spectrum of light emitted by solids at high temperatures. When a solid is heated to higher temperatures, the spectrum of light it emits includes radiation of higher and higher frequencies. If the temperature is high enough, the radiation will contain frequencies in the visible region, and the object will glow. For example, when iron is heated to 600 or 700°C, it glows with a red color—we say it is "red hot." Most of the spectrum of light it is emitting is still not visible, but a small amount is at the low-frequency end, the red end, of the visible region. As the iron is heated to still higher temperatures, it becomes orange. Finally, it becomes white, and at that point all the visible light frequencies are represented. Similarly, the whitish-yellow light given off by an ordinary incandescent light bulb is emitted from the white-hot tungsten filament, which is heated to over 2000°C by the electric current passing through it. At any particular temperature, the brightness (or intensity) of the emitted light at each frequency can be measured. Typical results are shown in Fig. 15–1(A).

Toward the end of the nineteenth century, physicists tried to use the concepts of the atom and of classical physics to explain the intensity of light as a function of frequency. According to those concepts, the atoms in a solid are oscillating, and therefore they should emit electromagnetic radiation (see Chapter 14). Although this explained why solids emit light, it did not successfully predict the observed spectrum of emitted light. In fact, classical theory predicted that the brightness of the light should be greatest at the highest frequencies; in other words, all bodies would glow with a blue color, if at all. Clearly, classical theory was not consistent with the experimental results; see Fig. 15–1(B). The German physicist Max Planck (1858–1947) tackled this problem. At a meeting of the German Physical Society during Christmas week of 1899, he presented his solution. He had developed a theory that was in accord with the experimental data, Fig. 15–1(B). But to do so he had had to make a startling assumption, one that seemed at variance with classical physics. He hypothesized that the vibrating atoms or molecules in a solid material can have only certain discrete amounts of energy. Now it had been believed for centuries that a vibrating object could have any amount of energy, large or small, depending on the amplitude of the vibrations. But Planck broke with this tradition and

Figure 15–1

(A) Spectrum of light emitted by a body, at two different temperatures. (B) Theoretical predictions (at 1000°C).

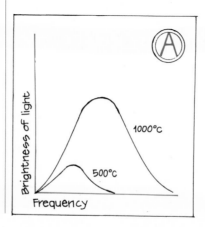

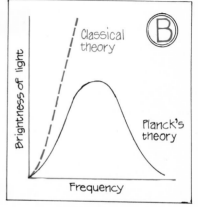

423

made the ad hoc assumption that the tiny molecular vibrations in a solid have a minimum energy, which is proportional to the frequency of vibration, f. That is,

$$E = hf,$$

where E is the energy of vibration, f is the vibration frequency, and h is a constant (later known as Planck's constant).[1] Furthermore, Planck hypothesized that the energy of vibration must be an integral multiple of this minimum energy; that is, the energy can only be: $1hf$, $2hf$, $3hf$, and so on. There would never be vibrations of molecules with an energy between these amounts. In other words, according to Planck, energy is not a continuous quantity. It is "quantized"—it exists only in certain "discrete" amounts. The smallest amount of energy that can exist is called the "quantum" of energy. This assumption that energy comes only in discrete amounts or quanta is known as **Planck's quantum hypothesis**.

Planck's quantum hypothesis was used successfully to predict the spectrum of light emitted by a heated solid, as indicated in Fig. 15–1(B). Yet the quantum hypothesis was not readily accepted by physicists. This is usually the case with any radically new idea. The concept of quantization, however, was not entirely unprecedented. For example, the mass of a block of iron is a whole number times the mass of one iron atom. This is because the block contains only an integral number of iron atoms. Nowhere will you find a piece of iron containing 506½ atoms of iron. We could call the mass of one atom of iron the "quantum of mass" for iron, and the mass of any piece of iron would then be an integral multiple of this quantum of mass.

Similarly, electric charge is quantized. The smallest electric charge found in nature is the charge on one electron. The electric charge on any macroscopic (ordinary-sized) body is due to an excess or a deficiency of electrons, and therefore the charge must be an integral multiple of the charge on one electron. The "quantum of electric charge" is thus the charge on one electron. If we call this charge "e," then any other charged body can have a charge of $1e$, $2e$, $3e$, or maybe $67,541e$. But it can never have a charge of $67,541.3e$.

We think of most other quantities, such as energy and velocity, as being continuous; that is, capable of taking on any value whatsoever. But Planck's hypothesis has forced us to recognize that these quantities are *not* necessarily continuous. According to Planck, energy at least is quantized.

[1]Planck's constant has been measured to be 6.6×10^{-34} joule•seconds.

Einstein and the
Photoelectric Effect

The quantum concept remained more or less in limbo for several years. Then Einstein extended it by introducing a new theory of light, which he published in 1905 soon after his famous article on special relativity. Einstein's theory was a bold extension of Planck's quantum hypothesis. Planck had assumed that the vibration energy of atoms and molecules in a solid is quantized. Einstein hypothesized that *light itself,* rather than being transmitted as a wave as previously thought, is instead transmitted as tiny particles, or "quanta." He proposed that each of these light quanta—or "photons," as they came to be called—would have an energy E directly proportional to the wave frequency f:

$$E = hf.$$

Again h is Planck's constant. According to Einstein's theory, then, a beam of light consists of a large number of tiny particles, or photons. And the greater the frequency of the light, the greater the energy of each photon.

Einstein pointed out that his theory could be tested by experiments on the "photoelectric effect." When these were performed a few years later they fully confirmed Einstein's ideas, as we shall now see. The photoelectric effect refers to the fact that when light shines on a metallic surface electrons are ejected from the metal, Fig. 15–2. The effect is used today in such devices as electric eyes and photographic light meters (which are discussed later in this chapter). The occurrence of the photoelectric effect was consistent with the classical idea that light is an electromagnetic wave consisting of electric and magnetic fields, since the electric field will exert a force on electrons and perhaps thrust some of them out of the metal. However, the wave theory gave quite different predictions on the details of this effect than Einstein's "particle," or photon, theory gave. Recall that the two main properties of a wave are its amplitude and its frequency (or wavelength); amplitude refers to the height of the crests of the wave, and frequency to the number of crests passing a given point per second. The amplitude of a light wave corresponds to the brightness of the light; and the frequency corresponds to the color. A brighter light would thus mean that the electric and magnetic fields are larger. So, according to the wave theory, we would expect that when a brighter light is shined on a metal the electrons should feel a greater force due to the greater electric field and therefore should be ejected with greater kinetic energy.

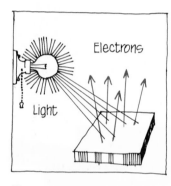

Figure 15–2

The photoelectric effect.

Einstein's photon theory, however, predicts that when a beam of light strikes a metal, each individual photon can collide with a single electron, causing it to be ejected from the metal. In the process, the electron absorbs all the energy of the photon, and the photon ceases to exist. Since electrons are held in the metal by attractive forces, some minimum energy (let's call it W) is required just to get an electron out of the metal. If the frequency f is so low that the photon energy hf is less than W, then the photons will not have enough energy to knock electrons out of the metal. This minimum frequency we call the "cut-off frequency," and designate it f_0. Furthermore, if the energy of the photons is greater than W, then a photon will have sufficient energy to knock an electron completely out of the metal. During the collision, energy is conserved; so part of the photon's energy goes into overcoming the forces holding the electron in the metal, and what is left over goes into the kinetic energy of the freed electron. Thus, according to the photon theory, if the frequency of the light is increased, the kinetic energy of the emitted electrons is increased. This is shown in the graph of Fig. 15–3. Finally, according to Einstein's theory, an increase in light intensity with no change in its frequency f means only that the light beam contains a greater number of photons, each still of energy hf. With the increase in number of photons, more collisions with electrons can occur and more electrons will be ejected; but since the energy of each photon has not been changed, the kinetic energy of the electrons is not changed.

In summary, Einstein's photon theory predicts that:

1. Increasing the intensity of the light increases the number of electrons ejected, but their energies remain the same.

Figure 15–3

Photoelectric effect. Kinetic energy of electrons increases with frequency. No electrons are emitted if the frequency is less than f_0.

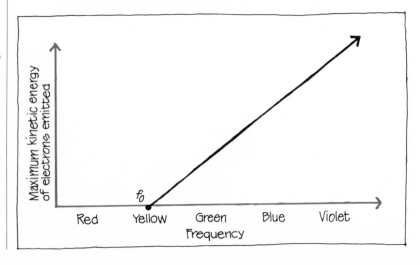

2. Light of higher frequency causes electrons to be ejected with greater kinetic energy.

3. No electrons are emitted if the frequency of the light is below a certain value.

These three predictions are very different from those of the wave theory, which predicts that (1) the energy of the ejected electrons should be greater for a greater intensity of light; (2) frequency would have no effect on the electron energy; (3) there would be no "cut-off" frequency—electrons would be ejected at any frequency.

When careful experiments were carried out in 1913–14, the results were fully in accord with Einstein's photon theory! For example, it was found that violet light readily ejects electrons from sodium; when yellow light is used instead, the electrons have less energy; and if red light is used, no electrons are ejected at all. And the intensity of the light was found to have no effect on the energies of the ejected electrons.

Einstein's brilliant theory was thus affirmed, and it was a crucial factor in the development of quantum theory.

2. THE WAVE–PARTICLE DUALITY

A number of additional experiments were carried out in the early part of the twentieth century that further confirmed Einstein's idea that light consists of particles, or photons. Thus, the particle nature of light rests on a firm experimental basis.

But what about the experiments performed by Young and others, which conclusively showed that light was a wave—in fact, an electromagnetic wave? These carefully performed experiments cannot be discounted. The wave nature of light also rests on a firm experimental basis!

So we are in a dilemma. In some experiments light behaves like a wave; in other experiments it behaves like particles. This is a rather unusual situation. Often in the history of science a physical theory has had to be discarded because it didn't agree with experimental results, or a theory has had to be modified to account for new experimental findings. But in the case of light, two incompatible theories have both been shown to have validity. Which theory is correct, the wave theory or the particle theory, seems to depend on the experiment being performed. Some experiments with light can only be explained by using a wave theory, whereas others can only be explained by using a particle, or "quantum," theory. Eventually physicists came to realize that this duality of light had to be accepted as a fact of life. It is often referred to as the "wave–particle duality."

Clearly, light is a more complex phenomenon than just a simple beam of particles or a simple wave.

The Principle of Complementarity

The great Danish physicist Niels Bohr (1885–1962) clarified the situation somewhat in a summary of the experimental data, which he called the "principle of complementarity." This principle states that in no single experiment does light show both wave and particle properties. Yet, both the wave and particle aspects of light are necessary for our understanding of light, and therefore these two aspects are "complementary." It is true that both the wave and the particle theories have some validity, but neither theory is complete in itself. Our knowledge of light is only partial unless we are aware of both its wave and its particle aspects.

It is impossible to "visualize" this duality. We cannot picture a combination of wave and particle. Hard as it may seem, we can only try to accept the experimental facts as they are and recognize the existence of both the wave and the particle properties of light.

Einstein's equation $E = hf$ itself relates the particle and wave properties of a light beam. E stands for a particle property, the energy of an individual photon, and f stands for a wave property, the frequency of the corresponding wave.

Matter Has a Wave Nature Too

In 1923, the French scientist Louis de Broglie (1892–   ) wrote a remarkable Ph.D. thesis in which he extended the idea of the wave–particle duality. De Broglie sensed the symmetry of nature. He argued that if light behaves sometimes like a wave and sometimes like a particle, then perhaps those things in nature that were previously considered to be particles (i.e., ordinary material objects, including electrons) might behave like waves in certain circumstances. De Broglie reasoned that the wavelength λ of a particle of mass m, which is moving with a velocity v, would be

$$\lambda = \frac{h}{mv},$$

where h is again Planck's constant. This equation also relates a wave property, the wavelength λ, and particle properties, the mass m and velocity v.

Two years later, de Broglie's hypothesis of "matter waves" was tested and was experimentally confirmed. The experiment was similar to Young's double slit experiment (see Chapter 12), but it was performed with electrons instead of light. A clear interference pattern was observed. Electrons behaved like waves! Shortly thereafter, neutrons and protons were shown to exhibit interference, and hence they too have wave properties. It was now apparent that the wave–particle duality applies to matter as well as to light.

The wave nature of matter is real. Why then are the particle properties of matter so obvious whereas its wave properties are not? For example, a baseball acts like a particle, not a wave. Wave properties are not observed because the wavelengths of most objects are extremely short. For example, the wavelengths of a 100-gram baseball moving at a velocity of 15 meters/second (50 feet/second) is about 10^{-34} meters, according to de Broglie's wavelength formula. Such an incredibly small wavelength is much too tiny to be detected. However, electrons have a very small mass, 9×10^{-31} kg, so, according to de Broglie's formula, they should have somewhat larger wavelengths. Indeed, an electron moving at a speed of 10^5 m/sec, which is an "ordinary" speed for an electron, has a wavelength of about 10^{-10} m; though small, this length can be dealt with, and it was with electrons that the wave properties of matter were first observed.

Just as in the case of light, the wave and particle aspects of matter are complementary. We must be aware of both the particle and the wave aspects in order to have a full understanding of matter. But again we must recognize that a "wave–particle" is impossible to visualize.

What Is an Electron?

At this point we might ask: What is the evidence that an electron is a *particle*? To answer this question we must go back to the early experiments that indicated that something, which came to be known as an electron, actually existed.

In the latter part of the nineteenth century it was observed that a gas kept at a low pressure in a glass tube glowed faintly when a high voltage was applied to it. If the gas was at a low enough pressure and the voltage was high enough, a bright fluorescent glow appeared at the end of the tube opposite the negative terminal, Fig. 15–4(A). In the 1890s an English physicist, J. J. Thomson (1856–1940), showed that the fluorescent glow could be deflected by an electric or a magnetic field, Fig. 15–4(B). This convinced Thomson that the glow was caused by charged particles coming from the negative terminal. From the direction of the bend he concluded that these particles possessed a negative charge, Fig. 15–4(C). Soon thereafter these theoretical particles were given the name "electrons."

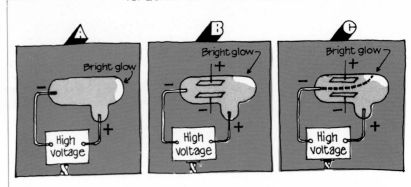

Figure 15–4

A gas is maintained at a low pressure in a glass tube into which two terminals intrude. (A) A high voltage causes the tube to glow at the end opposite the negative terminal. (B) Thomson showed that the electric field produced by two oppositely charged plates caused the glow to shift in position. (C) Thomson interpreted this phenomenon to be due to negatively charged particles (dotted line) that were moving away from the negative terminal.

Notice that in these and in subsequent experiments the existence of a negatively charged particle, called an electron, was inferred from the experimental data. That is, the only obvious explanation for the observations was that they were caused by particles that were negatively charged. No one actually *saw* an electron directly. Drawings that show an electron as a tiny sphere with a negative charge on it are really simplistic *depictions* of what we think an electron would look like; of course these are not very accurate pictures. For, as we saw above, an electron has wave properties as well as particle properties—that is, electrons are subject to the wave–particle duality—and the drawings suggest only their particle properties.

Part of the difficulty in understanding the wave–particle duality arises from the way we think. The pictures in our mind are all based on what we see in the world of everyday experience. We use the concepts of waves and particles because on the macroscopic scale energy is transferred from place to place only by the motion of particles or of waves. These two concepts, applied to the microscopic world of light and electrons, are really abstractions of the human mind. When we try to conceive of what light and electrons "really" are, our minds insist on some visual picture. Naturally we can form "pictures" only from our everyday experience. Yet there is no logical reason to expect that the fundamental entities, electrons and light, should conform to the visual images in our minds. Therefore we must try to understand light and matter through the experimental evidence. Since it is impossible to "see" an electron directly, our knowledge of it can be obtained only through indirect experiments. It is probably not even legitimate to ask, "What *is* an electron?" If one insists on an answer to that question, it is perhaps easiest to think of an electron as being something between a particle and a wave. But it is probably more accurate to say that an electron is merely the set of its properties that we can measure. Bertrand Russell once remarked that the electron is merely a "logical construction."

3. THE ATOM

It was not until the appearance of the quantum theory and the eventual acceptance of the wave–particle duality that the structure of the atom could at last be understood. It is to the study of the atom that we now turn.

Early Investigations of the Atom

Soon after the electron was discovered in the 1890s it was realized that it must be an essential constituent of atoms. J. J. Thomson, who is credited with discovering the electron, developed the first modern theory or "model" of the atom. In this model, negatively charged electrons are imbedded in a sphere of uniform positive charge, Fig. 15–5, much as plums are imbedded in a plum pudding.

A few years later, in 1909, Ernest Rutherford (1871–1937) and his colleagues performed some experiments that indicated that the "plum-pudding" model of the atom was not correct. In these experiments a beam of positively charged "alpha (α) particles" (which are helium nuclei emitted from a radioactive source; see Chapter 16) were directed at a thin sheet of gold foil, Fig. 15–6(A). The direction of many α particles was altered when they passed through the foil as a result of the attraction and repulsion of the charges in the atoms of the foil. Rutherford was astounded, however, to find that a few of the α particles actually rebounded, in nearly the same direction from which they had come! See Fig. 15–6(A). This could happen, he reasoned, only if the positively charged α particles were being repelled by a very strong positive charge. This positive charge would have to be concentrated in a very small region of space; no negative charge could be in the vicinity to nullify the strong repulsive effect, Fig. 15–6(B). Thus Rutherford concluded that the atom must contain a very tiny, but heavy, positively charged "nucleus," with the negatively charged electrons some distance away.

Since negative and positive charges attract one another, Rutherford had to explain why electrons do not fall into the nucleus. His explanation was that the electrons revolve around the nucleus at

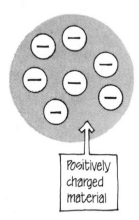

Figure 15–5

Plum-pudding model of the atom.

Positively charged material

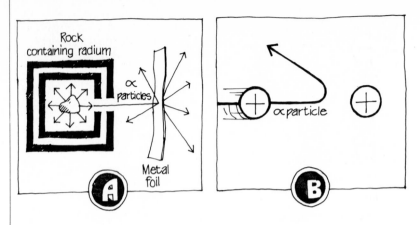

Figure 15–6

Rutherford's experiment.
(A) Experimental setup: α particles emitted by radium strike metallic foil; some rebound backward.
(B) Backward rebound of α particles explained as repulsion from positively charged heavy nucleus.

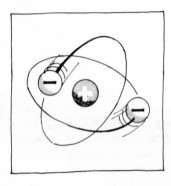

Figure 15–7

Rutherford's planetary model of the atom. The negative electrons revolve around the positive nucleus.

high speed, much as the planets revolve in orbits around the sun, Fig. 15–7. From his measurements Rutherford calculated that the nucleus must have a radius of only 10^{-12} to 10^{-13} cm. The electrons, on the other hand, revolve around the nucleus at a distance of about 10^{-8} cm, which is more than 10,000 times the radius of the nucleus itself. Rutherford determined that the nucleus contains nearly all the mass of the atom, more than 99.9 percent.

Rutherford's model of the "nuclear atom" is still the way we view the atom today, at least in broad outline. The details of atomic structure, however, still had to be worked out. For example, it was at once recognized that there was no reason for the hypothesized electron orbits to remain stable. Since the electrons are moving in orbits in which the direction of their velocity is changing, they are "accelerating"; and an accelerating electric charge must emit electromagnetic radiation, or light, as we saw in Chapter 14. When the electrons emit this light energy, they must lose kinetic energy to compensate—since energy is conserved—and so we would expect them to spiral into the nucleus, Fig. 15–8; that would mean the end of the atom. This obviously does not happen, since the matter all around us is made up of atoms and is quite stable. Rutherford's model of the atom clearly required some modification. By 1910 the necessary clues were at hand.

Atomic Spectra: The Key to the Structure of the Atom

Earlier in this chapter we saw that heated solids emit light with a continuous range of frequencies. The dominant frequency range depends on the temperature of the body. The radiation from solids is assumed to be due to the vibration of the atoms and molecules in the solid, and that motion is governed by the interaction of each

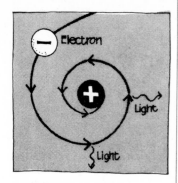

Figure 15–8

The orbiting electron is accelerating and thus must emit electromagnetic radiation. To compensate for this loss in energy, we would expect it to spiral into the nucleus.

Figure 15–9

Spectra of various elements in the gas state.

atom or molecule with its neighbors. In a gas, on the other hand, the atoms or molecules are assumed to be so far apart that they interact very little with each other, except during infrequent collisions. Thus, when the atoms in a gas emit light, it must be due to motion *within* the atom—that is, the motion of its *electrons*. Thus it is to the spectra of light emitted by gases that we must turn to gain knowledge of the internal structure of atoms.

The light emitted by heated or electrically excited gases had been observed early in the nineteenth century. This light was very strange. Instead of a continuous spectrum of wavelengths, like that emitted by heated solids, the light emitted by a gas consisted only of certain discrete wavelengths or frequencies; this is called a discontinuous, or "line," spectrum, Fig. 15–9.

The essential tool for observing and measuring the spectrum of light from a source is a spectroscope. And the essential part of a spectroscope is a prism, which, as we saw in Chapter 12, separates light into its component colors. The prism bends light of different wavelengths by different amounts, the shorter wavelengths of light being bent more than the longer wavelengths. For example, when white light passes through a prism, each of the component colors or wavelengths is bent by a different amount; all the colors of the rainbow from red to violet are seen. In the spectroscope, a viewing screen or photographic film is placed behind the prism, as shown in Fig. 15–10. Each component wavelength arrives at a different point, so the position on the screen or film is a measure of the wavelength. Film sensitive to ultraviolet or infrared light can be used to detect radiation in these nonvisible parts of the spectrum. Figure 15-9 shows the spectra of various elements in the gaseous state. Each is a "line spectrum"—that is, only certain distinct wavelengths are present and each appears as a line on the film. These discrete

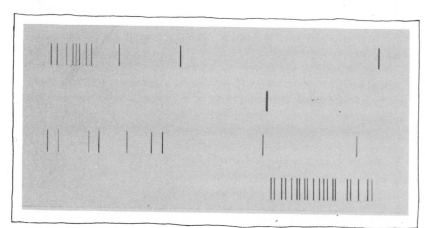

Atomic hydrogen

Sodium

Helium

Neon.

Blue wavelengths   Red wavelengths

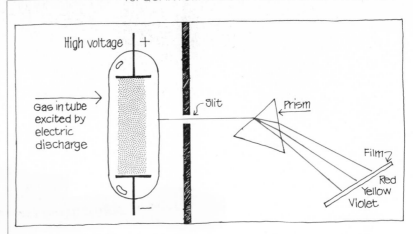

Figure 15–10

A spectroscope.

wavelengths are different for each element and are therefore characteristic of the element. In a sense, the spectrum of each kind of atom or molecule is a "fingerprint"—it is unique and can be used for identification. By examining the spectral lines emitted by a gaseous object—for example, a star, the atmosphere of a planet, or the flame of a burning object—scientists can determine what elements and compounds are present. Chemists have made great use of the spectroscope as a tool for chemical analysis, a field of research known as spectroscopy.

Now any reasonable theory of atomic structure will have to explain why atoms emit light of only certain discrete frequencies. Indeed, the theory should be able to predict just what these frequencies are for any particular atom.

Hydrogen is the simplest of all atoms—it has only one electron orbiting its nucleus. It also happens to have the simplest spectrum —there is a regular decrease in the spacing between the lines emitted by hydrogen, Fig. 15–9. Because of the simplicity of hydrogen and its spectrum, much of the early theoretical work centered on this single-electron atom. By the second decade of the twentieth century, the way had been paved for putting together a theory of atomic structure. The known facts were these:

1. Atomic spectra consist of discrete wavelengths. For hydrogen, the values of these wavelengths have a regularly decreasing spacing.
2. Rutherford had shown that the atom contains a tiny nucleus with electrons orbiting around it.
3. The "nuclear" or "planetary" model envisioned by Rutherford, while successful in many respects, did not account for the stability of electron orbits.

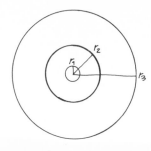

Figure 15–11

Bohr model of the hydrogen atom. Its one electron can revolve in any one of many circular orbits that can have only certain radii. The three smallest orbits are shown.

It was Niels Bohr who put all these pieces together and added one essential idea—the quantum hypothesis.

Bohr Develops a Theory of the Atom

After studying in Rutherford's lab for a few months in 1912, Bohr became convinced that the only way a line spectrum could be explained was to assume that the electrons revolving around the nucleus follow only particular circular orbits. Unlike the planets, which in principle could be orbiting at any arbitrary distance from the sun, the electrons in an atom can have only discrete "quantized" orbits (Fig. 15–11), according to Bohr's theory. Bohr did not know what caused the quantization of the orbits, but after working with the experimentally determined hydrogen spectrum, he deduced how the circular orbits must be laid out in order to obtain agreement with the experimental evidence. In hydrogen, the smallest orbit, called the "ground state," has a radius $r_1 = 0.53 \times 10^{-8}$ cm. The next largest possible orbit has a radius four times as large, $r_2 = 4r_1$; the third orbit's radius is nine times as large, $r_3 = 9r_1$; and so on.

An essential aspect of Bohr's theory is that the electrons not only revolve in quantized orbits; but in each of these orbits the electron has a discrete energy. Energy in the atom is quantized. Figure 15–12 shows the values for the energy, or the **energy levels**, that are possible for the single electron in the hydrogen atom. The $n = 1$ energy level is the energy when the electron is in the smallest orbit, or radius r_1; the $n = 2$ energy level corresponds to the next largest orbit; and so on. The integer n that labels the energy levels as well as the orbits is called the **quantum number** of the energy level or orbit. From Figs. 15–11 and 15–12 it is clear that the orbits farther from the nucleus have the greater energy.

Bohr was now able to explain the line spectrum emitted by hydrogen gas. According to Bohr, *an atom will emit a quantum of light whenever an electron jumps from an orbit of higher energy to one of lower energy.* In order to satisfy the law of the conservation of energy, the energy of the emitted photon, *hf,* will be exactly equal to the energy the electron loses when it jumps from the higher to the lower energy level. If we let E_h and E_l represent the energy of the electron in the higher and lower energy levels, respectively, the energy of the emitted quantum of light can be written as

$$hf = E_h - E_l.$$

The frequency of the emitted light is thus completely determined by the difference in energy of the two energy levels. Since only certain discrete energy levels exist, only certain frequencies for the

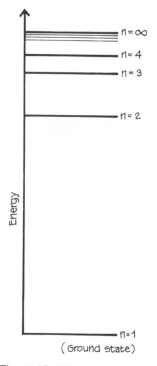

Figure 15–12

Energy levels for each orbit in hydrogen atom.

emitted light will exist. This explains why line spectra are observed instead of continuous spectra, Fig. 15–13. When all combinations of the energy levels E_h and E_l, as calculated from Bohr's theory, are put into the above equation, the frequencies predicted by that equation are in excellent agreement with the experimental spectrum of hydrogen, Fig. 15–9.

Bohr's theory assures the stability of the atom. When the electron is in the lowest orbit, it has no place to go. There is no lower energy level it can go to and emit more energy.

When the electron is in the lowest energy level, $n = 1$, the atom is said to be in its **ground state**. When the electron is in a higher energy level ($n = 2$, or more) and is thus capable of emitting light by jumping to a lower level, the atom is said to be in an **excited state**. Under normal conditions the majority of atoms in a gas are in the ground state. Collisions with other atoms, or the absorption of a photon of the right frequency, can cause an electron to be excited to a higher energy level. This is done on a large scale by heating the hydrogen gas or by passing an electric current through it

Figure 15–13

Some of the electron jumps that give rise to the emitted line spectrum of hydrogen.

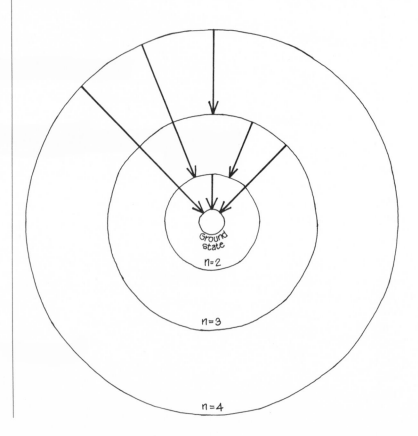

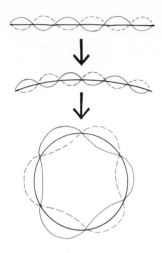

Figure 15–14

A circular standing wave compared to an ordinary standing wave.

Figure 15–15

A wave that does not close and hence interferes with itself. It rapidly dies out.

Figure 15–16

Standing circular waves for 2, 3, and 5 wavelengths on the circumference. n is the number of wavelengths and corresponds to the quantum number.

in a "discharge tube" (Fig. 15–10). In these circumstances, the atoms in the gas are excited to higher energy states and emit light in one or more steps as they return to the ground state. Thus the large collection of atoms in a gas will emit all the different possible spectral lines that are seen.

De Broglie Explains Bohr's Theory

But why should there be discrete orbits and energy levels? Why can't an electron move in just any orbit? Bohr had no answer for these questions, but a decade later de Broglie did. The answer, he pointed out, lies in the wave nature of matter.

De Broglie showed that Bohr's assumption of discrete orbits actually had a deeper basis in nature: each allowed electron orbit is actually a standing wave. Recall from Chapter 10 that when a violin string is plucked, a vast number of wavelengths are excited but only certain ones, those that have nodes at the ends, are sustained (Fig. 10–15); all other wavelengths interfere with themselves upon reflection and their amplitudes quickly drop to zero. Since electrons move in circles, de Broglie argued that the associated wave must be a circular standing wave that closes on itself, Fig. 15–14. If the wavelength is such that the wave does not close on itself (Fig. 15–15), then destructive interference takes place as the wave travels around the loop, and it rapidly dies out. Therefore, if the wave is to persist, the wavelength must be such that the wave closes on itself, as in Fig. 15–16. This can happen only if the circular orbit of the electron contains an integral number of wavelengths. The condition, then, for a stable orbit is that *the circumference of the orbit must be equal to an integral number of electron wavelengths.* The circumference of a circle of radius r is $2\pi r$, so the stable orbits are obtained only when $2\pi r_n = n\lambda$. In this equation n is an integer that is the quantum number of the electron's state, r_n is the radius of the

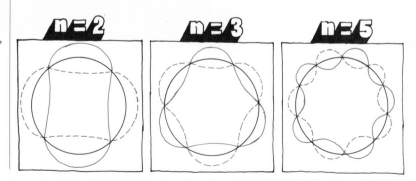

orbit that contains n electron wavelengths, and λ is the wavelength of the electron given by de Broglie's hypothesis, $\lambda = \dfrac{h}{mv}$. Making use of the above equation as well as Coulomb's law to express the force between the electron and the nucleus, de Broglie only needed a little algebra to calculate the possible orbital radii and energy levels in the hydrogen atom.

A word of caution: In viewing the circular electron waves given in Fig. 15–16 remember that the electron does not follow the wiggly wave pattern. The electron, considered as a particle, moves in a simple circle. The circular wave represents the *amplitude* of the electron "matter wave," and in Fig. 15–16 the wave amplitude is superimposed on the circular path of the orbit for convenience.

To summarize, we see that Bohr's theory was very successful in explaining the observed properties of hydrogen. Subsequently, de Broglie's wave concept of the electron gave the reason behind Bohr's successful theory and added a new and fundamental dimension to our understanding of the atom. The wave–particle duality was at the root of atomic structure.

4. QUANTUM MECHANICS

The Bohr theory of the atom provided a reasonably good picture of what an atom is like. But even with the theoretical basis provided by de Broglie's matter-wave hypothesis, there still were limitations. The Bohr theory successfully accounted for the spectrum of lines emitted by the hydrogen atom, but it was incapable of explaining the spectra of more complex atoms—atoms containing two or more electrons. Furthermore, the problem of the wave–particle duality was far from resolved. A more comprehensive theory was clearly needed. It was not long in coming. Within a year after de Broglie presented his matter-wave hypothesis, two physicists independently developed the desired theory: Erwin Schrödinger (1887–1961) and Werner Heisenberg (1901–1976).

The new theory, known as **quantum mechanics**, unified the wave–particle duality into a single consistent but highly mathematical theory. The wave and particle aspects no longer had to be applied on the basis of which worked best in a given situation. Instead, the wave and particle aspects were integrated unequivocally into one single formalism.

Quantum mechanics as a theory is accepted today by nearly all physicists as the fundamental theory underlying physical processes. A theory is as good as the experimental facts that it can confirm and predict, and quantum mechanics is very good indeed.

Although quantum mechanics deals mainly with the microscopic

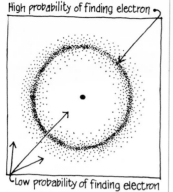

High probability of finding electron

Low probability of finding electron

Figure 15–17

Spherically symmetric electron cloud or "probability distribution."

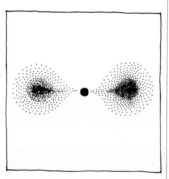

Figure 15–18

Dumbbell-shaped electron probability distribution.

439

world of atoms and light, in the macroscopic world of everyday life we do perceive light and we take for granted that ordinary objects are made up of atoms. Therefore if this new theory is really good, it must also account for the verified results of classical physics when it is applied to macroscopic phenomena; that is, when applied to macroscopic phenomena, the laws of quantum mechanics must yield the old classical laws. This is the "principle of correspondence" first enunciated by Bohr. Quantum mechanics fulfills this requirement as well. Quantum mechanics thus represents the latest and most complete description of the behavior of atoms.

Although we will not go into the difficult mathematics of quantum mechanics, we can discuss briefly many of its results and predictions as well as its tremendous influence on science and philosophy in the twentieth century.

Atoms: Energy Is Quantized but Orbits Are Diffuse

Quantum mechanics applied to atoms is a much more complete theory than the Bohr theory. But it does retain certain aspects of the older theory. It tells us that the electrons in an atom can exist only in discrete states that have certain discrete values for the energy. In fact, quantum mechanics predicts exactly the same energy levels for hydrogen as the older Bohr theory did, Fig. 15–12. But quantum mechanics is able to predict the energy levels for the more complex atoms as well, something the Bohr theory could not do. Thus the well-known line spectrum for each element, Fig. 15–9, is accurately predicted by quantum mechanics.

This new theory is not merely an extension of the Bohr theory, however. It is a much deeper theory, and it presents quite a different view of the atom. According to quantum mechanics, electron orbits are not the simple circular orbits of a particle depicted by the Bohr theory. Instead, the electron is spread out over a sizable region of space, as a sort of "cloud" of negative charge. For any given state of an atom, the size and shape of the electron cloud can be calculated. For the ground state of hydrogen, and for many other states as well, the electron cloud is much like that shown in Fig. 15–17, which is said to be spherically symmetric. An "orbit" is thus quite diffuse. The electron cloud indicates roughly the "size" of the atom; but just as a cloud does not have a distinct terminus or border, atoms do not have a precise and well-defined size or boundary. Not all electron clouds have a spherical shape; some are "dumbbell" shaped, as in Fig. 15–18. These latter are especially important in forming chemical bonds between atoms.

The electron cloud can be interpreted by using either the wave or the particle aspects of electrons. Notice first of all that what we mean by a particle is something that is *localized* in space—it will have a definite position at any given time. But a wave at any given time is *spread out* in space. Thus the electron cloud, spread out in space as it is (Figs. 15–17 and 15–18), reflects the wave nature of electrons. Electron clouds can also be interpreted as **probability distributions** for a "particle." If we were to make 1000 separate measurements of where an electron is (considering it a particle), we would most often find it in a region of space where the probability is high (dark area in Fig. 15–17) and only a few times would we find it in those regions where the probability is low. However, it is not possible to predict the actual path that an electron will follow. After measuring its position at one instant we cannot predict exactly where it will be in the next instant. We can only predict the *probability* of finding it in various different places. This is very different from classical Newtonian physics.

The fact that we cannot predict the exact position of an electron in an atom but can only give the probability distribution has deep philosophic consequences and leads to one of the most famous conclusions of quantum mechanics: the uncertainty principle.

The Uncertainty Principle

Whenever a measurement is made, some uncertainty or error is always involved. For example, if we want to measure the length of a board we cannot make the measurement exactly. Even with a ruler or a tape measure that has markings one millimeter apart, we cannot measure a distance with an accuracy of better than ½ mm or so. If we use more accurate instruments, we can make more accurate measurements. But there is always *some* uncertainty involved in any measurement, no matter how good the instrument. By using finer and finer instruments, we expect that we can reduce the uncertainty indefinitely.

But quantum mechanics tells us that there actually is a limit to the accuracy of certain measurements. This limit is not a restriction on how well instruments can be made. Instead, the limit is inherent in nature. This is the result of two factors: the unavoidable interaction between the thing observed and the observing instrument, and the wave–particle duality. Let's see how this comes about.

It is impossible to make a measurement on an object without somehow disturbing it, at least a little. Have you ever tried to locate a ping-pong ball in a dark room? In order to pick it up, you must first find its position. You grope around trying to find it; and just as you touch it, it bounces away from you. Frustrating! Now it's always true that whenever we measure the position of an object, be it a ping-pong ball or an electron, we must somehow touch it with something else that will carry the information about its position back to us. In locating a lost ping-pong ball in a dark room you could poke around with your hand or a stick until you touched it; or you might shine light on it and detect the reflected light with your eyes. If you search with your hand or a stick, you will find its position when you touch it. But in touching the ball you unavoidably move it, giving it some momentum; and thus you won't know its future position. The same is true if you try to observe the ping-pong ball visually, using light. In order for the object to be "seen," at least one photon must bounce off it, and the reflected photon must enter your eye or some optical instrument.[2] A photon striking an ordinary-sized object does not appreciably alter the motion or position of the object. But if the object is an electron, a single photon bouncing off it can give it considerable momentum and thus alter its motion and position quite radically—and in an unpredictable way. Thus we see that the mere act of measuring the position of an object at any one time renders inaccurate our knowledge of the position of the object at future times.

Where does the wave–particle duality come in? As we saw in Chapter 12, the position of an object can be measured with light to an accuracy no greater than the wavelength of the light used. So if we want to measure the position of an electron very accurately, we must use light of a short wavelength. But a short wavelength corresponds to photons of high frequency (since wavelength = wave velocity/frequency) and therefore of high energy (since $E = hf$); and the higher the energy of the photons, the harder they hit the electron and the more they disturb it. If we use photons of longer wavelength and correspondingly lower energy, the motion of the electron will not be altered as much; but the position measurement will be less accurate. The act of observing thus produces a significant uncertainty in either the *position* or the *velocity* of the electron! This is the essence of the **uncertainty principle** first enunciated by Heisenberg in 1927.

Heisenberg was actually able to give a quantitative statement of the magnitude of these uncertainties. He showed that the product of the uncertainty in momentum (mass × velocity) times the uncertainty in position can at best be equal to Planck's constant, h. Mathematically, the uncertainty principle can be written as

$$(\Delta mv)\,(\Delta x) \geq h.$$

Here, the symbol Δ means "uncertainty in"; so Δmv stands for the uncertainty in momentum, mv, and Δx for the uncertainty in position, x.

[2]Instruments have been developed that can detect single photons. Our eyes, on the other hand, are not sensitive to individual photons; so many photons will be required if we are to "see" the object with our eyes.

The uncertainty principle does not forbid single exact measurements, but it tells us that if we measure the position of an object exactly (so its uncertainty Δx equals zero), then its momentum (and hence its velocity) is completely unknown. Thus, while we could know the position of the object exactly at one instant, we would have no idea where it would be a moment later. Similarly we can in principle measure the velocity of an object exactly, but then we would know absolutely nothing about its position. If we wish to measure both the position and the velocity of an object at the same time, we can measure both, but only inexactly. The more accurate the measurement of one, the more inaccurate the measurement of the other.

Since Planck's constant is very small, the uncertainties prescribed by the uncertainty principle are usually negligible on the macroscopic level. But on the atomic level, those uncertainties are very significant. Since macroscopic objects are made up of atoms containing nuclei and electrons, the uncertainty principle has great relevance for our understanding of all of nature. In a way, the uncertainty principle most clearly expresses the probabilistic nature of quantum mechanics, and thus is often the stepping-off point for philosophic discussion.

Notice, however, that in this discussion of the uncertainty principle we have spoken of making measurements of the position and velocity of an electron *as if it were a particle*. But it isn't. The uncertainty principle is a reflection of the fact that an electron—and matter in general—is not of a purely particulate nature.

In a very real sense, what the uncertainty principle tells us is that if we wish to think of the electron as a particle, then there are specific limitations on this simplified view—namely, that the position and velocity cannot both be precisely known at the same time.

5. THE PHILOSOPHIC IMPLICATIONS OF QUANTUM MECHANICS

We Must Distinguish Our Description of Nature from Nature Itself

The model of an atom that shows electrons moving in orbits around a nucleus treats electrons as if they were actually particles—that is, as if an electron had a definite position at each point in time. Since electrons are not simply particles, they cannot be represented as following particular paths in space and time. This means that any description of matter in space and time cannot be precisely correct. This deep and far-reaching conclusion has been much discussed by philosophers during the past half-century. Perhaps the most influential philosopher of quantum mechanics was Niels Bohr. He was among the first to realize that a space–time description of actual atoms and electrons is not possible. Yet, he pointed out, a description of *experiments* performed on atoms or electrons *must* be given in terms familiar to our ordinary experience. Thus the description of any experiment must be given in terms of space and time, and in terms of other ordinary concepts such as waves and particles. But we should not let our descriptions of experiments lead us into believing that the electron itself actually exists in space and time as a particle. Nor should we be led into believing that the ultimate processes in nature actually occur in space and time. Quantum theory tells us that they do not. This distinction between our interpretation of experiments and what is "really" going on in nature is crucial.

This brings us back to our discussion in Chapter 1 of the similarities between the arts and sciences. The scientist is not always a logician—he is often much like a poet. This is no better illustrated than when a physicist talks about the ultimate constituents of matter. When a physicist says "an electron is like a particle," he is making a metaphorical comparison like the poet who says "love is like a rose." In both images a concrete object, a rose or a particle, is used to illuminate an abstract idea, love or electron.

Strict Determinism and Causality Are No Longer Valid

Until the beginning of the twentieth century the mechanistic description of nature embodied in Newtonian mechanics and in Maxwell's theory of electricity and magnetism was deeply embedded in scientific thought. The basic idea of classical physics is that once the position and the velocity of an object at a particular time are known, its future can be predicted if the forces on it are known. This meant, among other things (see Chapter 3), that the past and future unfolding of the universe, assumed to be made up of particulate bodies, was completely determined.

With the discovery that "particles" such as electrons have wave properties, physicists were forced to abandon this classical view. The deterministic view of the physical world, so deeply a part of classical physics, has been largely invalidated with the advent of quantum mechanics. Quantum mechanics tells us that we cannot accurately know both the position and the velocity of an object at the same time. Because of this, even if we were to know the force acting on an electron, we still could not predict precisely where the electron would go. Quantum mechanics does, however, allow us to calculate the probability that the electron will be observed (as if it were a particle) at various different places. According to quantum mechanics, then, only approximate predictions are possible. There is an inherent unpredictability in nature.

We should point out, however, that there are a few physicists who are not willing to give up the deterministic view of nature, a view in which only concrete particles exist. Thus they refuse to accept quantum mechanics as a complete theory. One of these few was Einstein, who insisted that "God does not play dice."

The vast majority of physicists accept quantum mechanics and the probabilistic view of nature, which is called the "Copenhagen Interpretation" of quantum mechanics. It is named after Bohr's home town, since it was developed there largely in discussions between Bohr and other prominent physicists. Although quantum mechanics has strongly influenced philosophers and other intellectuals, from poets to architects to artists, the reverse is true as well. For example, Bohr was strongly influenced by the nineteenth-century philosophers Søren Kierkegaard and William James. And the so-called "positivist" school of philosophy has also had a strong impact on twentieth-century science. Science is developed by humans, none of whom is an island.

What does quantum mechanics say about causality? In everyday life we observe that whenever we let go of a rock it falls to the ground. Since we have never seen a violation of this phenomenon—for example, a stone suddenly rising upward—we are confident that it will always fall downward. Similarly, whenever it is cold enough we observe that water freezes. It always does, at least within our experience. This repeatability of similar events is what gave rise to the idea of causality, as we discussed in Chapter 3. At the atomic level, however, causality does not seem to hold. A series of electrons that are treated alike in a given experiment will not all follow the same path and end up in the same place. Instead, certain probabilities exist that an electron will arrive at different places, Fig. 15–19.

Since matter is made up of atoms, even ordinary-sized objects do not obey a strict causality; they too are governed by chance and probability. For example, a finite probability exists that when you drop a rock it will suddenly fly up in the air; or that a lake will freeze over on a hot summer day. Of course the probability that such bizarre things will happen is incredibly small, but it is not zero. A quantum mechanical calculation might predict a probability that such an odd occurrence would happen maybe once in ten trillion years. Quantum mechanics predicts with very high probability that macroscopic objects will behave just as the classical laws of physics predict; but these predictions are probabilities, not certainties. That ordinary objects should behave with high probability in accordance with classical laws is a result of the large number of molecules in these objects. When large numbers of objects are con-

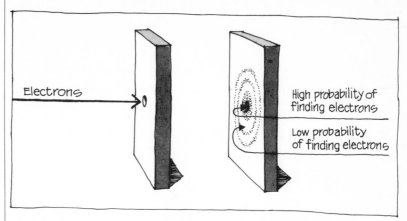

Figure 15–19

Electrons moving horizontally approach a slit. We cannot predict where they will go after passing through the slit. We can only predict the probability of finding them at various points. After many electrons have passed through the slit, the position of their arrival points on the screen will look like a diffraction pattern of waves passing through the slit.

sidered in any statistical situation, deviations from the average are negligible. It is the average configuration of vast numbers of molecules that follows the so-called "fixed laws" of classical physics with such high probability; this is what gives rise to an apparent "causality." Whenever small numbers of molecules are dealt with, deviations from classical law are readily observed.

The problem of human freedom and free will is somewhat alleviated by the new discoveries in physics. Since the world seems to be governed in part by chance instead of by precise fixed laws, a totally deterministic view of human beings and nature no longer seems reasonable. Although some thinkers feel that quantum mechanics allows human freedom to exist, it is unlikely that this question can be answered by science.

Many people feel that science should be able to answer all questions. However, it is probably a narrow view to expect that the "scientific method" is the only way of learning and knowing. In view of the changes within the field of science itself, a scientist must keep an open mind in these matters. There may well be other ways of knowing. After all, science was invented by human beings and is based on the *assumption* that there actually *is* a physical world out there beyond our senses. Eddington, as we saw in Chapter 1, said, "No one can deny that [one's own] mind is the first and most direct thing in our experience, and all else is remote inference." All that we know about this physical world comes in through our senses

445

and is "processed" and interpreted by our mind. Bohr in particular felt that humans always occupy the central position. It is, he said, impossible to distinguish sharply between the phenomena themselves and our conscious perception of them. To view science as having an existence independent of human beings is misleading. Science is a human creation, and it is a beautiful and admirable creation. It should be recognized for what it is, however; we should not expect too much of it.

Quantum Theory and Art

As we mentioned in Chapter 1, art and science—particularly in the twentieth century—show intriguing similarities. Just as modern physics has delved into the abstract world of the interior of atoms far removed from the everyday "commonsense" world, so too has art become abstract. Painters have reflected the new physics by showing us that matter is less than solid, that motion cannot be accurately rendered instant by instant. With Jackson Pollock and others in the 1940s, painting not only dispensed with representation but with anything resembling an image or even a geometrical shape (Fig. 15–20). The disappearance of the image from painting corresponds to the loss of the "object" in science (for example, we cannot visualize an electron). And the role of chance in contemporary painting also reflects nature as seen by the scientist.

The sculptor Alberto Giacometti has discussed the artist's role in trying to perceive and paint the real world. Like the scientist, the contemporary artist realizes that complete knowledge of the external world is impossible; and so, instead of trying to represent the external world in conventional modes, many modern artists seek to present an underlying reality or structure.

Figure 15–20

Detail from *Autumn Rhythm*, 1950, Jackson Pollock. Metropolitan Museum of Art.

6. COMPLEX ATOMS AND MOLECULES

The Exclusion Principle Clarifies the Structure of Complex Atoms

We have talked a great deal about the hydrogen atom, and the fact that its single electron can be in the ground state or in an excited state. What about more complex atoms, such as helium with 2 electrons, oxygen with 8, or gold with 79 electrons?

As in the case of hydrogen, every atom has its own characteristic set of possible energy levels. When an atom is in its ground state, we might expect that the electrons will all be in the lowest energy state. But this is not the case. Instead, as Wolfgang Pauli (1900–

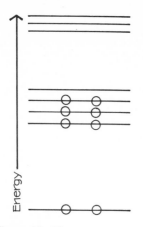

Figure 15–21

Schematic diagram of energy levels in oxygen showing those occupied by the eight electrons in the ground state.

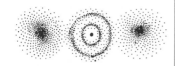

Figure 15–22

Electron probability distribution in oxygen.

1958) discovered in 1925, no more than two electrons can be in any given quantum level, a fact known as the **Pauli exclusion principle**. Thus only two electrons can be in the lowest energy state, two more can be in the next lowest, and so on. An atom is in the ground state when all the electrons in the atom are filling the lowest energy states possible. See for example Fig. 15–21. If one or more of the electrons is in a higher energy level so that a "vacancy" exists at a lower level, the atom is said to be in an excited state. When an electron drops down to fill in the vacancy, it emits a photon of light. The spacing of the energy levels is different in every atom and gives rise to the atom's characteristic spectrum, Fig. 15–9. Under normal conditions, an atom spends most of its time in the ground state.

The electrons that occupy the lowest energy levels in a complex atom are, on the average, closest to the nucleus. The electron clouds or probability distributions take on specific shapes and are the same for both electrons in a given state. For example, in oxygen the two electrons in the lowest energy level have spherically symmetric probability distributions. The next two electrons also have spherically shaped probability distributions, but on the average they are farther from the nucleus, Fig. 15–22. The next two electrons have dumbbell-shaped orbits. The final two electrons also have dumbbell-shaped orbits but around a different axis—that is, they are moving in and out of the paper in Fig. 15–22; one end of the dumbbell is above the paper and the other is below, so they are not shown in the diagram.

Molecules Are Formed from Atoms

The study of molecules and their formation is usually considered the subject of chemistry. The knowledge of how bonds form between atoms to make a molecule, however, is based in physics, and so we will discuss this subject briefly.

When two or more atoms come together to form a molecule, we say that a chemical "bond" has been formed. This bond usually comes about when one or more electrons are shared by both atoms, and it is the attraction of the nuclei of both atoms to these shared electrons that holds the two atoms together. Let's take a concrete example: the formation of magnesium oxide from atoms of magnesium and oxygen. A magnesium atom contains twelve electrons, but only the two outermost electrons—the so-called "valence" electrons—play a role in making a chemical bond. The other ten electrons make up a spherically symmetric probability distribution that shields the two outer electrons from the nucleus. Therefore the two valence electrons, instead of feeling the attraction of all twelve positive charges in the nucleus, feel a net attraction of only two plus charges. Since the two valence electrons are rather far from the

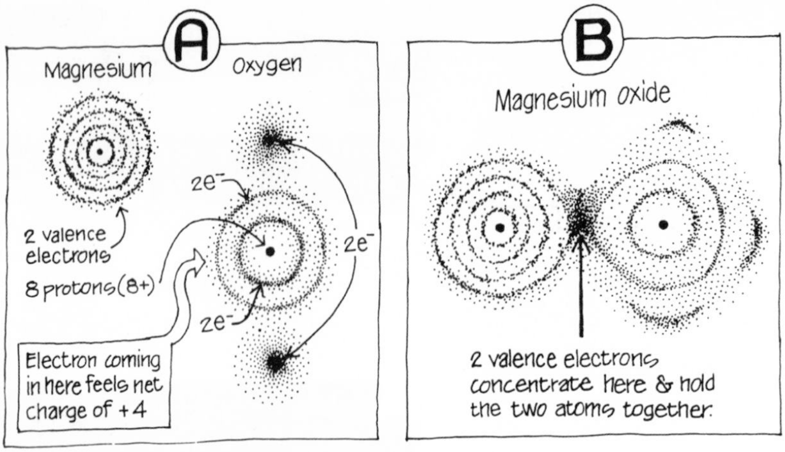

Figure 15–23

Electron probability distributions for (A) a magnesium atom and an oxygen atom, and (B) for a magnesium oxide molecule.

nucleus, the electric force holding them on the magnesium atom is not very strong. Consequently, when a magnesium atom approaches an oxygen atom, Fig. 15–23(A), the valence electrons of magnesium can be attracted to the oxygen atom if they happen to be in the correct orientation. The attraction occurs because the oxygen atom has two dumbbell-shaped orbits along two axes, but not along the third; so an electron that happens to move into this "opening" [see Fig. 15–23(A)] will feel the net attraction of four positive charges. This is true because in the particular region shown, the oxygen nucleus with its eight protons is shielded by only four electrons, the other four electrons being in dumbbell-shaped orbits off to the side.

Thus, the two valence electrons of magnesium are attracted to both their own nucleus and to the nucleus of the oxygen atom; they then can orbit both atoms but tend to spend most of their time between the two atoms, Fig. 15–23(B). The positively charged nuclei of the two atoms are attracted to this concentration of negative charge between them, and this electric attraction is the "bond" that holds the atoms together. This is known as a covalent bond. Most of the bonds that hold atoms together to form molecules are formed in a similar way. Again we see the great importance of the electric force.

7. PRACTICAL USES OF QUANTUM EFFECTS

In this chapter we have traced the history and development of the quantum theory from its beginnings at the turn of the century to its culmination in quantum mechanics. Along the way a number of related discoveries were made that have been put to practical use. We now discuss some of these.

X Rays

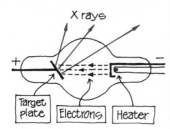

Figure 15–24

Diagram of an x ray tube.

In 1895 Wilhelm Roentgen (1845–1922) discovered that electrons striking a metal plate produced a mysterious kind of radiation, which he dubbed "x" radiation, or x rays, after the algebraic symbol "x" meaning an unknown quantity. He found that x rays were able to pass through various solid materials with greater or lesser ability, depending on the material. In particular he found that x rays passed through human flesh, but not readily through bone, and thus they would leave a photographic record of the body's bone structure on a photographic plate—a phenomenon we are all familiar with from our experiences in the doctor's or dentist's office.

Nearly two decades went by before it was determined that x rays are simply electromagnetic radiation of very high frequency. They differ from visible light only in frequency. Of course x rays too are subject to the wave–particle duality; individual x ray photons have much greater energy than visible light photons.

How are x rays produced? Figure 15–24 shows an x ray tube. Electrons are accelerated from the cathode to the plate by a voltage of many thousands of volts. When they strike the plate the electrons are rapidly decelerated, and the energy each electron loses in this process is emitted as a photon of very high energy—an x ray photon.

X ray machines used in medical laboratories throughout the world today make use of this x ray quantum phenomenon.

Phototubes—Light Meters, Sound Tracks, and "Electric Eyes"

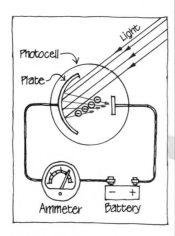

Figure 15–25

A photocell and circuit used in a light meter.

A **photoelectric cell** (or photocell) is an evacuated tube containing two metal electrodes or plates and making use of the photoelectric effect. It can be connected into a circuit containing a battery or other voltage source, Fig. 15–25. If no light shines on the plate, no electrons are emitted and no current flows in the circuit. When light does fall on the plate, electrons are emitted (by the photoelectric effect, Fig. 15–2) and they pass across to the other (positive) electrode. Thus a current flows in the whole circuit. When the light is brighter, more electrons are emitted and more current flows. Thus the current registered by an ammeter is a measure of the light intensity. This is how a photographic light meter works.

An "electric eye" that opens a door automatically or activates a burglar alarm is diagrammed in Fig. 15–26. Instead of an ammeter, the circuit contains a solenoid (see Chapter 14). In the normal situation, the iron rod is held in place inside the coil because a beam of light shines on the photocell causing a current to flow. But when someone steps into the path of the light beam, the current

Figure 15–26

Photoelectric effect used in a photocell to activate an automatic door or a burglar alarm.

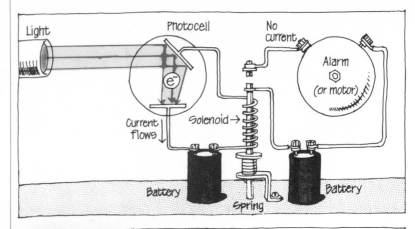

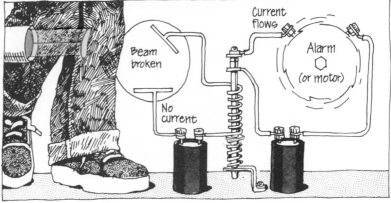

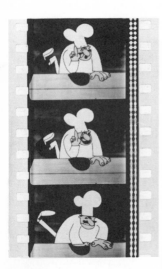

Figure 15–27

One type of sound track on moving picture film.

stops and the rod is released. A second circuit is completed when the rod moves to its new position and this starts a motor that opens the door or rings an alarm. Ultraviolet light is sometimes used in such devices, since it is less conspicuous than visible light.

The photoelectric effect is also used on the sound track of motion pictures. The variable shadings that appear on the side of the film, Fig. 15–27, are produced originally by electrical signals from the recording microphone, which control a shutter that in turn controls a light beam directed at this edge of the film. When the film is developed, this "sound track" has variations in shading that follow the original frequencies of the sound. When the film is projected, a small beam of light passes through the sound track to a photocell behind the film. The current flowing in the photocell circuit follows the shadings on the sound track; and when that current is amplified and fed into a speaker, it accurately reproduces the original sound.

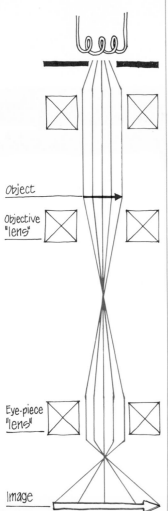

Figure 15–28

Diagram of an electron microscope. Squares represent magnetic field coils, with names for each lens that correspond to those of an ordinary light microscope.

Object

Objective "lens"

Eye-piece "lens"

Image

Figure 15–29

Scanning electron microscope photograph of surface of rat's tongue.

The Electron Microscope

Soon after the experimental confirmation that electrons have wave properties, it was suggested that electrons could be used to magnify tiny objects with much greater detail than was attainable with an ordinary light microscope. The resolution of a microscope—that is, the tiniest object that it can display clearly—is no smaller than the wavelength of the light used. So with visible light, a microscope cannot be used to observe anything smaller than about 400 nm (remember, 1 nm = 10^{-9} meter). A beam of electrons, however, can easily be produced with a wavelength of less than 1 nm. Thus, the electron microscope can increase our magnifying ability by a factor of a thousand or more.

An electron microscope is diagrammed in Fig. 15–28. Magnetic fields produced by currents in a coil of wire are used to focus the electron beam. These "magnetic lenses" play the same role that glass lenses do for visible light.

Recently a new kind of electron microscope, the "scanning electron microscope," has been developed that will produce photographs of tiny objects with a three-dimensional effect, see Fig. 15–29.

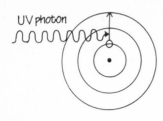

UV photon

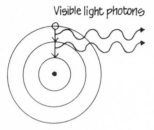

Visible light photons

Figure 15–30
Fluorescence.

Fluorescence and Phosphorescence

The materials used to coat the inside of fluorescent light tubes and the materials used on watch dials that glow in the dark have unique properties known as **fluorescence** and **phosphorescence**.

Materials that emit visible light when ultraviolet light is shined upon them are said to be **fluorescent**. Fluorescence can be readily explained as a quantum phenomenon at the atomic level. When a photon of ultraviolet light is absorbed by an atom of the fluorescent material, an electron is raised to a high-energy state. The electron, instead of jumping back to its original energy state in a single jump, may instead return in two or more jumps, as shown in Fig. 15–30. In this case, each of the emitted photons will have less energy than the energy of the original incoming ultraviolet photon. Therefore the emitted photons will have lower frequency and thus may be photons of visible light.

Fluorescent rocks are often seen in museums. When ultraviolet is shined on them, beautiful reds, greens, and violets are seen. The color emitted depends on the energy levels of the particular atoms that are excited. Fluorescent paints are also common today.

Fluorescent light bulbs work in a two-step process. An applied voltage accelerates electrons, which strike the atoms of the gas in the tube causing them to become excited. The excited electrons subsequently jump back down to their normal energy levels and emit ultraviolet photons in the process. The emitted photons strike the fluorescent coating on the inside of the tube, which, through the process of fluorescence just described, emits the visible light we see.

Objects such as luminous watch dials, which "glow in the dark," are said to be **phosphorescent**. Phosphorescence is much like fluorescence except that when the atoms of a phosphorescent material are excited, and this can be done even with visible light, the electrons remain in the excited state for a considerable length of time. Although many of them return to the ground state very soon after excitation, others remain in the excited state for as long as several hours. When they finally descend to the ground state they naturally emit a photon. When the source of the excitation is removed—for example, when the light is turned off—the material continues to emit light—it glows—as the many excited electrons drop one by one to the lower energy level.

Fluorescence and phosphorescence seem to us to be everyday phenomena. But they were both discovered by atomic physicists in their quest for an understanding of matter.

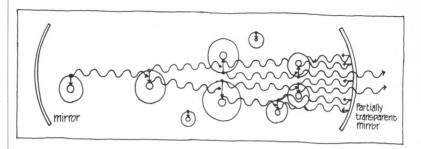

Figure 15–31

Schematic diagram of a laser showing excited atoms being stimulated to emit light.

The Laser and Holography

The laser is a device that can produce a very narrow, concentrated beam of light. The name **laser** is an acronym for "Light Amplification by Stimulated Emission of Radiation," which is a concise description of how a laser works. There are many types of laser, which use both solid and gaseous materials, but they all work in basically the same way. The material to be used is placed between two mirrors, one of which is partially transparent, Fig. 15–31. The atoms of the "lasing" material are periodically excited to high energy states by putting in a large amount of energy, often light energy from a photo flash tube. Just as in the case of phosphorescence, many of the atoms remain in the excited state, although some drop almost immediately to the ground state, emitting a photon. An atom that has just dropped to the ground state is shown on the left in Fig. 15–31. If the emitted photon strikes another atom that is still in the same excited state, it *stimulates* the second atom to drop to the lower energy level with the emission of a photon, which then moves in the same direction as the first photon; so there are now two photons. These two photons may strike two more excited atoms, and two new photons will be emitted. Now there is a total of four photons, all moving in the same direction. As the process continues, the number of photons multiplies. When the photons strike the end mirrors, most of them are reflected; as they move in the opposite direction, they continue to stimulate other excited atoms to emit photons. As the photons move back and forth between the reflecting mirrors, a small percentage of them pass through the partially transparent mirror at one end. It is these photons that make up the usable laser beam.

The light emitted from a laser is unique in several ways. Since it is produced by one or more atomic transitions, it consists of a single frequency or a series of distinct frequencies. Furthermore,

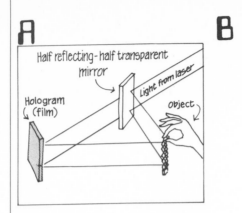

Figure 15–32

(A) Making a hologram. The laser light is split by the mirror, half going directly to the film, half to the object from which it is reflected to the film. The two beams interfere at the film, producing the hologram.

(B) Photograph of a holographic image. This hand, offering a diamond ring and bracelet, appeared to protrude from the window of a New York jewelry store. One passerby assailed it with an umbrella.

unlike most light sources in which the light is emitted in all directions, laser light comes out parallel in a straight, narrow beam.

It is the high concentration of light energy over a very small area that makes the laser so useful. A laser is not a *source* of energy—energy must be put in, and the laser *converts* this energy into an intense narrow beam of light energy. Lasers have many practical uses. They have become a useful surgical tool; the intense beam can be used to destroy tissue in a localized area or to "weld" broken tissues together, such as the reattachment of a detached retina. Lasers have been used in biological research to destroy tiny organelles within a living cell in order to study the behavior of the cell without that organelle. Lasers are used for welding and machining metals, because their narrow intense beams can heat a tiny area to very high temperatures.[3]

Because laser beams diverge so little, when they are directed at the moon the reflected beam will still be strong when it arrives back on earth. By means of lasers, the distance from the earth to the moon has been very accurately measured.

One of the most interesting uses of the laser is to produce three-dimensional pictures known as **holograms**. In an ordinary pho-

[3]Science fiction stories and movies lead us to believe that hand-held lasers may someday be used as weapons to destroy ordinary-sized objects. This is highly unrealistic. Only a mammoth-sized laser could possibly transform enough energy to destroy an ordinary-sized object.

tograph the film simply records the intensity of light reaching it at each point. When the photograph or transparency is viewed, light is shined on or through it and the intensity of light is reproduced, giving us a two-dimensional picture. When a laser hologram is made on film, the laser light is split into two parts by a mirror. One part goes directly to the film; the other part is directed to the scene to be photographed and is reflected back to the film. The interference of these two beams when they reach the film allows the film to record both the *intensity* and the relative *phase* of the light waves at each point, Fig. 15–32. It is the additional information contained in the phase that allows a three-dimensional picture to be projected when laser light is shown on or through the film.

Viewing a hologram is an uncanny experience. The hologram image can be projected into the middle of a room and one can walk around it, viewing it from all sides as if one were viewing the actual object. But there is nothing material there.

The principles of holography are currently being applied to moving pictures, and it may not be too long before we have truly three-dimensional movies and television.

REVIEW QUESTIONS

1. Planck's quantum hypothesis was
   (a) that electric charge is quantized
   (b) that light consists of particles called photons
   (c) that the energy of molecular vibration in solids is quantized
   (d) stated in 1940

2. Einstein hypothesized the existence of
   (a) electrons
   (b) atomic nuclei
   (c) particles of light
   (d) the uncertainty principle

3. When light is directed at a metal surface, the emitted electrons
   (a) are really waves
   (b) have nearly zero energy
   (c) have energies that depend on the frequency of the light
   (d) have energies that depend on the intensity of the light

4. A photon of which color light has the most energy?
   (a) red
   (b) green
   (c) yellow
   (d) violet

5. The "principle of complementarity" states that
   (a) we must complement our experiments with theories
   (b) we must complement our principles with action
   (c) matter has a wave nature

(d) the wave and particle aspects of light and matter are complementary, and both aspects must be considered in order to gain a complete understanding

6. We say that electrons have wave properties because
   (a) they exhibit interference and diffraction
   (b) they move along wavelike paths
   (c) they have the same frequencies as sound waves
   (d) they move from crest to trough

7. Rutherford's nuclear model of the atom hypothesized that
   (a) electrons are embedded in a sphere of uniform positive charge
   (b) electrons are attached to the nucleus
   (c) an atom is mostly empty space with electrons orbiting the nucleus
   (d) there can be no nuclei without electrons

8. The study of line spectra emitted by atoms
   (a) led to an explanation of light emitted by heated solids
   (b) led to a model of the atom in which electrons can be only in certain energy states
   (c) explained de Broglie's hypothesis of the wave nature of matter
   (d) explained the photoelectric effect

9. Light emitted from excited atoms has frequencies that are related to
   (a) the color of the atom
   (b) the size of the atom
   (c) the size of orbits in the atom
   (d) the difference in energy levels in the atom

10. According to the modern view of the atom,
    (a) electrons revolve around the nucleus in well-defined orbits
    (b) electrons do not have a localized orbit but are spread out in space, like a cloud
    (c) electrons are essentially identical to photons
    (d) electrons do not actually exist

11. The uncertainty principle states that
    (a) it is impossible to know exactly both the position and the momentum of a particle at the same time
    (b) the position of an electron can never be measured precisely
    (c) there is uncertainty in the momentum of an electron
    (d) there is uncertainty as to how an electron will behave in a magnetic field

12. According to quantum mechanics, if we think of the electron as a particle, we can determine its position
    (a) precisely at all times
    (b) only when its speed is precisely known
    (c) only on a probabilistic basis
    (d) whenever it is outside its cloud

13. Quantum mechanics
    (a) has led to a revival of determinism
    (b) emphasizes causality
    (c) says that causality and determinism are not strictly valid
    (d) has few philosophic implications

14. According to the Pauli exclusion principle,
    (a) electrons are excluded from nuclei
    (b) electrons cannot be removed from atoms
    (c) electrons must be in particular energy states
    (d) no more than two electrons can be in the same energy state

15. Atoms are held together to form molecules by
    (a) gravity
    (b) the attraction of electrons for other electrons
    (c) the attraction of nuclei for shared electrons between them
    (d) valence electrons that operate via the nuclear force

16. When high-speed electrons strike a metal plate, which of the following is emitted?
    (a) x rays   (c) molecules
    (b) atoms   (d) beta rays

17. The advantage of an electron microscope over an ordinary light microscope is that
    (a) it can take pictures more quickly
    (b) it employs quantum mechanics
    (c) it uses less energy
    (d) it can magnify with greater detail

18. One important feature of a laser is that it
    (a) can excite atoms
    (b) can be used in television sets
    (c) produces a high concentration of energy over a very small area
    (d) generates large amounts of energy

EXERCISES

1. What can you say about the relative temperatures of reddish, bluish, and whitish-yellow stars?
2. If all objects radiate energy, why can't we see them in the dark?
3. Describe what we mean by a "particle"; by a "wave."
4. Why do we say that light has wave properties? Why do we say that light has particle properties?
5. Why do we say that electrons have wave properties? Why do we say that electrons have particle properties?
6. What is the difference between a photon and an electron? Be specific.
7. Calculate the wavelength of a 2-kg ball traveling at 5 m/sec.
8. Calculate the wavelength of an electron (mass = 9.1×10^{-31} kg) moving at a speed of 2×10^5 m/sec.

457

9. What is the lowest frequency of light that will bring about the emission of electrons from a metal if the minimum energy W required is 4×10^{-19} joules? (Planck's constant $h = 6.6 \times 10^{-34}$ J•sec.) What wavelength and color is this?

10. Why do you suppose the particle nature of the electron was "discovered" before its wave nature?

11. An electron has a wavelength of 5.3×10^{-7} m. This is the wavelength of green light. Will the electron appear green? Explain why or why not.

12. What must be the electron's speed in the above question?

13. How can you tell if there is oxygen on the sun?

14. Some materials produce colors other than yellow when thrown into a fire. Why?

15. Compare the light emitted by solids with the light emitted by gases.

16. In Rutherford's planetary model of the atom, what keeps the electrons from flying off into space?

17. Discuss the differences between Rutherford's and Bohr's theories of the atom.

18. In a helium atom, which contains two electrons, do you think that on the average the electrons are closer to the nucleus or farther away than in a hydrogen atom? Why?

19. In an atom, why can't an electron orbit exist whose circumference is equal to 2½ wavelengths?

20. How can the spectrum of hydrogen contain many lines when a hydrogen atom contains only one electron?

21. What happens to an atom when it emits light?

22. Is energy conserved when an atom emits a photon of light? Explain.

23. What do we mean when we say an atom is "excited"?

24. Does an atom have more energy when it is excited? If so, what kind of energy is it?

25. When an atom jumps from the first excited state in hydrogen to the ground state, light of frequency 2.5×10^{15} Hz is emitted. What is the difference in energy between these two states?

26. Discuss the differences between Bohr's view of the atom and the quantum mechanical view.

27. Explain why Bohr's theory of the atom is not compatible with quantum mechanics, particularly the uncertainty principle.

28. What does the uncertainty principle make us uncertain about?

29. Explain why it is that the more massive an object is the easier it becomes to predict its future position.

30. In view of the uncertainty principle, why does a baseball seem to have a well-defined position and speed whereas an electron does not?

31. Discuss whether something analogous to the uncertainty principle operates in a public opinion survey. That is, do we alter what we are trying to measure when we take such a survey?

32. A cold thermometer is placed in a hot bowl of soup. Will the temperature reading of the thermometer be the same as the temperature of the hot soup before the measurement was made?

33. If Planck's constant were much larger than it is, how would this affect our everyday life?

34. The sun rose this morning. It rose yesterday morning, too. What evidence do we have that it will rise tomorrow? Do you *know* it will rise tomorrow?

35. On numerous occasions it has been observed that a stone thrown up in the air comes back down again. Can you therefore predict with certainty that this will happen again?

36. Compare the Newtonian and modern quantum views of physical reality.

37. How could you distinguish between an electron and an x ray that have the same wavelength?

38. Photographs have been made with electron microscopes that purport to show the outline of large molecules and even larger atoms like gold. Does this prove that atoms and molecules exist? Discuss in view of the fact that our understanding of the operation of an electron microscope and the photographs it produces is based on the assumption that atoms and electrons exist. You might consider if circular reasoning is involved; or if the atomic theory is self-consistent.

39. Why do fluorescent colors seem so bright?

16

NUCLEAR PHYSICS

1. THE NUCLEUS

As we discussed in the last chapter, Rutherford's experiments at the beginning of this century revealed the existence of a very tiny but massive nucleus at the center of the atom. While some physicists were exploring the nature of the atom and developing the quantum theory, others had already begun to investigate the nucleus itself. A number of surprises awaited them as they delved into the mysterious world of the nucleus, surprises that have affected all humanity.

Structure of the Nucleus

One of the questions physicists asked was whether the nucleus itself has a structure. Experiments indicated that indeed it does. However, the nucleus turned out to be a complicated entity that is still not thoroughly understood. It is always useful to formulate a *model* (or a "picture") when dealing with difficult concepts, and the simplest model of the nucleus pictures it as made up of tiny particles. Of course these particles also have wave properties, but for ease of visualization and language we will refer to them simply as particles.

By the early 1930s, researchers had ascertained that the nucleus of any atom contains two types of particles: **protons** and **neutrons**. Different kinds of atom were believed to contain different numbers of protons and neutrons; this is what ultimately distinguished one kind of atom from another. Although the proton had been detected earlier, it was clearly identified by Rutherford in 1919. Protons were found to have a positive charge exactly equal in magnitude, but opposite in sign, to the charge on an electron. The neutron was discovered in 1932 by another English physicist, James Chadwick (1891–1974). As its name suggests, the neutron is electrically neutral; it has no net charge. Its mass is almost identical to the mass of the proton; so it differs from a proton mainly in its lack of electric charge. These two constituents of the nucleus—protons and neutrons—are collectively referred to as **nucleons**.

The number of protons in a nucleus is designated by the symbol Z and is called the **atomic number** of the nucleus. The total number of nucleons (protons plus neutrons) is designated by the symbol A and is called the **atomic mass number**. This name is used because the mass of a given nucleus is essentially A times the mass of one nucleon. Thus a nucleus having 6 protons and 8 neutrons has an atomic number $Z = 6$ and atomic mass number $A = 6 + 8 = 14$.

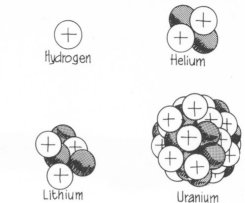

Figure 16–1

Different nuclei are made up
of different numbers of
protons and neutrons.

The number of electrons in a neutral atom determines the properties of that atom and, in fact, determines the kind of atom it is: carbon, oxygen, iron, or whatever. Since the charge on the proton is the same as the charge on the electron, but of opposite sign, the number of protons in the nucleus is the same as the number of "orbiting" electrons, Fig. 16–1. Therefore, the atomic number Z is equal to the number of electrons in a neutral atom and therefore determines the kind of atom it is (see Table 8–2). Because neutrons are neutral, they do not affect the number of electrons surrounding the nucleus; thus the number of neutrons has little effect on the properties and behavior of the atom. However, neutrons do have an effect on the stability of the nucleus as a whole, as we shall see shortly.

Isotopes

In order to specify a given nucleus, we need give only the two numbers Z and A. It is common practice to use a special symbol to represent this information:

$$_Z^A X,$$

where X is the chemical symbol for the element (e.g., C for carbon, H for hydrogen, etc.; see Table 8–1), Z is the atomic number (the number of protons), and A is the atomic mass number (the total number of nucleons). For example, $_6^{14}C$ means a carbon nucleus that has six protons and eight neutrons for a total of fourteen nucleons.

Since the atomic number defines the type of atom, it is redundant to give both the symbol for the element and its atomic number. For example, carbon always has six protons—by definition of what

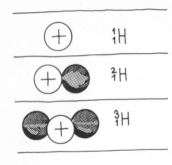

Figure 16–2

The three isotopes of hydrogen.

carbon is—so the subscript 6 is not really necessary. Consequently $^{14}_{6}C$ is sometimes written simply as ^{14}C; in words, we say "carbon fourteen." Similarly, $^{238}_{92}U$ may be written ^{238}U, and we say "uranium two-thirty-eight."

For a particular kind of atom, nuclei are found with different numbers of neutrons. For example, a hydrogen nucleus, in addition to its single proton, may contain no neutrons, or one neutron, or, very occasionally, two neutrons, Fig. 16–2. Similarly, carbon atoms may have 5, 6, 7, 8, 9, or 10 neutrons. Nuclei that contain the same number of protons but different numbers of neutrons are referred to as **isotopes**. Thus ^{11}C, ^{12}C, ^{13}C, ^{14}C, ^{15}C, ^{16}C are all isotopes of carbon.

Similarly, the known isotopes of hydrogen are ^{1}H, ^{2}H, ^{3}H. Some atoms have 10 or more isotopes.

For each element, a mixture of the various isotopes is found in nature, although one isotope is usually predominant. For example, 98.9 percent of naturally occurring carbon is the isotope ^{12}C, and 1 percent is ^{13}C, with other isotopes present only in very small amounts. Similarly, naturally occurring chlorine contains 75 percent of the isotope ^{35}Cl and 25 percent of ^{37}Cl.

Nuclear Forces

An interesting question arises when we consider that nuclei contain protons—often in large numbers—and these protons repel each other since they are positively charged. Since the average distance between protons is approximately 10^{-13} cm, Coulomb's law tells us that the repulsive forces must be extremely large. What, then, holds the nucleus together? Why doesn't it fly apart? There must be a force of some kind that holds the nucleons together in the nucleus. This force cannot be the electric or magnetic force, nor can it be gravity; these would all be much too weak. Some new force that is far stronger than previously known forces must be involved. We call this new force the **nuclear force**.

Evidently the nuclear force is an attractive force that acts between all nucleons. Two protons attract each other via the nuclear force at the same time as they repel each other via the electric force. Neutrons, since they are electrically neutral, interact only via the attractive nuclear force. It is the attraction of each nucleon for other nucleons via the nuclear force that holds them all together in the nucleus.

Sometimes this nuclear force is called the **strong nuclear force** to distinguish it from a second, much weaker, nuclear force called the **weak nuclear force**. The latter is detected in certain types of radioactivity. The two nuclear forces, along with the electromagnetic force and the gravitational force, are the four known forces in nature.

Nuclear forces are far more complex than the other known forces. Although we know a great deal about them, we still do not fully understand them. Because of their complexity, the nuclear forces, unlike the gravitational and electromagnetic forces, cannot yet be described precisely by simple mathematical equations.

Although the strong nuclear force is the strongest force we know, we are unaware of it in everyday life. The reason is that the nuclear force has a very short *range*—it is very strong between two nucleons 10^{-13} cm apart, but it is practically zero if the distance between them is only slightly greater than that. The forces of gravity and electricity, by contrast, are "long-range" forces; they decrease with the square of the distance, and so they can be felt even at fairly large distances. Thus we are aware of them in everyday life.

Nuclear forces, however, occasionally do reveal themselves in a dramatic, though not always pleasant, fashion. We answered the question of what holds the nucleus together by introducing the strong nuclear force. Yet because of the strange nature of this force it is not always able to do its job. If a nucleus has too many protons relative to its number of neutrons, or too many neutrons relative to its number of protons, the nuclear force is weakened. Nuclei that are unbalanced in this fashion are not stable. The effects of this instability had already been observed by the turn of the century, although the nucleus itself had not yet been discovered.

2. RADIOACTIVITY

Becquerel Discovers Radioactivity

The story of the nucleus has its beginnings in 1896. In that year the French physicist Henri Becquerel (1852–1908), while studying the phenomena of fluorescence and phosphorescence, began to wonder whether phosphorescent materials might emit the newly discovered and mysterious x rays. He exposed various phosphorescent minerals to bright sunlight in order to excite them and then placed them next to a photographic plate wrapped in opaque paper to exclude light. If x rays were emitted, they would penetrate the paper and blacken the photographic plate. He met with no success until he tried an obscure mineral labeled "pitchblende from Bohemia." Pitchblende, which is uranium ore, did blacken the plates. Further experiments by Becquerel showed that even when the uranium was kept in the dark for many weeks, by which time all phosphorescence, including x ray emission, would certainly have stopped, the photographic plates were darkened just as much! A new and different kind of radiation was apparently being emitted, and the phenomenon was called **radioactivity**.

Shortly after Becquerel's discovery, the brilliant Polish-born scientist Marie Sklodowska Curie (1867–1934) set to work on this new kind of radiation. In the course of her work, Madame Curie isolated two new elements, polonium and radium, both of which were highly radioactive. Soon many other radioactive elements were discovered. In every case the radioactivity was found to be unaffected by the strongest physical and chemical treatments. Excessive heating or cooling, and the action of powerful chemical reagents, had no effect on the activity of these radioactive elements. It soon became clear that the source of radioactivity must be deep inside the atom, far from the orbiting electrons. Indeed, a few years later Ernest Rutherford found that radioactivity was accompanied by the transformation of one element into another. It then became apparent that radioactivity is the result of the disintegration of an unstable nucleus: isotopes of certain elements are not stable under the action of the nuclear force, and these unstable nuclei "decay" by emitting radiation or "rays."

Three Types of Radiation: Alpha, Beta, and Gamma Rays

In 1898 Rutherford and others began to study the emitted rays and found them to be of three distinct types. They were first classified by their penetrating power. One kind of radiation could barely penetrate a piece of paper. The second kind was able to pass through as much as $1/8$ inch of aluminum. The third was able to pass through as much as several inches of lead. These three kinds of radiation were named alpha (α), beta (β), and gamma (γ) radiation, after the first three letters of the Greek alphabet. Each is found to have a different electric charge and hence is bent differently in a magnetic field, Fig. 16–3; alpha rays are positively charged, beta rays are negatively charged, and gamma rays are neutral. Some radioactive nuclei, bismuth-212 for example, emit all three types of radiation, whereas others emit only one or two types.

Further investigation showed that all three kinds of radiation consisted of particles that were already well known. Gamma rays are very high-energy photons of light. Beta rays are ordinary electrons, identical to those that orbit the nucleus. Finally, alpha rays are identical to the nuclei of helium atoms. Thus an alpha particle is made up of two protons and two neutrons, Table 16–1.

Figure 16–3

Alpha and beta rays are bent in opposite directions by a magnetic field. Gamma rays are not bent at all.

	Table 16–1	
The Three Types of Radiation		
ORIGINAL NAME	CHARGE	ACTUAL PARTICLE EMITTED
α (alpha)	$+2e$	^4_2He nucleus
β (beta)	$-1e$	Electron*
γ (gamma)	0	High-energy photon

*Some radioactive nuclei emit positive β rays, which are identical to electrons except that they are positively charged ($+1e$). These are known as *positrons* and are "antiparticles" to the electron; we will discuss them later in this chapter.

Radioactive Decay Involves a Change Within the Nucleus

Clearly, something important is happening in the nucleus when it emits an alpha, beta, or gamma ray.

The emission of gamma rays is the simplest case. It is very much like the emission of photons by excited atoms. Like the atom, the nucleus itself can exist in a number of different energy states. A nucleus in an excited state is not stable. It jumps down to its ground state, where it *is* stable, and in the process emits a photon of light. Since the possible energy levels in a nucleus are much farther apart than are atomic energy levels, the photons emitted have much higher energy. These high-energy photons are called gamma rays. How does a nucleus get into an excited state? It may get that way because of a violent collision with another particle, or as the result of a previous radioactive decay.

Alpha or beta decay occurs in certain nuclear isotopes because the nuclear force is unable to keep those isotopes stable. When an alpha or beta particle is emitted by a nucleus, the nucleus actually changes character. A new element is formed. Consider first the emission of an alpha particle. In emitting the alpha particle the original "parent" nucleus loses two of its neutrons and two of its protons. Therefore the remaining "daughter" nucleus has two fewer protons and two fewer neutrons; it is the nucleus of a different element. This changing of one element into another is called **transmutation**.

As an example, consider a uranium nucleus, $^{238}_{92}\text{U}$, which "decays" by emitting an alpha particle. Since four nucleons leave the

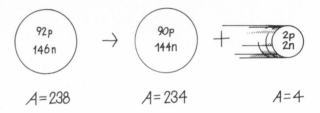

$A=238$ $A=234$ $A=4$

Figure 16–4

Uranium-238 decays with the emission of an alpha particle (^{4}He nucleus) to thorium-234.

^{238}U nucleus, only 234 nucleons remain; and, since two of the displaced nucleons were protons, the atomic number decreases from 92 to 90, Fig. 16–4. Looking at the periodic table (Table 8–2), we see that the element with atomic number 90 is thorium. We can write this transmutation using nuclear symbols as

$$^{238}_{92}U \rightarrow \, ^{234}_{90}Th + \, ^{4}_{2}He.$$

The arrow indicates that the parent nucleus $^{238}_{92}U$ changes into the daughter nucleus $^{234}_{90}Th$ with the emission of the helium nucleus (an alpha particle).

Transmutation also occurs when a nucleus emits a beta particle (an electron), Fig. 16–5. The electron is formed within the nucleus at the instant it is emitted. The electron has such a small mass that the atomic mass of the daughter nucleus is essentially the same as that of the parent. However, since electric charge is conserved, the daughter nucleus is left with an additional positive charge when the emitted electron leaves the parent nucleus. For example $^{14}_{6}C$ decays with the emission of an electron, and the daughter nucleus must

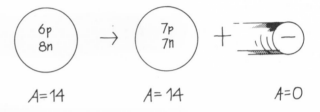

$A=14$ $A=14$ $A=0$

Figure 16–5

Carbon-14 decays to nitrogen-14 when it emits a beta particle (an electron).

then have a net charge of +7 (that is, + 7 − 1 = +6), and this is a nucleus of nitrogen:

$$^{14}_{6}C \rightarrow \,^{14}_{7}N + e^-.$$

It is just as if one of the neutrons in the carbon nucleus emitted an electron and became a proton: $n \rightarrow p + e^-$. Indeed, free neutrons outside a nucleus actually do decay in this fashion. So we can think of beta decay as the transformation of a neutron into a proton within the nucleus, accompanied by the creation and emission of an electron.

Detailed measurements in beta-decay experiments indicated that an anomaly existed. Neither the total energy nor the total momentum seemed to be conserved in the process. The possibility that these two conservation laws might prove to be invalid was very troubling to physicists. They could not be sure, however, whether energy and momentum were indeed not being conserved in beta decay or whether they had overlooked some aspect of the process.

A resolution of the problem came in the early 1930s when theoretical physicists postulated that, in addition to the electron, a new kind of particle was also being emitted in beta decay. This hypothesized particle was assumed to carry off just enough energy and momentum so that the total energy and momentum would be conserved. The new particle apparently has no measurable rest mass but does have energy just as a photon does. It is electrically neutral and was dubbed the **neutrino**—meaning "little neutral one"—by the great Italian physicist Enrico Fermi (1901–1954), who worked out a detailed theory of beta decay in 1934. It has proved to be very difficult to detect the neutrino directly, which indicates that it interacts with matter only very weakly. However, in 1956, complicated experiments produced further evidence of the neutrino, but by then most physicists had come to take its existence for granted.

Notice that in all three types of decay the *total* atomic number Z, which represents the electric charge, does not change; nor does the total number of nucleons, A (Figs. 16–4 and 16–5). This is a result of the conservation of electric charge and of a new conservation law, the conservation of nucleons: *the total number of nucleons always remains constant, although one type can be changed into the other type*.

Half-Life

Any macroscopic sample of radioactive material, for example uranium, contains a vast number of radioactive nuclei. These nuclei do not all decay at once. They decay one by one over a long period

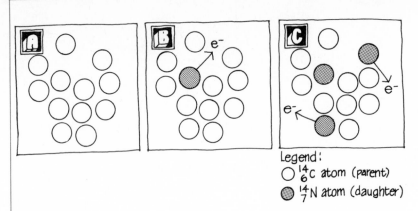

Legend:
○ $^{14}_{6}C$ atom (parent)
◉ $^{14}_{7}N$ atom (daughter)

Figure 16–6

In a sample of radioactive nuclei, the nuclei decay one by one. Thus the number of parent nuclei is continually decreasing. As soon as a ^{14}C nucleus emits the electron, the nucleus has become a ^{14}N nucleus. In (A), all the nuclei are ^{14}C. In (B), one has decayed to ^{14}N, and in (C) two more have decayed.

of time. This is a random process; it is impossible to predict just when a given nucleus will decay. However, it is possible to determine on the basis of probabilities approximately how many nuclei in a given sample will decay in a given time period. This is another example of the probabilistic nature of matter.

Some radioactive isotopes are much more unstable than others. The more unstable an isotope is, the more quickly a sample of many such nuclei will decay. A highly unstable isotope is said to be very radioactive, because a macroscopic sample of many such nuclei will emit alpha, beta, or gamma rays at a relatively high rate.

Radioactive decay is a "one-shot" process. That is, once a particular parent nucleus has decayed into the daughter nucleus, it cannot do it again; there is one less parent nucleus to decay. The number of parent nuclei is thus continually decreasing, Fig. 16–6. A quantitative measure of the longevity of a particular isotope is its **half-life**. The half-life of an isotope is the time it takes for half the original amount of the isotope in a given sample to decay. For example, the half-life of ^{14}C is 5700 years. If at some time a piece of petrified wood contains, say, 10 g of ^{14}C, then after 5700 years it will contain only 5 g. After 5700 more years (a total of 11,400 years) half of the remaining 5 g will have decayed, so that only 2.5 g remain; and so on, Fig. 16–7.

The half-lives of known radioactive isotopes vary from as short as 10^{-22} seconds to as long as many billions of years, Table 16–2. The half-life of very long-lived isotopes, e.g., ^{238}U with a half-life of

Table 16-2
Some Commonly Occurring Radioactive Isotopes

ISOTOPE	PARTICLES EMITTED	HALF-LIFE
Carbon 14 (^{14}C)	β^-	5700 years
Cobalt 60 (^{60}Co)	β^-, γ	5.24 years
Strontium 90 (^{90}Sr)	β^-	29 years
Iodine 131 (^{131}I)	β^-, γ	8 days
Polonium 214 (^{214}Po)	α, γ	1.6×10^{-4} seconds
Radium 226 (^{226}Ra)	α, γ	1620 years
Uranium 238 (^{238}U)	α, γ	4.5×10^9 years

4½ billion years, is determined not by waiting for 4½ billion years but by measuring only the beginning of a curve such as that of Fig. 16–7.

The longer the half-life of an isotope, the more slowly it decays, and the lower its "activity" or rate of emission of alpha, beta, or gamma rays. Conversely, very active isotopes have short half-lives. In any radioactive sample the activity depends not only on the half-life but on the number of nuclei present. For a given radioactive isotope, the more nuclei that are present, the more nuclei that can decay. As decay proceeds and the number of parent nuclei decreases, the rate at which alpha, beta, or gamma rays are emitted decreases as well. Thus the number of particles emitted by a radioactive sample decreases in time and follows a curve like that of Fig. 16–7.

Figure 16–7

The amount of ^{14}C remaining decreases by half every 5700 years.

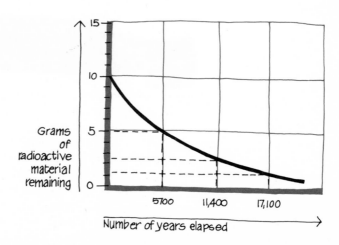

Detection of Radiation

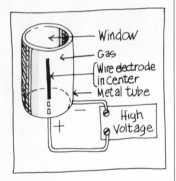

Figure 16–8

Diagram of a Geiger counter.

Individual "elementary" particles, such as neutrons, protons, electrons, alpha particles, and gamma rays, cannot be detected directly by our senses. Therefore a variety of instruments has been developed to detect them.

One of the most common is the **Geiger counter**, which is a common tool in the laboratory as well as for uranium prospectors. (Since uranium is one of the most common naturally radioactive substances, its presence can often be determined by detecting emitted alpha particles.) The inside of a Geiger counter consists of a cylindrical metal tube with an electrical wire (electrode) running down its center, Fig. 16–8. The tube is filled with a gas, and the end of the tube is a very thin "window" through which charged particles can enter. The electrode is maintained at a high voltage with respect to the walls of the tube. The voltage is slightly less than that required to pull the electrons off the atoms in the gas. However, if a charged particle enters the tube it may strike an atom and knock an electron out of the atom (this is called "ionizing" the atom). The freed electron is attracted to the positive wire, and as it moves toward the wire it strikes other atoms, ionizing them. Soon an avalanche of electrons strikes the electrode, causing an electric pulse; this pulse can be amplified, counted by a counting device, or sent to a loudspeaker. In the latter case, the minor explosion within the tube caused by the passage of a single particle is heard as a click in the speaker. A faster clicking rate signifies a greater number of particles and a greater amount of radioactivity.

A **scintillation counter** makes use of a solid, liquid, or gas known as a "scintillator." The atoms of the scintillator are easily excited when they are struck by an incoming particle and emit light when they drop down to their ground states. The emitted photons strike a metal that emits electrons (the photoelectric effect), which are multiplied by a special "photomultiplier tube" into detectable electric signals. Scintillation counters can be used not only to detect charged particles—to which Geiger counters are mainly limited—but also to detect neutral particles such as gamma rays and neutrons.

With a "cloud chamber" and the more recent and more effective "bubble chamber," the *paths* of charged particles can be seen. In a cloud chamber, a gas is cooled to a temperature slightly below its usual condensation point. The gas molecules do not condense immediately unless dust or an ionized molecule is present for them to condense on. However, when a charged particle passes through such a "supercooled" gas, it ionizes the molecules along its path, and tiny bubbles form on the resulting ions; the row of bubbles indicates the path of the charged particle, Fig. 16–9. In the bubble

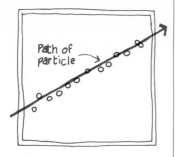

Figure 16–9

Bubbles formed by the passage of a charged particle in a cloud or bubble chamber (magnified).

chamber, invented in 1952 by D. A. Glaser, the path of a particle is seen as a series of gaseous bubbles within a "superheated" liquid: The temperature of the liquid is slightly above its boiling point, and the bubbles characteristic of boiling will form around the ions produced by the passage of a charged particle, Fig. 16–9. Because the bubble chamber makes use of a liquid—often liquid hydrogen—the density of the atoms in the chamber is much greater than in a cloud chamber; thus the chance of an encounter between a charged particle and the nucleus of one of the atoms of the fluid is much greater. The bubble chamber is thus very useful for studying collisions between charged particles and nuclei, Fig. 16–10. A study of such "nuclear reactions" provides information about the nature of the nuclear force.

Figure 16–10

(A) A bubble chamber.
(B) Bubble chamber photo of particle tracks.

Radioactive Dating

Radioactive decay is an extremely interesting phenomenon that has helped physicists unravel the intricacies of the atomic nucleus. The phenomenon of radioactivity has many useful applications as well. In this section we discuss how it can be used to find the age of ancient materials; in the next section we will study some of its medical applications.

The age of any object made from once-living matter, such as wood, can be determined by using the natural radioactivity of ^{14}C. All living plants absorb carbon dioxide (CO_2) from the air, using the carbon and expelling the oxygen. The vast majority of these carbon atoms are ^{12}C, but a small percentage, one out of 100 million, is the radioactive isotope ^{14}C. In spite of the fact that ^{14}C decays with a half-life of 5700 years, the ratio of ^{14}C to ^{12}C in the atmosphere has remained nearly constant over many thousands of years. This happens because neutrons in the cosmic radiation that impinges on the earth from outer space collide with atoms of the atmosphere. In particular, collisions with nitrogen produce the following nuclear transformations: $n + {}^{14}_{7}N \rightarrow {}^{14}_{6}C + p$. That is, a neutron strikes and is absorbed by a ^{14}N nucleus, and a proton is knocked out in the process. The remaining nucleus is ^{14}C. This continual production of ^{14}C in the atmosphere balances the loss of ^{14}C by radioactive decay.

As long as a plant or tree is alive, it continually uses the carbon from carbon dioxide in the air to build new tissue and to replace old. Animals eat plants, so they too are continually receiving a fresh supply of carbon for their tissues. An organism cannot distinguish ^{14}C from ^{12}C and, since the ratio of ^{14}C to ^{12}C in the atmosphere remains constant, the ratio of ^{14}C to ^{12}C within the living organism remains constant as well. But when the organism dies, carbon dioxide is no longer absorbed; and because the ^{14}C decays radioactively, the ratio of ^{14}C to ^{12}C in the dead organism decreases in time. Since the half-life of ^{14}C is 5700 years, the $^{14}C/^{12}C$ ratio decreases by half every 5700 years. Thus, if archeologists find an ancient bone, an artifact, or a tool made from wood or other organic matter, they can measure the $^{14}C/^{12}C$ ratio and determine the age of the object. For example, if the $^{14}C/^{12}C$ ratio of an ancient wooden tool is half of what it is in living trees, then the object must have been made from a tree that was felled 5700 years ago.

Carbon dating is useful only for determining the age of objects less than about 20,000 years old. The amount of ^{14}C remaining in older objects is usually too small to measure accurately. However, radioactive isotopes with longer half-lives can be used in certain circumstances to obtain the age of older objects. For example, the decay of ^{238}U, because of its long half-life of 4.5×10^9 years, is useful in determining the age of rocks on a geologic time scale. When molten material solidifies into rock, the uranium present in the material becomes fixed in its position, and the daughter nuclei that result from the decay of uranium will necessarily be fixed in the same position. Thus, by measuring the amount of ^{238}U remaining in the material relative to the amount of daughter nuclei in the same place, the time of solidification (when the rock was formed) can be determined.

These radioactive dating methods have been used to measure the age of the earth as approximately 5 billion years. The age of rocks in which the oldest fossilized shells of sea animals are embedded indicates that these animals first appeared about 500 million years ago. The earliest fossilized remains of mammals are found in rocks 200 million years old. Radioactive dating has similarly revealed that creatures akin to man first appeared on earth about 2 million years ago. Radioactive dating has thus been indispensable in determining the history of the earth and the evolution of biological organisms living on it.

Biological Uses and Dangers of Radioactivity

A number of radioactive isotopes have found use as radioactive "tracers." Radioactive isotopes such as ^{14}C can be incorporated into almost any organic molecule. The details of how food molecules are digested, and to what parts of the body they are diverted, can be "traced" by following the movement of radioactively labeled molecules through the body using a Geiger counter or a scintillation counter. Radioactive tracers have been used to determine how amino acids and other essential food molecules are metabolized by various organisms from humans to microorganisms.

A radioactive isotope of iodine, ^{131}I, is very useful for detecting thyroid disorders. Since the thyroid gland accumulates iodine, ^{131}I injected into the body will also accumulate there. The rate of accumulation of ^{131}I can be measured with radioactivity detectors, and this yields information on the functioning of this gland. Also, if the thyroid gland is overactive or cancerous, huge amounts of ^{131}I injected into the body accumulate in the abnormal region of the gland; the intense radioactivity emitted by this iodine isotope can destroy the diseased tissue.

The intense radiation given off by ^{60}Co (cobalt-60) is a more general therapeutic agent, which is used to destroy malignant cells in cancerous tumors. Cancer therapy is also accomplished by focusing a beam of high-energy particles—for example, protons accelerated by a high-energy accelerator—on the malignant tumor,

Fig. 16–11. In either case, the collision of a high-energy particle with an organic molecule causes the atoms within the molecule to be reoriented or even detached, and the disrupted molecule no longer functions properly. When the radiation is intense, many molecules within the cells are altered, and the cells are unable to function properly and they die. Even if only a few molecules essential to the life of the cell are damaged, the cell dies. In this way cancerous tissue can be destroyed.

But excess radiation striking healthy cells can damage or destroy them in the same way. Often the genetic material is altered and a **mutation** results. Most mutations are detrimental and are carried on to the next generation. Some mutations cause a cell to begin multiplying indiscriminately—the beginning of a cancerous growth. Radiation can thus cause cancer as well as help cure it. In any kind of radiation treatment special precautions must be taken to avoid exposing normal tissue; some exposure may be unavoidable, but it must be kept to a minimum.

A certain amount of radiation is tolerable. In fact, our bodies are constantly subjected to the low-level radiation of cosmic rays. But our bodies can take only so much. If the level of radiation from all sources is too high, the normal repair mechanisms of the body will not be able to repair the damage. For this reason the government sets maximum levels of radiation to which individuals can be safely exposed. But these levels are only approximate and are somewhat arbitrary. We should all be aware of what radiation is, and what its effects are, so that we can intelligently express our concern over its uses.

Radiation damage is cumulative. Many small doses are almost as harmful as a single large dose. X rays are also damaging and must be considered when the safe dosage for humans is calculated. Increased concern has developed in recent years over the routine use of medical x rays, and it is generally agreed that their use should be avoided whenever possible. For example, skin tests, rather than chest x rays, are often used now for detection of tuberculosis.

Radiation is an extremely useful tool with a great variety of applications. Used with caution, it can be of great benefit to humanity.

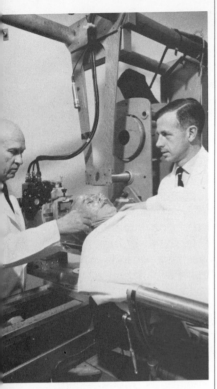

Figure 16–11
Cancer patient being exposed to a beam of charged particles from a high-energy accelerator.

3. FISSION AND FUSION: NUCLEAR ENERGY

Artificial Transmutation of Elements

Whenever a nucleus undergoes alpha or beta decay, the remaining daughter nucleus corresponds to a different element. The transformation of one element into another, called transmutation, also occurs as a result

Figure 16–12

Enrico Fermi (1901–1954) was a twentieth-century anomaly, for he contributed significantly to both experimental and theoretical physics. Among his many important accomplishments, besides the artificial production of new elements, were the first detailed theory of beta decay, a theory describing the statistical behavior of certain kinds of particles (known as "Fermi-statistics"), and a great amount of experimental work.

of **nuclear reactions**. For example, we saw earlier that ^{14}C is produced in the atmosphere by the collision of neutrons with nitrogen nuclei. The first transmutation to be observed in the laboratory via a nuclear reaction was detected by Ernest Rutherford in 1919. He observed that when alpha particles pass through nitrogen gas, some of the alpha particles are absorbed and protons are emitted. He concluded that nitrogen nuclei are transformed into oxygen nuclei by the reaction

$$\,^4_2He + \,^{14}_7N \rightarrow \,^{17}_8O + \,^1_1H.$$

In the decade that followed, a number of other nuclear reactions were observed. The artificial transformation of elements (we say "artificial" meaning it is done in the laboratory rather than occurring naturally) took a great leap forward in the 1930s when Enrico Fermi (Fig. 16–12) realized that neutrons would be the most effective projectiles for producing new elements. Since neutrons have no electric charge, they are not repelled by positively charged nuclei as alpha particles are. Therefore, even slow-moving neutrons can enter a nucleus and bring about a transformation.

Between 1934 and 1936, Fermi and his coworkers in Rome produced many new isotopes by bombarding different elements with neutrons. Fermi soon realized that by bombarding the heaviest known element, uranium, it might be possible to produce heretofore unknown elements whose atomic numbers were greater than that of uranium! The experiments were carried out and, after several years of hard work, it was suspected that two new elements, neptunium ($Z = 93$) and plutonium ($Z = 94$), had been produced, Fig. 16–13. (The elements were produced unequivocally several years later at the University of California, Berkeley.) But, as was shown shortly thereafter, what Fermi had actually observed was an even stranger process—a process that was destined to play an extraordinary role in the world. Fermi's work was interrupted at the very

Figure 16–13

Series of reactions in which neptunium and plutonium are produced after bombardment of uranium-238 by neutrons.

$n + \,^{238}_{92}U \longrightarrow \,^{239}_{92}U$	Neutron captured by $^{238}_{92}U$
$\,^{239}_{92}U \longrightarrow \,^{239}_{93}Np + e^-$	$^{239}_{92}U$ decays by beta decay to neptunium-239
$\,^{239}_{93}Np \longrightarrow \,^{239}_{94}Pu + e^-$	$^{239}_{93}Np$ itself decays by beta decay to plutonium-239

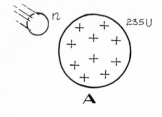

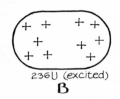

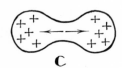

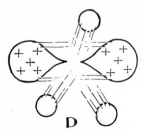

Figure 16–14

The fission of a ^{235}U nucleus induced by capture of a neutron.

threshold of this discovery. In 1938 he was awarded the Nobel Prize, and he chose to use the trip to Sweden to escape the fascist oppression of Mussolini.

The Discovery of Fission

During the late 1930s German scientists repeated and advanced some of Fermi's work. They performed extensive chemical analyses of the elements produced when uranium was bombarded by neutrons. In 1939 two chemists, Otto Hahn and Fritz Strassman, made an amazing discovery: uranium bombarded by neutrons produced not only the new elements neptunium and plutonium, but also smaller nuclei such as that of ^{141}Ba. (barium-141) Two refugees from nazism working in Sweden, Lise Meitner and Otto Frisch, quickly realized what had happened. The uranium nucleus, they said, after absorbing a neutron, actually split into two almost equal pieces! This was quite startling in view of the fact that in previously known nuclear reactions only tiny fragments were knocked out of a nucleus. But now, in certain circumstances, it appeared that a nucleus could actually be split in half. It was soon discovered that this happens much more readily with the rarer uranium isotope, ^{235}U, than with the more common ^{238}U.

Lise Meitner named this new phenomenon **nuclear fission**. The process by which it is assumed to occur is shown in Fig. 16–14. The neutron absorbed by the ^{235}U nucleus gives the nucleus (now ^{236}U because of the additional neutron) extra internal energy; the resulting increased motion of the nucleons causes the nucleus to take on abnormal, elongated shapes [Figs. 16–14(B) and (C)]. When the nucleus elongates into the shape shown in Fig. 16–14(C), it just keeps going and splits roughly in two, Fig. 16–14(D). The two resulting nuclei, called **fission fragments**, are not always the same; they can be different for different individual fissions. A common fission reaction is one in which barium is produced:

$$n + {}^{235}_{92}U \rightarrow {}^{236}_{92}U \rightarrow {}^{141}_{56}Ba + {}^{92}_{36}Kr + 3\,n.$$

Whenever a uranium nucleus fissions, several neutrons are released along with the two fission fragments. Fermi, who settled in the United States in 1939, realized that the neutrons released in a fission could be used to generate a **chain reaction**. That is, one neutron initially causes the fission of one uranium nucleus, and in the process two or three more neutrons are released. These in turn can cause the fission of additional uranium nuclei; each of these new fissions releases more neutrons capable of causing more fissions. Thus it multiplies into a chain reaction, Fig. 16–15.

Figure 16–15
A chain reaction.

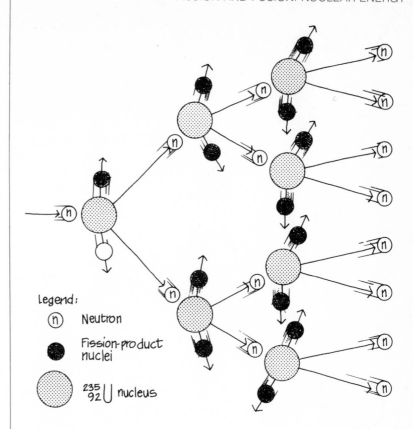

Legend:

(n)   Neutron

●   Fission-product nuclei

⬤ $^{235}_{92}U$ nucleus

Fission Releases Energy

In any fission reaction, when we calculate the sum of the masses of the resulting fission fragments and free neutrons, we find that the total is less than the mass of the original uranium nucleus. Mass is lost during the fission process. Where has it gone?

The answer lies in Einstein's famous statement of the equivalence of mass and energy, $E = mc^2$. The mass that disappears in a fission reaction is changed into energy! This energy appears as kinetic energy of the fissioned nuclei and neutrons; it is transferred to other atoms and molecules through collisions, and thus is ultimately manifested as heat.

The amount of energy released in a single fission reaction can be calculated by using $E = mc^2$. In this situation, m represents the difference in mass between the original ^{236}U nucleus and the final nuclei plus freed neutrons. This energy is about 3×10^{-17} joules, which does not seem like a lot, but on an atomic scale it is enor-

479

mous. In a chain reaction, in which a vast number of fissions occur, the total energy released on a macroscopic scale can be tremendous. For example, if only one gram of uranium were to undergo fission, it would release more than 3,000,000 times as much energy as burning the same amount of natural gas or coal would release.

This fantastic source of energy could be a great boon for humanity. But alas, history played one of its cruelest tricks. At the moment of this great discovery, the world was collapsing. Instead of aiding humanity, nuclear energy would first be used for destructive purposes.

The Bomb

By early 1940, all research into the fission process was suddenly enshrouded in secrecy, both by the Western powers and by Germany. Hitler banned the sale of uranium from the Czechoslovakian uranium mines he had recently taken over. Physicists in the United States became alarmed. They were acutely aware of what would happen if the enormous energy available in the nuclear fission process should be used in a bomb. There was little question in their minds that Hitler would order his scientists to develop such a bomb. A group of eminent physicists approached Albert Einstein and convinced him that the president of the United States must be made aware of the facts about nuclear fission.

Although Einstein was a passionate pacifist, he signed a now-famous letter to President Roosevelt informing him of the possibility of a fission bomb and citing evidence that Germany might already be on the road to producing one.

President Roosevelt recognized the significance of Einstein's words. But at first, government support was limited. Finally, in 1942 an all-out program, known as the Manhattan Project, was begun. At that point it had still not been proved that energy from fission could be obtained on a large scale; that is, it had not been demonstrated that a large-scale chain reaction could actually be made to occur. It is true that every time a uranium nucleus fissions it releases two or three additional neutrons, which in turn can potentially cause further fissions and a chain reaction (Fig. 16–15)—that is, if the neutrons don't escape from the piece of uranium before they are captured by a uranium nucleus. If the piece of uranium is too small, most of the neutrons will escape through the surface before they can cause another fission, and thus the chain reaction will not be sustained. Therefore a large enough mass of uranium must be used so that more neutrons will be captured (and produce further fissions) than will escape. The minimum mass of uranium necessary to produce a "self-sustaining" chain reaction is known as the **critical mass**. In early 1942 it was not known how big the critical mass would have to be. It was necessary to determine what that critical mass was and then to demonstrate that a self-sustaining chain reaction could actually be made to occur.

After months of planning, the first nuclear reactor (Fig. 16–16) was constructed beneath the stands of Stagg Field at the University of Chicago under the direction of Enrico Fermi. Fermi's reactor was equipped with cadmium control rods whose function was to absorb neutrons. With the rods fully in place, most of the neutrons produced in the fissionable material would be absorbed before they could produce further fissions. Thus a runaway chain reaction, and subsequent explosion, could be prevented. As the control rods were pulled out, fewer neutrons would be absorbed and more would be available to cause fission. And, if everything went well, it was hoped that a sustained chain reaction would occur. On December 2, 1942, Fermi was ready to make the critical test. That afternoon, as Fermi slowly withdrew the cadmium control rods, the reactor went "critical": the first self-sustaining chain reaction had been produced.

With this (perhaps dubious) success, work on the nuclear bomb (or atomic bomb, as it is popularly called) began in earnest. A great many technical problems remained, however. In 1943 a secret new laboratory was constructed on the site of a boys' school in New Mexico known as Los Alamos. Under the directorship of J. Robert Oppenheimer (1904–1967), that laboratory became the home of scientists from all over the United States and Europe. For the next two years they worked together and in mid-1945 produced the first bombs. The roster of names at Los Alamos in those years is astonishing; it seems that nearly all of the world's great physicists and chemists were there for at least a part of those two years.

After the war, these scientists were criticized by some for having participated in the development of this destructive weapon. Many

Figure 16–16

The first nuclear reactor, which was built under the grandstand of Stagg Field at the University of Chicago. (A) Photograph taken during the addition of the nineteenth layer of graphite. (B) Scale model of the reactor.

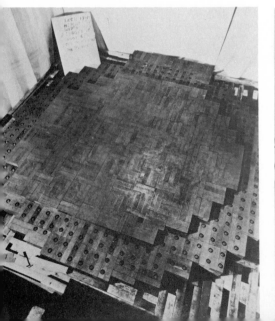
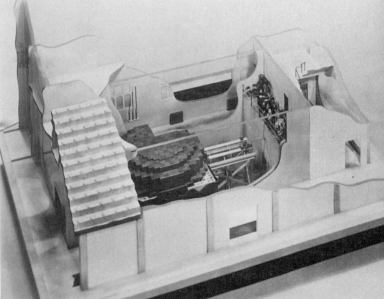

of the scientists themselves regretted that they had helped. Yet, if we examine their motivations and the times in which they worked, criticism may seem out of place. The German military under Hitler was overrunning Europe, and German scientists were presumably trying to build an atomic bomb. In addition, Japan was waging a successful war against the United States. Many of the scientists at Los Alamos were perfectly aware that what they were producing posed a great danger to humankind. But they had to make a choice. Stopping Hitler and Japan was uppermost in their minds.

We can be thankful for one irony. It seems that Hitler did not realize that a practical bomb could be built and so did not push hard for it. Furthermore, German scientists may not have been sufficiently sympathetic to Hitler's aims to give themselves wholeheartedly to this project. In fact, Germany did not even come close to producing an atomic bomb.

Nuclear Reactors

The technology developed during World War II was not all wasted on building weapons. Emphasis on peaceful uses of nuclear energy began after the war, and before long nuclear reactors were built that could supply cities with large amounts of electricity.

The principles behind a nuclear reactor are quite simple. The heat energy from fission is used to boil water, changing it to steam. The steam then drives a turbine, which in turn drives an electric generator, Fig. 16–17.

The essential difference between a nuclear power reactor and a nuclear bomb is the rate at which the energy is released. In a bomb, since few neutrons are allowed to escape, the chain reaction occurs very swiftly and an explosion takes place. In a nuclear reactor, the multiplication of neutrons is limited by control rods, which, just as in Fermi's first reactor, absorb some of the neutrons so that they cannot cause fission. The control rods are adjusted so that an average of only one neutron per fission is allowed to cause another fission, thus keeping the reactor just barely "critical."

Since ^{235}U is easily fissionable and ^{238}U is not, the fact that only about 0.7 percent of naturally occurring uranium is ^{235}U means that the world's supply of fissionable fuel is limited. However, during the war years it was found that plutonium-239 is also easily fissionable; in fact, one of the first two atomic bombs used plutonium. But, since plutonium does not occur naturally, it must be produced artificially by bombarding the common uranium isotope, ^{238}U, with neutrons, as we saw earlier. So-called "breeder reactors" would use this fact to "breed," or produce, ^{239}Pu from ^{238}U. Because ^{238}U is much more

Figure 16–17

Diagram of a nuclear reactor. Heat generated by the fission process in the fuel is carried off by hot water or liquid sodium and is used to boil water to steam in the heat exchanger. The steam drives a turbine to generate electricity and then is cooled in the condenser.

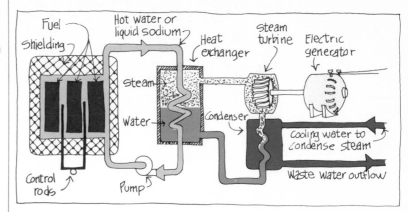

abundant than ^{235}U, the amount of available fissionable fuel in the world would be enhanced.

Nuclear power reactors have several advantages over conventional power plants that burn coal or natural gas. The latter produce considerable air pollution through the release of the combusted gases, whereas nuclear plants give off essentially no air pollutants. Furthermore, our accessible reserves of the fossil fuels coal, oil, and gas are being depleted, and nuclear power is a viable alternative.

Nuclear power plants have disadvantages as well. Like fossil fuel plants, they give rise to several forms of pollution. Nuclear power plants require cooling water, which is ejected at a higher than normal temperature, usually into a river, the ocean, or the air. Thermal pollution may upset the ecology of marine life in the immediate vicinity, or in the case of air cooling towers it may affect the weather over land areas. Even more important is the fact that the nuclei produced in fission are very highly radioactive, and the safe disposal of these radioactive wastes is a serious problem.[1] Most often, the highly radioactive material remaining after the fissionable fuel has been used up is sealed in heavy concrete casings and dropped in the ocean or buried. Leakage of the radioactive wastes during transport or disposal is unlikely, but it is possible and has in fact occurred. Eventually, because of its finite size, the earth may not be able to hold safely all the waste products, nuclear or otherwise, produced by human beings.

---

[1] The radioactive fission products produced by a bomb explosion are referred to as radioactive fallout. Like any kind of radioactivity, fallout can be dangerous to living organisms. It cannot be controlled during a bomb explosion and spreads throughout the atmosphere, often coming down in rain. Nuclear testing in the atmosphere is thus highly dangerous to human life. This is the reason behind the treaty of 1963, signed by more than 100 nations, that bans the testing of nuclear weapons in the atmosphere.

There is also the remote possibility that an accident will occur in a power plant permitting the escape of radioactive material into the atmosphere. Great precautions are taken to prevent such accidents, and safeguards are built in to prevent the escape of radioactivity in case of accident. Nonetheless, a slight chance of accident exists, and nuclear power plants are usually located far from urban centers.

Breeder reactors, if developed, will present an additional risk, because plutonium interacts with the human body in such a way that it is highly poisonous even in microscopic amounts. Extreme measures would have to be taken to prevent its escape. Because the risk is so high, many scientists and environmentalists believe that a breeder reactor should not be developed.

Fusion

The mass of every nucleus is less than the sum of the masses of its constituent protons and neutrons. For example, the mass of the nucleus of the helium isotope $_2^4$He is less than the mass of two protons plus the mass of two neutrons. Thus, when two protons and two neutrons come together to form a helium nucleus, there is a loss of mass. This loss is manifested in the release of a large amount of energy. The process of building up nuclei by bringing together individual protons and neutrons, or building larger nuclei by combining small nuclei, is called **fusion**. It is believed that all the elements in the universe were originally formed through the process of fusion. Today, fusion is continually taking place within the stars, including our sun, and it is these fusion reactions that produce the intense light energy emitted by the stars. Life on earth depends on the energy we receive from the fusion reactions occurring in the sun.

The simplest fusion reactions of practical importance involve the two isotopes of hydrogen: $_1^2$H, which contains one neutron as well as the proton and is called **deuterium**; and $_1^3$H, which contains two neutrons and a proton and is called **tritium**:

$$_1^2\text{H} + {_1^2\text{H}} \rightarrow {_1^3\text{H}} + {_1^1\text{H}}$$
$$_1^2\text{H} + {_1^2\text{H}} \rightarrow {_2^3\text{He}} + \text{n}$$
$$_1^2\text{H} + {_1^3\text{H}} \rightarrow {_2^4\text{He}} + \text{n}$$

Deuterium, the main "fuel" for each of these reactions, can itself be formed by the fusion of a proton and neutron, although not so readily. Each of these fusion reactions yields a tremendous amount of energy due to the fact that the resulting nucleus or nuclei on the right side of the above equations have less mass than the original ones on the left.

Physicists have been trying to produce a power reactor that uses fusion ever since the Second World War. The fusion reactions most likely to succeed in a reactor are those listed above using deuterium and tritium. It would be a great boon to mankind to be able to use fusion as a source of energy. Gram for gram, fusion releases more energy than fission does. Fusion produces almost no radioactive waste products. In fact, fusion is a "cleaner" energy-producing process than others presently known. For fuel, fusion requires only deuterium, and deuterium is readily available. In ordinary water (H_2O), one out of every 3500 molecules contains a deuterium isotope (these molecules are called "heavy water"); that means about 1 gram of deuterium per 8 gallons of water. And the oceans contain enough water to supply us with deuterium as fusion fuel for a very long time.

Unfortunately, considerable difficulties still exist, and a usable fusion reactor is not yet a reality. The problems are associated with the fact that all nuclei have a positive charge and thus repel each other. However, if they can be brought close enough together so that the short-range attractive nuclear force can come into play, that force will pull the nuclei together and fusion will occur. In order for the two nuclei to get close enough together, they must have very high speeds, Fig. 16–18. Since high speed of atoms corresponds to high temperature, very high temperatures are required for fusion to occur; hence fusion reactors are referred to as "thermonuclear" reactors. The sun and other stars are very hot, many millions of degrees, so the nuclei are moving fast enough there for fusion to take

Figure 16–18

(A) Low-speed ^2_1H nucleus is deflected by a second ^2_1H nucleus. (B) High-speed ^2_1H nucleus strikes a second ^2_1H nucleus and fusion occurs, producing ^3_2He and a neutron.

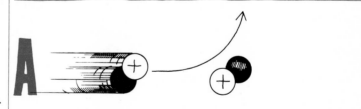

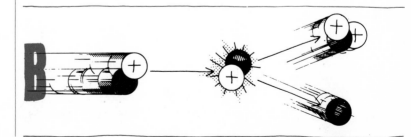

place readily. And the energy released by the fusion reactions themselves keeps the temperature high so that further reactions can occur. Thus the sun and the stars represent self-sustaining thermonuclear reactors. But on earth such high temperatures are not easily attained, at least not in a controlled manner.

After the Second World War, however, when it was realized that the temperature produced within a fission (or "atomic") bomb was close to 100 million degrees, the development of a thermonuclear (or "hydrogen") bomb was undertaken. The hydrogen bomb used an ordinary atomic bomb as a fuse. The explosion of the atomic bomb generates the high temperatures necessary for fusion to occur, and the rapid release of fusion energy results in a devastating explosion.

The uncontrolled release of thermonuclear energy in a bomb was fairly easy to obtain; but to realize usable energy from fusion reactions in a slow and controlled manner turned out to be extremely difficult. The problem has not yet been solved. The high temperatures necessary for controlled fusion can now be obtained. One way is to use a high-powered laser. However, difficulties still exist in containing the atoms that are ionized at temperatures of many millions of degrees. This ionized gas, nuclei plus electrons, is known as a **plasma**. All known materials vaporize at temperatures of only a few thousand degrees, and so cannot be used as containers. In an effort to overcome this problem, physicists have used a magnetic field, Fig. 16–19. Charged particles encircle magnetic field lines (see Chapter 14); when they encounter a region where the magnetic field lines are very close together, they are reflected. The configuration of Fig. 16–19 is known as a "magnetic bottle." Unfortunately, this bottle develops "leaks" and the charged particles at high fusion temperatures escape after a short time. Other

Figure 16–19
A magnetic bottle.

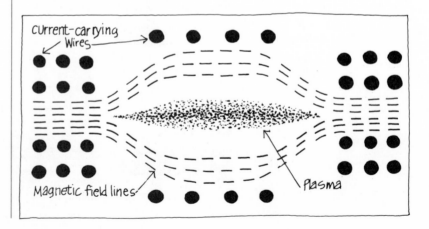

current-carrying wires

Magnetic field lines

Plasma

configurations of magnetic fields are being tried, and the problem of leaks is slowly being overcome.

At a time when the supply of conventional fuels, both fossil fuels and uranium, is rapidly being depleted, achieving controlled fusion seems especially important. The practically inexhaustible supply of deuterium fuel for fusion, and its slight environmental polluting effect, suggest that fusion might be a great help in solving our energy problems. Furthermore, it might be possible to obtain electric energy directly by passing the plasma into an MHD generator (see Chapter 14) rather than by using the less efficient heat engine.

4. ELEMENTARY PARTICLES

We have already met several fundamental or "elementary" particles, as they are called —namely protons, neutrons, electrons, neutrinos, and gamma rays, or photons. In recent years a large number of other elementary particles have been discovered whose role in the universe is not so clear as that of these more familiar ones.

Discovery and Production of New Particles

The field of elementary-particle physics dates back at least to 1937, when the Japanese theoretical physicist Hideki Yukawa (1907–   ) postulated the existence of a new particle. Yukawa reasoned that just as the photon is a quantum of light related to electromagnetic forces, so should there be a corresponding quantum or particle for the strong nuclear force. He calculated that the mass of this particle should be intermediate between that of the electron and that of the proton; hence it is called a meson (Greek for "middle"). The particle predicted by Yukawa was finally discovered in cosmic radiation in 1947. It is known as the "π," or pi, meson.

Apparently the nuclear forces are very complicated, because a great many other mesons have been discovered since, some of which are listed in Table 16–3. The mesons are all unstable and decay into lighter particles, usually other mesons, electrons, or gamma rays. Mesons can be positively or negatively charged, or neutral. For example, the pi meson can have a charge +e (e = charge on one electron), −e, or zero; the eta meson, on the other hand, can only be neutral.

Elementary particles heavier than nucleons but sharing some of the latters' properties have also been discovered and are collectively called "baryons" (meaning "heavy"). The first baryon, other than nucleons, was discovered by Fermi in 1952. Lighter particles,

Table 16–3
Elementary Particles
(Partial List)

FAMILY	PARTICLE NAME AND SYMBOL	RELATIVE MASS*	ELECTRIC CHARGE	NAME OF ANTI-PARTICLE
Electro-magnetic	Photon (γ)	0	0	Itself
Leptons	Electron (e)	1	–	Positron
	Neutrino (ν)	0	0	Antineutrino
	Muon (μ)	207	–	Positive muon
Mesons	Pion, or Pi meson (π)	$\begin{cases}273\\264\end{cases}$	+ 0	Negative pion Itself
	K meson (K)	990	+, 0	Anti-K meson
	Eta (η)	1070	0	Itself
Baryons	Proton (p)	1836	+	Antiproton
	Neutron (n)	1839	0	Antineutron
	Lambda (Λ)	2183	0	Anti-lambda
	Sigma (Σ)	2340	+, 0, –	Anti-sigma
	Xi (Ξ)	2590	–, 0	Anti-xi
	Omega (Ω)	3280	–	Anti-omega

*Relative to mass of electron.

such as electrons, are known as "leptons." Many of the known baryons and leptons are listed in Table 16–3 along with the mesons.

Where do all these elementary particles come from? They are produced in nuclear reactions, but only when there is sufficient energy to create the additional mass of the new particle. For example, if a proton of great enough kinetic energy strikes another proton, some of the proton's kinetic energy is changed into mass, the mass of a pi meson:

$$p + p \rightarrow p + p + \pi^0$$
$$p + p \rightarrow p + n + \pi^+$$

Because $E = mc^2$, the incoming proton must have enough kinetic energy to create the mass of a meson. If the proton does not have enough energy, the protons merely bounce off each other and no pi meson is produced. To obtain the necessary high energy, large

Figure 16–20

Photograph of the Fermilab high-energy accelerator at Batavia, Illinois. The particles move in a nearly circular path 2 km in diameter.

accelerators or "atom smashers" (Fig. 16–20) have been built that can accelerate protons and other charged particles to very high energies. These accelerators can then be used to produce and study the many kinds of elementary particles, as well as to study the nature of nuclei and nuclear forces.

The roles of these elementary particles, and how these particles are related to nuclear forces, are being actively investigated. A great deal remains to be learned about them.

Antiparticles

In 1932 a particle was discovered that had the same mass as an electron and that behaved like an electron in every way except that it had a positive charge. It was a positive electron, or **positron**, and is said to be the **antiparticle** to the electron. Antiparticles corresponding to the other known particles also exist. In each case the antiparticle has the same mass as the particle, and many of its other properties are also the same except its charge, which is exactly opposite. For example, the antiproton has the same mass as a proton but has a negative charge.

Antiparticles are produced in nuclear reactions. However, anti-

particles do not exist very long in the presence of ordinary particles.

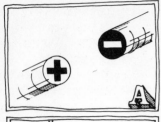
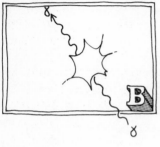

Figure 16–21

(A) An electron and a positron approach each other. (B) They collide and are annihilated, creating two gamma rays.

For example, when a positron encounters an electron, the two vanish—they "annihilate" each other—and gamma ray photons are given off in the process, Fig. 16–21. Since energy must be conserved, the masses of the original electron and positron appear as the energy of the gamma rays. The annihilation of the masses thus appears as electromagnetic energy. This is another example of Einstein's $E = mc^2$.

It is possible that far out in space there could exist "antimatter": matter whose atoms are made of antiparticles. In such a world the nuclei would consist of antiprotons and antineutrons; the nuclei would be negatively charged, and positrons would be orbiting them. These "anti-atoms" would be much like ordinary atoms. But should an antiworld made of antimatter approach our world made of ordinary matter, annihilation would take place, resulting in a huge explosion with the release of a vast amount of energy.

The New J/ψ Particles and the Nature of Matter

Since beginning the search for the basic constituents of matter, scientists have discovered smaller and smaller entities. They found that ordinary matter is composed of molecules and that molecules are made up of atoms. Atoms themselves are made up of nuclei surrounded by orbiting electrons; nuclei, in turn, consist of neutrons and protons. And protons and neutrons are not alone—a vast number of other elementary particles exist (Table 16–3); well over 200 are already known. Where will it end? Are all these "elementary" particles truly fundamental? Are there perhaps particles within the known particles that are even more basic? These are the questions that physicists are now seeking to answer.

Some hint as to the reason behind the vast number of "elementary" particles comes to us from the equivalence of mass and energy. Just as an atom has many different energy levels or excited states, perhaps the fundamental particles also have excited states of different energies; these higher energy levels might then appear to us as different particles of greater mass. Many of the "elementary" particles may thus be considered as excited energy states of the more basic particles.

Various theories have been developed recently that combine this idea with the idea that there are even smaller, more basic particles than those now known. One of these theories proposes that mesons and baryons are composed of three simpler, more fundamental particles known as "quarks." The great proliferation of elementary particles is then simply a manifestation of the various excited states of combinations of quarks.

The discovery in 1974 of a new class of particles called J/ψ ("psi") seems to confirm a theory that postulates the existence of a fourth quark, called the "charmed" quark. The J/ψ particles are believed to be made of a charmed quark and its antiquark.

Experimental physicists are working intensely on these problems. Newly developed super-high-energy accelerators may help to reveal the more deeply held secrets of nature. And of course in all of this we must remember that, as we mentioned in Chapter 15, the basic entities we have been calling elementary "particles" are not really particles at all, for they also have wave properties. We call them particles only for convenience.

Today, the questions facing physicists are among the most fundamental ever posed. They are deeply intellectual questions— questions about the nature of matter and energy, questions about the nature of physical reality itself.

And So . . .

The following exchange is taken from congressional testimony given by a leading physicist, Robert R. Wilson, concerning the value of building the high-energy particle accelerator research facility at Batavia, Illinois.

Senator Pastore: Is the accelerator connected in any way with the security of our country?

Dr. Wilson: No, sir, I do not believe so.

Senator Pastore: It has no value in this respect?

Dr. Wilson: It only has to do with the respect with which we regard one another, the dignity of men, our love of culture. It has to do with those things. It has nothing to do with the military, I am sorry.

Senator Pastore: Don't be sorry for it.

Dr. Wilson: I am not but I cannot in honesty say it has such applications; but it has to do with whether we are good painters, good sculptors, great poets, I mean all the things that we really venerate and are patriotic about in our country. In that sense, this new knowledge has everything to do with honor and country but it has nothing to do with defending our country except to help make it worth defending.

* * * *

REVIEW QUESTIONS

1. Nuclei normally contain
   (a) protons
   (b) electrons
   (c) neutrons
   (d) both electrons and protons
   (e) both neutrons and protons

2. Atomic number refers to
   (a) the number of neutrons in a nucleus
   (b) the number of protons
   (c) the total number of neutrons and protons
   (d) the number of neutrons minus the number of protons

3. Atomic mass number refers to
   (a) the number of neutrons in a nucleus
   (b) the number of protons
   (c) the total number of neutrons and protons
   (d) none of these

4. Isotopes have in common
   (a) the same number of protons
   (b) the same number of neutrons
   (c) the same atomic weight
   (d) the same number of neutrons and protons

5. $^{235}_{92}U$ contains
   (a) 92 protons and 235 neutrons
   (b) 92 protons and 143 neutrons
   (c) 235 protons and 92 neutrons
   (d) 143 protons and 327 neutrons

6. A nuclear force is assumed to exist in order to
   (a) keep electrons orbiting the nucleus
   (b) explain why, in spite of electric repulsion of protons, the nucleus doesn't break apart
   (c) explain radioactivity
   (d) allow fission to occur

7. Helium nuclei emitted in radioactive processes are known as
   (a) alpha particles
   (b) beta particles
   (c) gamma rays
   (d) quarks

8. A $^{60}_{27}Co$ nucleus decays by emitting an electron to become the nucleus
   (a) $^{60}_{27}Co$
   (b) $^{60}_{26}Fe$
   (c) $^{60}_{28}Ni$
   (d) $^{59}_{27}Co$

9. A $^{238}_{92}U$ nucleus decays by alpha decay to a nucleus containing how many protons?
   (a) 88
   (b) 90
   (c) 91
   (d) 92
   (e) 93
   (f) 94

492

10. The half-life of a radioactive isotope is one day. How much of the original material will be left after two days?
    (a) none
    (b) one-half
    (c) one-fourth
    (d) all

11. Uranium is found to be accompanied by an equal number of daughter nuclei in some ancient rocks. If the half-life of uranium is 4.5×10^9 years, how old are the rocks?
    (a) 2.25×10^9 years
    (b) 4.5×10^9 years
    (c) 9×10^9 years
    (d) none of these

12. Particles released in radioactive decay
    (a) can cause considerable biological damage
    (b) cannot cause biological damage
    (c) can cause biological damage only in sick people
    (d) are very slow moving

13. Complete the following nuclear reaction: $^4_2\text{He} + \,^{16}_8\text{O} \rightarrow \,^{19}_9\text{F} + ?$:
    (a) p
    (b) n
    (c) ^4_2He
    (d) γ
    (e) e^-

14. Nuclear fission is the process whereby
    (a) small nuclei are combined to form a larger nucleus
    (b) nuclei are caused to boil
    (c) radioactivity was discovered
    (d) large nuclei, like that of uranium, split apart

15. A chain reaction refers to
    (a) radioactive decay
    (b) alpha particles creating nuclear reactions
    (c) the fusion process
    (d) neutrons produced in one fission going on to produce more fissions

16. The difference between a nuclear reactor and a nuclear bomb is
    (a) the total amount of energy released
    (b) the rate at which the energy is released
    (c) the number of neutrons released
    (d) the kind of fuel used

17. The combining of two light nuclei to form a heavier one is called
    (a) fission
    (b) fusion
    (c) radioactive decay
    (d) a chain reaction

18. $E = mc^2$ does *not* apply in the case of
    (a) production of mesons
    (b) fission
    (c) fusion
    (d) it applies in all three cases

EXERCISES

1. How many protons and how many neutrons are contained in each of the following nuclei: $^{16}_{8}O$, $^{17}_{8}O$, $^{3}_{2}He$, $^{208}_{82}Pb$, $^{238}_{92}U$.

2. How many protons and how many neutrons are there in the following nuclei: $^{35}_{17}Cl$, $^{197}_{79}Au$, $^{1}_{1}H$, $^{247}_{97}Bk$.

3. What are the atomic number and the atomic mass number of a nucleus with 20 protons and 24 neutrons?

4. How do we know that there is such a thing as a strong nuclear force?

5. The $^{3}_{1}H$ isotope of hydrogen decays by beta emission into what nucleus?

6. Fill in the missing nucleus or particle in the following decays:

$$^{210}_{84}Po \rightarrow {}^{206}_{82}Pb + ?$$
$$^{60}_{27}Co \rightarrow {}^{60}_{28}Ni + ?$$
$$^{60}_{27}Co \rightarrow ? + \gamma$$
$$^{12}_{7}N \rightarrow {}^{12}_{6}C + ?$$

7. Fill in the missing particle or nucleus in the following decays:

$$^{226}_{88}Ra \rightarrow ? + \alpha$$
$$^{45}_{20}Ca \rightarrow ? + e^{-}$$
$$^{28}_{15}P \rightarrow {}^{28}_{14}Si + ?$$
$$^{238}_{92}U \rightarrow {}^{238}_{92}U + ?$$

8. Which is most like an x ray: an alpha, beta, or gamma ray?

9. Why are many artificially produced radioactive isotopes rare in nature?

10. Can hydrogen or deuterium emit an alpha particle?

11. If only 1/16 of a radioactive sample remains after one year, what is the half-life of the material?

12. A nucleus of radium ($^{224}_{88}Ra$) decays in several steps to lead ($^{208}_{82}Pb$). How many alpha particles and how many (negative) beta particles are emitted in the process?

13. How might radioactive tracers be used to find a leak in an underground pipe?

14. Can carbon-14 dating be used to measure the age of stone walls and tablets from ancient civilizations?

15 How does the process of radioactive decay differ from processes that follow Newton's laws?

16. Immediately after a $^{238}_{92}U$ nucleus decays to $^{234}_{90}Th + {}^{4}_{2}He$, the daughter thorium nucleus still has 92 electrons circling it. Since thorium normally holds only 90 electrons, what do you suppose happens to the two extra ones?

17. Because of the very short range of the nuclear force, a proton or a neutron is attracted by means of the nuclear force only to other nucleons that are very close. Therefore, in a large nucleus like uranium, nucleons on one side of the nucleus do not exert an attractive nuclear force on nucleons on the opposite edge of the nucleus; yet the electrical repulsion is readily felt over this distance. Explain why the nu-

clear force cannot hold very large nuclei together and therefore why nuclei heavier than uranium do not exist naturally.

18. Carbon extracted from an ancient wooden club found by archeologists in Mexico contains a ratio of ^{14}C to ^{12}C that is only one-fourth that found in living trees. How old is the club?

19. A proton strikes a $^{20}_{10}Ne$ nucleus, and an alpha particle is emitted. What is the remaining nucleus? Write down the reaction equation.

20. Why is it easier for a neutron than for a proton to penetrate a nucleus?

21. Can matter be created or destroyed?

22. If ^{235}U released only 1.5 neutrons per fission on the average, would a chain reaction be possible?

23. ^{235}U releases an average of 2.5 neutrons per fission compared to 2.7 for ^{239}Pu. Which of these two nuclei would you think has the smaller critical mass?

24. Why don't chain reactions occur in natural deposits of uranium ore?

25. Discuss how the course of history might have been changed if during the Second World War scientists had refused to work on developing a nuclear bomb. Do you think it would have been possible to delay the building of such a bomb indefinitely?

26. Researchers in molecular biology are moving toward an ability to perform genetic manipulations on humans. The moral implications of future discoveries along these lines has led to a warning that this may be the molecular biologist's "Hiroshima." Discuss.

27. The energy produced by a nuclear reactor appears in what form?

28. Discuss the relative merits and disadvantages, including pollution, of power generation by fossil fuels, nuclear fission, and nuclear fusion.

29. What is the basic difference between fission and fusion?

30. Why can't fusionable material be confined in an ordinary container rather than in a complex magnetic bottle?

31. Why is a "critical mass" needed before a chain reaction can occur?

32. In what ways would fusion reactors be preferable to fission reactors?

33. What is the advantage of using neutrons to bombard nuclei in producing nuclear reactions?

34. Where does the energy come from that is produced in the fission process? In the fusion process?

35. Does the electric force play a role in either radioactivity or fission? Explain.

36. What breeds what in a breeder reactor?

37. The light energy emitted by stars (including the sun) comes from the fusion process. What conditions in the interior of stars make this possible?

38. Why must the fission process release neutrons if it is to be useful?

39. Fill in the missing particles or nuclei:

$$^{4}_{2}He + {}^{16}_{8}O \rightarrow {}^{19}_{10}Ne + ?$$
$$^{4}_{2}He + {}^{16}_{8}O \rightarrow {}^{8}_{4}Be + ?$$

$$^{2}_{1}H + ^{2}_{1}H \rightarrow ^{4}_{2}He + ?$$
$$n + ^{239}_{94}Pu \rightarrow ^{141}_{54}Xe + ^{97}_{40}Zr + ?$$

40. Fill in the blanks for the following reactions:

$$n + ^{133}_{55}Cs \rightarrow ? + \gamma$$
$$n + ^{133}_{55}Cs \rightarrow ^{132}_{54}Xe + ?$$
$$^{6}_{3}Li + ^{7}_{4}Be \rightarrow ^{4}_{2}He + ?$$
$$p + n \rightarrow p + ? + \pi^{-}$$

41. Two protons approach each other at the same speed. How much kinetic energy must each have if a pion (mass = 2.5×10^{-29} kg) is to be produced according to the reaction $p + p \rightarrow p + p + \pi^{0}$? (Hint: Use $E = mc^2$.)

42. How much energy is released when an electron and a positron annihilate each other? (Mass of each is 9.1×10^{-31} kg.)

43. If a large quantity of antimatter exists in the universe, we expect that it would be quite far away from our galaxy. Could it be close to us? Why?

CHAPTER SUMMARIES FOR REVIEW

1

1. PHYSICS: A HUMAN ENDEAVOR

The aim of science is to find order in the physical world that surrounds us. The **scientific method**, which consists of (1) observation of the physical world, (2) the formulation of theories, or hypotheses, and (3) careful testing of those theories, is sometimes viewed as the basis for scientific development. A clearer view of scientific progress reveals that it is achieved not in a routine, mindless fashion but rather with creativity and perception, as in the arts.

The emphasis of this book is the *ideas* of physics, but some mathematical tools are useful in grasping those ideas. Two quantities are **proportional** if a change in one implies or produces a change in the other. The two are **directly proportional** if an increase in one leads to a corresponding increase in the other. They are **inversely proportional** if an increase in one leads to a corresponding decrease in the other. A **proportionality constant** is a number (or constant) used to express a proportionality as an equality.

Two **systems of measurement** are ordinarily used today. The **British system**, commonly used in the United States, is the system using such units as inches, feet, and pounds. In the **metric system**, more widely used throughout the world, the different-sized units are related to each other by simple powers of ten. Conversion between the two systems is accomplished by use of the appropriate proportionality constant.

2. MOTION: IN THE HEAVENS AND ON THE EARTH

The ancients viewed motion as being of two distinct types: celestial (motion of heavenly bodies) and terrestrial (motion of bodies on the earth). Observations of heavenly bodies showed that they appear to move from east to west. The sun, the moon, and the planets, which move at different rates from the stars, appear to move relative to them. Early views of the universe, culminating in that of **Ptolemy** about A.D. 150, were **geocentric**—i.e., earth-centered. Later, **Copernicus** proposed a **heliocentric**, or sun-centered, view of the universe in which the earth and the other planets moved in circular orbits about the sun. Using the observations of Tycho Brahe, **Kepler** refined the heliocentric view by hypothesizing that the planets move in elliptical orbits and worked out their relative speeds. **Galileo**, who made many great contributions to scientific thought, used a telescope to make observations of the heavenly bodies, which further supported the heliocentric viewpoint.

Motion, or change of position of an object with time, is described by giving the speed or velocity and the acceleration. **Speed** is the rate of change of position; the term **velocity** refers to the speed and also the direction. **Acceleration** is the rate at which velocity changes. Velocity and acceleration are examples of quantities called **vectors**. Vector quantities are defined by both a magnitude and a direction. **Scalar** quantities, such as time and temperature, are defined only by a magnitude (an amount).

Force, which is a vector quantity, is a push or a pull and may distort the shape of objects as well as change their motion. **Friction** is a commonly observed force that confused early thinkers about the relationship between force and motion. Galileo reasoned that a body moving horizontally would continue moving at constant velocity if no friction or other forces acted on it. He concluded furthermore that falling bodies all fall with the same acceleration when the air friction is absent or can be neglected.

3. MOTION: NEWTON'S SYNTHESIS

Sir Isaac Newton formulated three basic laws that describe the principles of motion. **Newton's first law of motion** states that if the *net* force on an object is zero, an object at rest will remain at rest and an object in motion will continue in motion at constant velocity. This tendency to resist a change in motion is called **inertia**. **Mass** is a quantitative measure of the inertia of a body. **Weight**, on the other hand, is defined as the gravitational force (or pull) on an object.

Newton's second law of motion states that the net force acting on a body will cause it to accelerate—i.e., change its motion—directly proportional to the net force acting on it and inversely proportional to the mass ($F = ma$).

Newton's third law of motion states that whenever one body exerts a force on another body, the second body exerts an equal and opposite force on the first.

The net force is the **vector sum** of all the forces acting on an object. Vector quantities can be added graphically by such methods as the parallelogram method and the tail-to-tip method. In the latter method, the tail of the arrow representing each successive vector is placed at the tip of the previous one. The sum is the vector drawn from the tail of the first vector to the tip of the last one. A vector can also be **resolved** into its **components**, which are two vectors (usually perpendicular to each other) that add to give the original vector. When the net force on an object is zero, the object is said to be in equilibrium.

The motion of a **projectile**—i.e., an object moving through the air—may be analyzed by examining the horizontal and vertical components of the motion separately.

An object moving at constant speed in a circle has an acceleration toward the center of the circle, or a **centripetal acceleration**. A force on the object acting toward the center of the circle produces this acceleration, thereby keeping the object moving in its circular path.

Newton's **law of universal gravitation** states that every body in the universe attracts every other body with a force proportional to their masses and inversely proportional to the square of the distance between them.

4. ROTATIONAL MOTION AND BODIES IN EQUILIBRIUM

Rotational motion—the motion of a body rotating about an axis—can be understood by extending Newton's laws of motion. A rotating body possesses **rotational inertia**, which depends on the mass of the body and on how far from the axis of rotation the mass is distributed. The **torque** about an axis due to an applied force is the product of the applied force and the **lever arm**, which is defined as the perpendicular distance from the axis to the line along which the force acts. Torque produces changes in rotational motion. The **angular acceleration** of a body is directly proportional to the net torque and inversely proportional to its rotational inertia.

The **center of gravity** of a body is that point at which the force of gravity on the entire body can be considered to act. The motion of any object can be considered the sum of two motions: translational motion of the center of gravity and rotation about the center of gravity.

If a body is to remain at rest, or in a state of uniform motion, the sum of all the forces on it must be zero and the sum of all the torques on it must be zero. It is then said to be in **equilibrium**. These two conditions for equilibrium can be used to analyze the forces and torques acting on a body.

The nature of the equilibrium of a body at rest can be determined by displacing the body slightly. The body is said to be in **stable**, **unstable**, or **neutral** equilibrium, depending on whether it subsequently returns to its original position, continues to move, or remains in the new position. A body whose center of gravity (c.g.) is higher than its base of support will be stable only if the c.g. is not beyond the base of its support.

All objects are **elastic** to some extent—i.e., their shape changes when a force is exerted on them and when the force is removed, they tend to return to their original shape. **Hooke's law** states that the amount of deformation is proportional to the applied force. If the force is sufficiently great, the **elastic limit** is exceeded and the object will not return to its original shape when the force is removed. If the **ultimate strength** is exceeded, the object will break. A body is under **tension**, **compression**, or **shear** depending on whether the forces on it tend to increase its length, decrease its length, or distort it crosswise.

5. ENERGY AND MOMENTUM

The **work** done by a force is defined as the product of the magnitude of the force and the distance through which the force acts. **Simple machines** such as a lever do not save us work but they offer a **mechanical advantage**, which is the ratio of the output force to the input force.

   Energy is the ability to do work. **Kinetic energy** is the energy of motion. **Potential energy** is the energy an object has because of its shape or position. Other forms of energy include thermal, electrical, and nuclear energy. Although energy can be transformed from one kind to another or transferred from one object to another, the total amount of energy remains constant, as stated by the **law of conservation of energy**.

   Power is defined as the rate at which work is done or at which energy is transformed.

   The **momentum** of an object is defined as the product of its mass and its velocity. The **law of conservation of momentum** states that the total momentum of a group of objects remains constant if no external forces act on the group. The **angular momentum** of a body depends on the body's mass, its rotational velocity, and how far from the axis of rotation the mass is distributed. The **law of conservation of angular momentum** states that the total angular momentum of an object remains constant if no external torques act.

6. RELATIVITY

The **theory of relativity** examines how we observe physical events. All observations and measurements are made in some frame of reference. The velocity of an object depends on the frame of reference in which it is measured, so we say that velocity is a **relative concept**. Position and time are also relative concepts.

   The theory of relativity is based on two principles. The first is the **relativity principle**. It states that the laws of physics are the same in all uniformly moving reference frames—i.e., there is no preferred reference frame that can be considered at absolute rest. The second principle, which was supported by the famous experiment of **Michelson and Morley**, states that the speed of light has the same value in all reference frames.

The theory of relativity predicts many unusual results. Two events that are **simultaneous** in one frame of reference may not be simultaneous in another frame of reference. Moving clocks run slowly (**time dilation**). Moving objects appear shorter (**length contraction**). Moving objects have **increased masses**. Velocities must be added in a special way (no longer just a simple sum). The speed of light is a maximum limit on speed that no ordinary object can reach or exceed. Energy and matter are equivalent and can be changed into one another as described by the relationship $E = mc^2$. The ideas of relativity led to the concept of a four-dimensional "space-time," where **time** is the "**fourth dimension**."

7. FLUIDS

There are three common states of matter—solid, liquid, and gas. **Solids** have a fixed shape and volume, **liquids** have a fixed volume but take on the shape of their container, and a **gas** does not have a fixed shape or volume. Because liquids and gases have the ability to flow, they are both referred to as **fluids**.

The **density** of a substance is defined as the mass per unit volume. The **specific gravity** is the ratio of the density of the substance to the density of water.

Pressure is defined as the force per unit area. In a liquid, the pressure is exerted in all directions and is proportional to the density and the depth in the liquid. **Archimedes' principle** states that the buoyant force on a body in a fluid is equal to the weight of the fluid displaced by the body. Whether a body sinks or floats in a liquid depends on whether the density of the body is greater or less than the density of the liquid. **Pascal's principle** says that the pressure applied to an enclosed fluid is transmitted throughout the fluid.

The atmosphere of the earth is made up primarily of the gases oxygen and nitrogen, with small amounts of other materials. The pressure, which may be measured with a **barometer**, and density of the earth's atmosphere decrease with altitude.

Fluid flow can be classified as either **streamline** (smooth) or **turbulent**. **Viscosity** describes the resistance of fluids to flow and results from internal friction in the fluid. Pumps, including the heart, cause fluids to flow by creating differences in pressure. In a continuous tube or channel, fluids flow fastest in those places where the cross-sectional area is smallest. **Bernoulli's principle** states that where a fluid flows fastest, the pressure is lowest.

8. ATOMS AND TEMPERATURE: KINETIC THEORY

Matter is believed to be made up of tiny particles, about 10^{-8} cm in diameter, called **atoms**. Atoms can combine to form **molecules**. An **element** is a substance made up of only one kind of atom, whereas a **compound** contains different kinds of atoms, generally combined as molecules. **Brownian movement** provided the most direct early evidence in favor of atomic theory. Atoms are considered to consist of a very tiny, but heavy, **nucleus** about which even smaller **electrons** orbit. Microscopically the differences between solids, liquids, and gases can be attributed to the strength of the forces existing between the atoms or molecules.

The **temperature** of a body is a measure of how hot or cold it is. A **thermometer** commonly measures temperature on either the **Celsius** (or centigrade) scale or the **Fahrenheit** scale. The absolute, or **Kelvin**, temperature scale begins at absolute zero (0 K = $-273°C$), but it is believed that this temperature cannot actually be reached.

Some laws have been formulated that describe the behavior of gases. **Boyle's law** states that, at constant temperature, the volume (V) of a fixed amount of gas is inversely proportional to the pressure (p) applied to it. **Charles's law** states that, at constant pressure, the volume of a fixed amount of gas is directly proportional to the absolute temperature (T). **Gay-Lussac's law** states that, at constant volume, the pressure in a fixed amount of gas is directly proportional to the absolute temperature. **Avogadro's law** states that equal volumes of gases at the same pressure and volume contain equal numbers (N) of molecules. All these laws can be combined into one relationship known as the **ideal gas law**, which is given in symbols as $pV = NkT$, where k is a universal constant—i.e., the same constant for all gases. It is an "ideal" law because real gases follow it closely only when the pressure is not too high and when the gas is not too dense.

Kinetic theory, which refers to the idea that matter is made up of molecules, which are always in motion, describes the absolute temperature of a body as a measure of the average kinetic energy of its molecules.

9. HEAT

The **thermal energy** of a body is the total energy of all the molecules in the body. The term **heat** describes the transfer of thermal energy from one place to another or from one body to another, because of a temperature difference. The **first law of thermodynamics** assumes heat is a kind of energy and states that the total amount of energy remains constant.

Thermal energy, or heat, can be transferred in three ways. **Conduction** is the transfer of energy through the collision of molecules. **Convection** is the transfer of energy through the mass movement of molecules from one place to another. **Radiation**, which does not require the presence of matter, is the transfer of energy in the form of electromagnetic waves.

Heat is involved in **changing the state** of a substance even though there may be no change in temperature. At the **boiling point**, a liquid may change to a gas without a change in temperature. At the **melting point** a solid can change to a liquid. The heat, or energy input, is needed because a change of state corresponds to a change in the potential energy of the molecules.

A **heat engine** is a device to change thermal energy into useful work. The **efficiency** of a heat engine is defined as the ratio of the work done to the heat input. One way of stating the **second law of thermodynamics** is that no heat engine can transform a given amount of thermal energy completely into work; some heat is always given off as exhaust. A more general statement of the second law of thermodynamics is that natural processes tend toward a state of greater disorder. **Entropy** is the quantitative measure of disorder.

10. WAVE MOTION AND SOUND

A **wave** is a vibration that moves and carries energy, but not matter, from point to point. There are two kinds of waves: **transverse**, in which the particles of the medium move perpendicularly to the direction of wave motion, and **longitudinal**, in which the particles of the medium vibrate parallel to the direction of wave motion. The distance between successive crests, or any two successive identical points on a wave, is called the **wavelength**. The **frequency** is the number of crests that pass a given point per unit time. The **amplitude** is the maximum displacement of a particle in the medium from its normal position. **Wave velocity** is defined as the velocity at which the wave crest or wave shape appears to move and can be found by multiplying the wavelength by the frequency.

Waves can **reflect** from obstacles and they can also **diffract** (bend) around obstacles. Waves can also **interfere** with one another, resulting in increased amplitude (**constructive interference**) at some points and reduced amplitude (**destructive interference**) at other points. If waves traveling in opposite directions interfere in such a way that the regions of constructive and destructive interference occur at fixed positions, a **standing wave** results. The oppositely traveling waves may be an incident wave and its subsequent reflection. The frequencies at which standing waves may occur in an object are called the natural or **resonant frequencies** of the object.

Sound travels as a longitudinal wave. **Pitch** refers to the frequency of the sound wave, and **loudness** is related to the amplitude. The **intensity** of a sound refers to the amount of energy the sound wave is carrying. The quality, or **timbre**, of a sound depends on the mixture of frequencies present.

The **Doppler effect** describes the observed change in the pitch of a sound if either the source of sound or the listener is moving. If they are approaching each other, the pitch is higher; if they are moving apart, the pitch is lower. **Shock waves** and **sonic booms** are large-amplitude **wave pulses** that occur when the speed of the source of vibration exceeds the speed of sound in the medium in which it is moving.

11. LIGHT AND LENSES

Light is a form of energy that travels with a speed of 300,000 km/sec in a vacuum or in air. For many purposes, light can be considered to travel in straight lines called **rays**. When light is **reflected** from a flat surface, the angle of reflection equals the angle of incidence. **Refraction** refers to the change of direction that light undergoes when it passes from one transparent medium into another. The relative sizes of the angle of refraction and the angle of incidence depend on the relative speeds of light in the two media. If the speed of light is greater in the second medium, there is a **critical angle** beyond which **total internal reflection** occurs. In this case all the light is reflected from the surface and none passes through.

Both **mirrors** and **lenses** form **images** of objects. In the case of a flat mirror, the image is as far behind the mirror as the object is in front. The **focal point** of a lens is that point at which parallel rays (approximately parallel rays come from an object that is very far away) come to a focus. The distance from the center of the lens to the focal point is called the **focal length**. Cameras, microscopes, telescopes, and the human eye are common examples of the use of lenses to form images.

12. THE WAVE NATURE OF LIGHT

Light travels as a transverse wave. Young's double-slit experiment showed that light exhibits **interference**: light from two slits interacts to produce greater amplitude in some places and less amplitude in others. Light also exhibits **diffraction**, i.e., it bends around objects. Light can also be polarized. **Polarization**, which occurs only in transverse waves, means that the vibrations are restricted to only one plane. The wave theory of light also explains the phenomenon of

refraction, where light bends as it passes from one medium to another.

Different **colors** correspond to different wavelengths. Red has the longest wavelength and violet the shortest. The **speed** of light in most transparent materials depends on the color or wavelength of the light. Since most objects do not emit light, their color is due to the wavelengths of light they reflect. Colors occurring because of reflection, such as in pigments, combine **subtractively** to produce other colors. The resulting color of the mixture is due only to those wavelengths that are not absorbed (and are therefore reflected) by the pigments in the mixture. On the other hand, colors occurring because of the emission of light, as in colored lights, combine **additively**: the resulting color is due to the sum of all the wavelengths emitted by the sources.

The eye contains two kinds of light receptors, rods and cones. The **rods** are more sensitive to light, but the **cones** distinguish colors.

13. ELECTRICITY

There are two kinds of electric charge: **positive** and **negative**. Like charges repel each other and opposite charges attract. The law of **conservation of electric charge** states that electric charge is neither created nor destroyed. Positive and negative charges may be separated, but such a separation always produces equal amounts of each, so that the *net* charge is unchanged.

Coulomb's law states that the magnitude of the force between two charged bodies is directly proportional to the product of the charges on the bodies and inversely proportional to the square of the distance between them.

Materials in which electric charge flows relatively freely are called **conductors**. If the charge cannot move freely, the material is called an **insulator**. There are a few intermediate materials called **semiconductors**.

The field concept is useful in understanding forces that *act at a distance*, such as gravitational and electrical forces. According to this concept, the **electric field** produced by a charge permeates all of space. The force that one charge exerts on another can be described as the interaction of the one charge with the field of the other. The magnitude of the electric field at any point in space is defined as the force per unit charge that would be exerted on a tiny test charge if it were placed at that point.

A charged body in an electric field possesses potential energy by virtue of its position in that field. The potential energy per unit charge is called the **electric potential** (V).

Electric **current** (*I*) is the rate of flow of electric charge. A continuous conducting path between the terminals of a source of electrical energy such as a battery is called a **circuit**. The difference in electric potential, or **voltage**, produced by batteries or generators can produce electric currents. **Ohm's law** states that the current in a circuit is directly proportional to the difference in potential ($I = V/R$). The proportionality constant *R* is called the **resistance**. **Electric power** is proportional to the product of potential difference and the current ($P = IV$). Currents can be either direct, **DC** (the current always flows in the same direction), or alternating, **AC** (the current direction periodically changes, or alternates).

Vacuum tubes operate by the flow of electrons through an evacuated tube from one electrode to another. **Semiconductor devices** operate by the flow of **electrons** (negative) and **holes** (positive) through solid material. If electron flow predominates, the material is called **n-type**; if hole flow predominates, it is **p-type**. **Diodes**, which have only two electrodes or two types of semiconductor material, allow current to flow in only one direction and can therefore be used to change AC to DC. A **triode** vacuum tube, with three electrodes, and its counterpart, the **transistor**, with three sections of semiconductor material, are used primarily to amplify small signals.

The picture tube of a TV set is a large vacuum tube called a **CRT**, or cathode ray tube, in which electrons are accelerated to high speed, focused, and directed to various regions of the fluorescent screen at the front by electric and magnetic fields.

14. MAGNETISM AND ELECTROMAGNETISM

Every magnet has two *poles:* **north** and **south**. Like poles of two magnets repel each other; unlike poles attract each other. Using the field concept on magnets, we can speak of the **magnetic field** surrounding a magnet. The force between two magnets is said to be an interaction of one magnet and the magnetic field of the other. Besides acting on other magnets, magnetic fields exert forces on moving charged particles and on electric currents. The force between a magnetic field and an electric current is the driving force in such devices as motors, loudspeakers, galvanometers, and solenoids.

Magnetic materials contain small regions known as **domains**, which act as tiny magnets. Permanent magnets are made by aligning the domains. Electric currents also produce magnetic fields.

The earth possesses a magnetic field, which causes a compass needle—which is a small magnet—to point "north." The magnetic field of the earth interacts with charged particles streaming in from space to protect us from the radiation of those particles.

A voltage and current can be produced in a circuit if the magnetic field through the circuit changes. This is called **induction** and is described by **Faraday's law**, which states that a changing magnetic field gives rise to an electric field. Transformers and generators are based on this principle of induction.

Electrical energy is usually produced by a turbine that turns a generator. The turbine may be driven by wind (as in a windmill), by water (over a dam or in the tides), or by steam. The steam may be made by using the energy from such diverse sources as fossil fuels, nuclear fuels, the sun, or the heat of the earth itself (geothermal). Electricity may be made without turbines by direct conversion in solar cells and by magnetohydrodynamics (MHD).

Maxwell postulated that the inverse of Faraday's law is also true: a changing electric field gives rise to a magnetic field. This enabled Maxwell to tie all of electricity and magnetism together into one theory. Maxwell's theory predicted that accelerating electric charges (or, equivalently, changing electric currents) give rise to electric and magnetic fields that move through space at the speed of light. These moving fields are called **electromagnetic waves**. Visible light is composed of electromagnetic waves in a particular range of frequencies. Other ranges of frequencies of electromagnetic waves include infrared, ultraviolet, microwaves, x rays, and radio waves.

15. QUANTUM THEORY AND ATOMIC STRUCTURE

The spectrum of electromagnetic radiation emitted by a heated solid can be explained by **Planck's quantum hypothesis**, which states that the vibrational energy of molecules in a solid is quantized—i.e., it exists only in discrete amounts. **Einstein's quantum theory of light** states that light travels as tiny particles, or quanta, called **photons**. The **photoelectric effect**, in which electrons are emitted from a metal surface when it is illuminated by light of the appropriate frequency, confirmed this theory of Einstein's. According to the **wave–particle duality**, light exhibits both wave and particle properties. **De Broglie** postulated that matter exhibits a similar duality. Electrons, for example, have been shown to exhibit both wave and particle properties.

Rutherford contributed to our understanding of the atom by his discovery of the **nucleus**, a heavy, positively charged core in each atom. The electron orbits around the atom were explained by the Bohr theory. Atoms and molecules in gases are observed to emit only certain frequencies of light. This pattern of frequencies is called a **line spectrum**, as opposed to the continuous spectrum emitted by hot solids. According to the **Bohr theory** of the atom, electrons orbit the nucleus of an atom only in certain discrete orbits, or states. According to de Broglie, these discrete states arise from the wave nature of electrons. When an electron jumps from one state to a state of lower energy, the atom emits a photon of light corresponding to the energy difference in the two states.

Quantum mechanics refers to our present theory of matter, in which electrons in an atom are visualized as a wave or a **probability cloud** surrounding the nucleus. The **uncertainty principle** states that there is a limit on the accuracy with which particle properties such as position and momentum may be determined. Quantum theory involves a **probabilistic view** of matter at the atomic level rather than a **deterministic view**, in which all occurrences are the result of specific causes.

The **Pauli exclusion principle** specified that no more than two electrons can be in any given quantum level of an atom. The resulting knowledge of the electron structure of atoms aids in understanding the bonds formed between atoms to make molecules.

X rays are high-energy photons emitted when high-speed electrons are suddenly decelerated by striking a metal target. **Photoelectric cells** are a practical application of the photoelectric effect. An **electron microscope** uses the wave properties of electrons to form a greatly magnified image of an object. In **fluorescence**, atoms are excited from their lowest, or ground, energy state to higher, excited states by ultraviolet or visible light. The atoms then emit visible light of a lower frequency when they return in several steps to the ground state. In **phosphorescence**, which is very similar, the atoms continue emitting visible light over a period of time after the exciting source is removed. A **laser** is a device that produces an intense, narrow beam of light. **Holograms** are three-dimensional images produced with laser light.

16. NUCLEAR PHYSICS

The nucleus of an atom contains **protons** (with a positive charge) and **neutrons** (with no charge—i.e., neutral), which are held together by the **nuclear force**. **Isotopes** are nuclei with the same number of protons (and therefore nuclei of the same element) but with different numbers of neutrons.

Unstable nuclei that transform themselves into other nuclei by emitting particles are called **radioactive**. Rutherford observed three types of emission from radioactive nuclei: **alpha**, in which a helium nucleus is given off; **beta**, in which an electron is given off; and **gamma**, in which a high-energy photon is given off. The **half-life** of a radioactive substance is the time required for half of the substance to decay. The decay of radioactive elements in an object can be used to date the object. Nuclei can also be artificially transformed by **nuclear reactions**, which involve collisions between nuclei and other particles, such as neutrons.

Fission is the disintegration of a heavy nucleus, such as a uranium nucleus, into two smaller nuclei plus several neutrons, with an accompanying release of energy. If enough fissionable material is present (the **critical mass**), a **chain reaction** can occur, in which the neutrons produced in one fission can cause the fissions of other nuclei, which, in turn, cause the fissions of still other nuclei, and so on. The process may be self-sustaining and releases a large amount of energy. A **nuclear reactor** uses the energy produced by a chain reaction to produce electrical power in a controlled fashion.

Fusion is the combination of small nuclei, such as hydrogen nuclei, into larger nuclei, such as helium nuclei, with an accompanying release of energy. The high temperatures required for fusion to occur cause the atoms to become completely ionized. This ionized gas, nuclei plus electrons, is known as a **plasma**.

Besides protons, neutrons, and electrons, a great many other fundamental or "elementary" **particles** have been discovered. The **antiparticle** corresponding to each particle has the same mass but the opposite charge. For example, a positron, or antielectron, has the same mass as an electron but has a positive charge. Particles and antiparticles completely annihilate each other when they collide, and their mass is converted to energy and released. A great deal remains to be learned about the "elementary" particles and their antiparticles.

POWERS
OF
TEN
NOTATION

In physics we often have to deal with very large or very small numbers. For example, the diameter of an atom is about 0.00000001 centimeter. On the other hand, the number of water molecules in a *liter* of water is about 33,000,000,000,000,-000,000,000,000. And the distance from the earth to the nearest star is about 40,000,000,000,000 kilometers. Dealing with numbers with this many zeros is difficult, and it is easy to make a mistake by dropping a zero or two when writing them down. To avoid problems, we commonly use the **powers of ten notation**.

You are undoubtedly aware that ten squared is a hundred; that is, $10 \times 10 = 10^2 = 100$. Similarly, ten cubed, or ten to the third power, is a thousand: $10 \times 10 \times 10 = 10^3 = 1000$. Similarly, $50,000 = 5 \times 10 \times 10 \times 10 \times 10 = 5 \times 10^4$. For large numbers, it is easier to use the powers of ten notation, 5×10^4, instead of writing the number out as 50,000. The number of water molecules in a quart of water given above is written as 3.3×10^{25} molecules.

In general, a number is written in powers of ten notation as the product of a simple number times ten to some power. The exponent attached to the ten is just the number of places the decimal point is moved to the right to obtain the fully written out number. Thus 5.0×10^4 means moving the decimal point four places to the right to give 5.0000. (see if this works for the number 3.3×10^{25}, the number of water molecules in a quart of water).

Powers of Ten		
10^9	=	1,000,000,000	one billion
10^6	=	1,000,000	one million
10^5	=	100,000	one hundred thousand
10^4	=	10,000	ten thousand
10^3	=	1,000	one thousand
10^2	=	100	one hundred
10^1	=	10	ten
10^0	=	1	one
10^{-1}	=	0.1	one-tenth
10^{-2}	=	0.01	one one-hundredth
10^{-3}	=	0.001	one one-thousandth
10^{-6}	=	0.000001	one one-millionth

When the number is less than 1, say 0.001, the exponent or the power of ten is written with a negative sign; thus $0.001 = \frac{1}{10 \times 10 \times 10} = \frac{1}{10^3} = 1 \times 10^{-3}$ Similarly, $0.002 = 2 \times 10^{-3}$. A negative exponent on a power of ten means that the decimal point is moved that number of places to the left: $2.0 \times 10^{-3} = 0.002$.; the diameter of atoms is thus about 1×10^{-8} cm. The table contains a brief summary of powers of ten notation.

ANSWERS TO MULTIPLE CHOICE QUESTIONS

CHAPTER 1

1(b), 2(b), 3(a), 4(b), 5(a), 6(b).

CHAPTER 2

1(a), 2(c), 3(c), 4(d), 5(d), 6(c), 7(b), 8(a), 9(c), 10(d), 11(c), 12(a), 13(a), 14(b).

CHAPTER 3

1(a), 2(b), 3(d), 4(d), 5(d), 6(a), 7(a), 8(a), 9(b), 10(a), 11(d), 12(c), 13(c), 14(a), 15(d).

CHAPTER 4

1(d), 2(a), 3(d), 4(c), 5(a), 6(c), 7(b), 8(a), 9(d), 10(b), 11(a), 12(a), 13(b), 14(d), 15(b).

CHAPTER 5

1(b), 2(b), 3(b), 4(d), 5(d), 6(b), 7(d), 8(c), 9(a), 10(d), 11(c), 12(c), 13(d), 14(a), 15(a), 16(b), 17(b).

CHAPTER 6

1(b), 2(c), 3(a), 4(a), 5(c), 6(c), 7(c), 8(a), 9(b), 10(b), 11(a), 12(b), 13(a), 14(c).

CHAPTER 7

1(c), 2(d), 3(a), 4(b), 5(b), 6(c), 7(d), 8(b), 9(a), 10(c), 11(c), 12(d), 13(b), 14(d).

CHAPTER 8

1(b), 2(c), 3(a), 4(b), 5(d), 6(c), 7(d), 8(e), 9(b), 10(a), 11(d), 12(d), 13(c), 14(c), 15(c).

CHAPTER 9

1(b), 2(b), 3(c), 4(c), 5(b), 6(b), 7(c), 8(a), 9(b), 10(d), 11(a), 12(a), 13(d), 14(b), 15(b).

CHAPTER 10

1(d), 2(b), 3(c), 4(a), 5(b), 6(a), 7(b), 8(d), 9(b), 10(a), 11(b), 12(c), 13(b), 14(a), 15(a), 16(c), 17(b), 18(c).

CHAPTER 11

1(a), 2(d), 3(c), 4(c), 5(b), 6(b), 7(b), 8(b), 9(d), 10(b), 11(a), 12(b).

CHAPTER 12

1(b), 2(b), 3(d), 4(b), 5(a), 6(c), 7(d), 8(b), 9(b), 10(d), 11(d).

CHAPTER 13

1(a), 2(b), 3(b), 4(a), 5(c), 6(c), 7(b), 8(c), 9(a), 10(c), 11(c), 12(a), 13(a), 14(b), 15(a), 16(b), 17(c), 18(a).

CHAPTER 14

1(b), 2(c), 3(d), 4(c), 5(a), 6(c), 7(b), 8(d), 9(a), 10(d), 11(c), 12(d), 13(c), 14(d), 15(a), 16(b).

CHAPTER 15

1(c), 2(c), 3(c), 4(d), 5(d), 6(a), 7(c), 8(b), 9(d), 10(b), 11(a), 12(c), 13(c), 14(d), 15(c), 16(a), 17(d), 18(c).

CHAPTER 16

1(e), 2(b), 3(c), 4(a), 5(b), 6(b), 7(a), 8(c), 9(b), 10(c), 11(b), 12(a), 13(a), 14(d), 15(d), 16(b), 17(b), 18(d).

INDEX

8
B 9
C 0
D 1
E 2
F 3
G 4
H 5
I 6
J 7